Topics in Medicinal Chemistry

Volume 40

Topics in Medicinal Chemistry (TMC) covers all relevant aspects of medicinal chemistry research, e.g. pathobiochemistry of diseases, identification and validation of (emerging) drug targets, structural biology, drugability of targets, drug design approaches, chemogenomics, synthetic chemistry including combinatorial methods, bioorganic chemistry, natural compounds, high-throughput screening, pharmacological in vitro and in vivo investigations, drug-receptor interactions on the molecular level, structure-activity relationships, drug absorption, distribution, metabolism, elimination, toxicology and pharmacogenomics. Drug research requires interdisciplinary team-work at the interface between chemistry, biology and medicine. To fulfil this need, TMC is intended for researchers and experts working in academia and in the pharmaceutical industry, and also for graduates that look for a carefully selected collection of high quality review articles on their respective field of expertise.

Medicinal chemistry is both science and art. The science of medicinal chemistry offers mankind one of its best hopes for improving the quality of life. The art of medicinal chemistry continues to challenge its practitioners with the need for both intuition and experience to discover new drugs. Hence sharing the experience of drug research is uniquely beneficial to the field of medicinal chemistry.

All chapters from Topics in Medicinal Chemistry are published OnlineFirst with an individual DOI. In references, Topics in Medicinal Chemistry is abbreviated as Top Med Chem and cited as a journal.

Andreas Brunschweiger • Damian W. Young
Editors

DNA-Encoded Libraries

With contributions by

M. J. Anderson · A. Brunschweiger · I. F. S. F. Castan ·
E. Detta · R. M. Franzini · J. S. Graham · Y. Gui ·
H. Hassan · T. Kodadek · A. Lessing · X. Li ·
L. Marcaurelle · J. I. Montgomery · B. A. Morgan ·
J. Odger · C. L. A. Salvini · A. Satz · J. Scheuermann ·
R. V. Stanton · C. Taylor · W. Tear · M. J. Waring ·
G. Yao · Y. Zhang · H. Zhu

Springer

Editors
Andreas Brunschweiger
Department of Chemistry and Chemical Biology
TU Dortmund University
Dortmund, Germany

Damian W. Young
Department of Pharmacology and Chemical Biology, Department of Pathology and Immunology, Center for Drug Discovery
Baylor College of Medicine
Houston, TX, USA

ISSN 1862-2461 ISSN 1862-247X (electronic)
Topics in Medicinal Chemistry
ISBN 978-3-031-18628-8 ISBN 978-3-031-18629-5 (eBook)
https://doi.org/10.1007/978-3-031-18629-5

This Springer imprint is published by the registered company Springer Nature Switzerland AG
The registered company address is: Gewerbestrasse 11, 6330 Cham, Switzerland

Preface

This book on DNA-Encoded Libraries (DELs) comes 30 years after the idea was first proposed by Lerner and Brenner. Their concept of encoding synthetic chemical information with oligonucleotide sequences was initially viewed as more of an academic curiosity. However, several influential publications that emerged in the mid-2000s illuminated the unique and powerful drug discovery potential for DELs, catapulting the technology into a spotlight that has not diminished since. The current DEL landscape is vital and it is continuing to expand. To date, hundreds of research papers on the topic of DEL technology have been published. Small biotechs to large pharmaceutical companies alike are increasingly relying on DELs as their principal discovery platform. Academic scientists are entering the DEL arena at a rapid pace, both as a means to supply small molecules for their newly identified disease targets and as innovators to DEL technology itself. All of these aspects point to the fact that DEL technology has now become a pillar in therapeutic discovery.

We believe that the timing of this Topics in Medicinal Chemistry volume devoted to DNA-Encoded Libraries is quite appropriate. Books of this type provide both the bird's-eye view and a detailed glimpse into time of a given field and can have great influence on its future direction. For example, A Handbook of DNA-Encoded Libraries was published in 2014 and was a thoughtful compilation of information related to the general technologies required for the synthesis and selection of DELs. This book undeniably contributed to the tremendous increase in DEL activity that has been observed over the past 8 years and is responsible for bringing many newcomers to the area. However, the DEL field has matured considerably over the 8 years since the Handbook of DNA-Encoded Libraries was published. The general scientific areas that DEL was built around have now evolved into finely tuned DEL-specific versions. In this Topics in Medicinal Chemistry volume, leaders within the DEL-field review these discrete areas of modern DEL science. We hope that the assembly of these topics will continue to motivate new scientists to utilize DELs as a discovery engine. At the same time, we believe that this book will be highly relevant to experienced DEL practitioners who may have less familiarity with some components of the DEL process.

The chapters that comprise this volume are roughly intended to follow the flow of the DEL pipeline from "concept to clinic." The astrophysicist Carl Sagan said: "Without understanding the past you cannot understand the future." Thus, Chap. 1 by Kodadek and Satz provides insight into the pre-DEL era defined by combinatorial chemistry and pre-DEL barcoding strategies for combinatorial libraries. They detail some of the challenges associated with combinatorial libraries that provided a major driving force for the conceptualization of DELs.

The encoding of chemical information is the hallmark of DEL technology. Encoding can be deployed in two basic ways: to solely record the synthesis information or additionally as a means of driving the chemical reactions to produce the encoded products. In Chap. 2, Brunschweiger and Scheuermann describe the various encoding strategies that have been developed for DELs.

Synthetic chemistry has always been a focal point of DEL technology given that the ability to generate large numbers ($\sim 10^9$) of structurally diverse compounds is correlated to the capability to modulate a wide spectrum of biological targets. However, the reactions used to construct DELs are constrained by the requirement for mostly aqueous conditions and the presence of a DNA oligonucleotide which can diminish reaction efficiency. Rather than providing an exhaustive list of DEL-compatible chemical reactions, which previous reviews have achieved, in Chap. 3, Waring and coworkers describe the development of general chemical platforms and mechanistic genres that can enable the synthesis of novel DELs with high fidelity

The preceding chapters on encoding strategies and DEL-chemical reactions address questions surrounding how to construct DELs but an equally important set of questions to consider concern which DELs should be made. Based on the analysis of years of previous DEL syntheses and selection outcomes, in Chap. 4, Franzini and Zhang turn a lens toward the generation of DELs having more drug-like properties. They point out that high-quality libraries are brought into existence based on parameterized decisions around physicochemical properties of building blocks as well as considerations of their structural features. The judicious selection of chemical reactions to assemble the building blocks is also an important aspect.

Computational science has long had an important role in chemistry and drug discovery; however, DELs generate unique opportunities for computational deployment. In Chap. 5, Zhu et al. describe the various aspects of computational science that have been applied to DEL technology within Pfizer. They point out how cheminformatics can enable the selection of appropriate building blocks to generate the best library outcomes. Furthermore, they provide deep insight into computational workflows that allow for best practices for analyzing sequencing data from selections leading to high-quality off-DNA hits.

In the early development of DEL, there was a major emphasis on innovation in the areas related to constructing the libraries (i.e., chemistry and encoding). While concerted efforts continue in these areas, recent years have brought about an increase in new technologies for selecting DELs against important targets. In Chap. 6, Gui and Li review the topic of selection strategies. While classical DEL selection strategies have focused mainly on binding affinity, newer strategies which can reveal functional binders have now emerged.

For therapeutic research, the ultimate measure of the utility of a new screening platform is whether it leads to clinical candidates. In Chap. 7, Marcaurelle and coworkers from GSK document the continuum in going from DEL selection to clinical candidate. While the preceding chapters in this book provide depth in critical DEL-specific components, the chapter by Marcaurelle et al. ties all of the chapters together to demonstrate the breadth and rigor that is required for clinical lead development. They demonstrate that DEL technology passes the ultimate test leading to clinical candidates. Moreover, they suggest that the assembly of technologies outlined in this book are poised to generate scores of novel drugs in the future.

Finally, given the many impressive advancements outlined in this book, it would be hard to imagine that DEL technology has reached a plateau. This is underscored by the last chapter which is an account by Barry Morgan of HitGen. Morgan, a key driver of DEL technology as it is widely practiced today, provides a historical account of the field and a perspective on where it may go in the future.

In closing, this volume on Topics in Medicinal Chemistry on DNA-Encoded Libraries is intended to encourage the application of DEL technology toward the interrogation of new biological targets. Furthermore, the volume should serve to identify areas where further innovation is warranted toward increasing the efficiency of DEL technology. The fulfillment of either of these objectives will undoubtably lead to new medicines for treating human disease.

Dortmund, Germany
Houston, TX, USA
August 25, 2022

Andreas Brunschweiger
Damian W. Young

Contents

Top Med Chem (2022) 40: 1–40
https://doi.org/10.1007/7355_2022_145

Published online: 11 October 2022

A History of Selection-Based High-Throughput Screening Technologies for Hit Identification

Thomas Kodadek and Alexander Satz

Contents

Abstract DNA-encoded libraries (DELs) are now used extensively in both academic and commercial laboratories for the discovery of protein ligands. DEL technology is attractive because it allows very large compound libraries to be screened at a low cost. In this chapter, we consider the origins of DELs in the broader picture of combinatorial chemistry and trace many important developments in the pre-DEL era that shaped current thinking in this area. The strengths and weaknesses of traditional high-throughput screening (HTS), the previously dominant technology for hit finding, are considered. A brief history of the development of both encoded and non-encoded combinatorial libraries is presented, with a particular focus on the critical issue of the "chemical space" covered by the libraries made in

T. Kodadek (✉)
Department of Chemistry, UF Scripps Biomedical Research, Jupiter, FL, USA
e-mail: kodadek@ufl.edu

A. Satz
WuXi AppTec, WuXi HitS, Planegg, Germany

the pre-DEL era and how technological limitations shaped their design. We also describe genetically encoded peptide libraries and methods to screen them.

Keywords Combinatorial chemistry, DNA-encoded library, High-throughput screening, Ligand, Phage display

The basic ideas underpinning DNA-encoded library (DEL) technology were first enunciated by Brenner and Lerner in 1992 [1]. Soon thereafter came the first experimental demonstration of the synthesis and screening of a DEL by Gallop and co-workers [2]. While these seminal papers are properly hailed as the dawn of the DEL era, it is important to recognize that they were part of a wider explosion of interest in combinatorial chemistry in the late 1980s and early 1990s. Indeed, some of the key concepts that made DEL technology possible were developed for libraries not encoded by DNA. In this chapter we revisit some of these advances and attempt to place the development of DNA-encoded compound library (DECL) technology in the context of the broader effort to develop effective methods to screen large numbers of compounds for protein (and now RNA) ligands. (Note the term DECL is narrower than DEL, as it solely describes DNA-encoded libraries of small molecules and not peptides).

1 1990–1991: Combinatorial Chemistry's "Big Bang"

One could argue that the combinatorial chemistry revolution began in earnest in 1990. That year saw the first reports of the synthesis of massive combinatorial libraries of peptides and methods to screen them for ligands to a target protein (TP).

One set of papers reported the creation and screening of phage-displayed combinatorial peptide libraries [3–5]. Five years earlier, Smith had described the insertion of foreign DNA sequences into the gene encoding the gpIII coat protein of bacteriophage M13, resulting in the display of the polypeptides encoded by these DNA sequences on the surface of the phage (five copies of gpIII are displayed on the "nose" of M13) (Fig. 1). He further demonstrated that the displayed polypeptide could be recognized by an antibody, providing a way to enrich the phage displaying this protein from a mixture of viruses. This suggested that it should be possible to create large libraries of peptides displayed at the N-terminus of gpIII by cloning into the gene a mixture of oligonucleotides with the general formula $(NNK)_n$, where N was any of the four deoxynucleotides and K represents G or T (to reduce the frequency of stop codons in the mixed sequence). The protocols worked out by the groups of Smith, Devlin, and Dower indeed created vast libraries of phage displaying a single peptide sequence at the N-terminus of gpIII. They also showed that by panning the phage library against immobilized protein targets (antibodies or streptavidin), phage displaying peptide ligands for these targets could be highly enriched. By carrying out multiple rounds of panning and amplification through

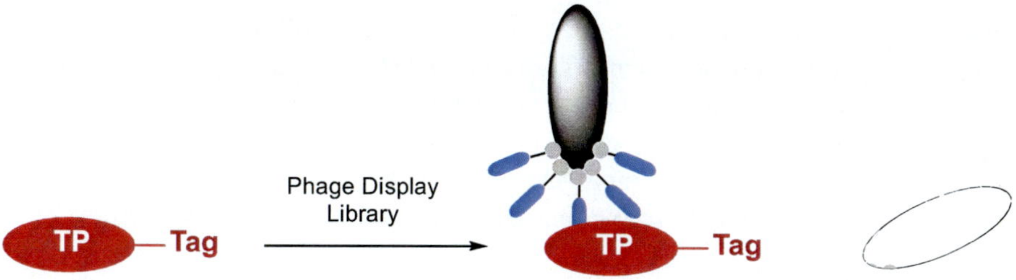

Fig. 1 Peptide ligand discovery using phage display. Bacteriophages that display five copies of a peptide library (colored ovals) fused to the terminus of the gene V protein are incubated with a tagged target protein (TP). The TP is pulled down and the precipitate washed thoroughly. Bound phage are released and mixed with E. coli, resulting in amplification of the bound phage. The process is repeated until phage display ligands to the TP are highly enriched

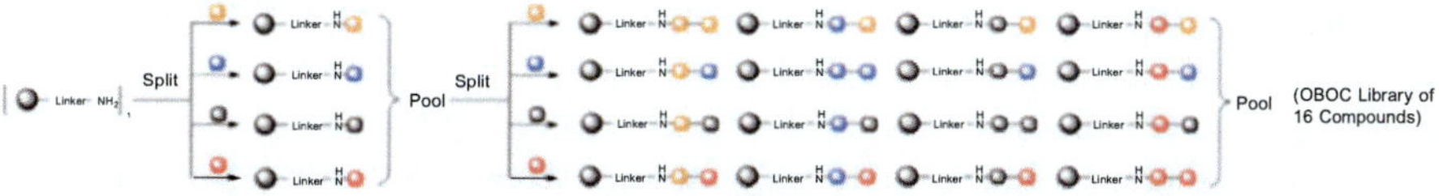

Fig. 2 Schematic representation of the split and pool strategy for combinatorial library synthesis. A simple exam is shown where all possible combinations of a two-mer containing four possible building blocks at each position are synthesized. The colored squares represent building blocks, for example amino acids for synthesis of peptide libraries

infection of E. coli, a modest number of distinct clones were obtained that displayed peptide ligands for the target. The sequences of these peptides were deduced by Sanger sequencing of the relevant gene region in individual phage colonies.

These phage-displayed peptide libraries were arguably the first DELs. They introduced the critical concept that one could screen a vast mixture of molecules against an immobilized target protein in a single tube, so long as the molecule retains a stable physical linkage to an encoding DNA. In phage display, the structure of the virus supplies this linkage.

The other major advance, reported the following year in papers from Houghten et al. [6] and Lam et al. [7] was the application of the split and pool strategy by Furka et al. [8] to the synthesis of combinatorial libraries. Figure 2 shows the basic concept of the split and pool strategy in the context of solid-phase synthesis for the simple example of creating all 16 combinations of a dimeric molecule built from four different building blocks. A collection of beads displaying an amino group is split into four tubes. For peptide synthesis, the amine is acylated with a different N-protected amino acid in each tube. Afterwards, the beads are pooled, and the protecting group is removed. The beads are then split again, and the process is repeated. After a final pooling and deprotection, the population consists of beads displaying 16 different peptides. Critically, each bead displays many copies of a single peptide and thus these are called one bead one compound (OBOC) libraries. To make an analogy to phage display, each bead is the functional equivalent of a single phage clone. If 10 different amino acids are employed at each split and six rounds of split and pool are carried out, the result will be 1,000,000 (10^6) hexapeptides. Split and pool synthesis is unrivaled in its ability to create huge molecular diversity with only a modest synthetic investment.

Of greatest relevance to the eventual development of DEL technology, Lam demonstrated that an OBOC library can be screened by incubating it with a labeled target protein, then manually recovering the beads that retain the label [7].

2 Exploring Encoded Library Strategies for Synthetic Libraries

After the Big Bang of 1990–1991, it was tempting to suggest that the tidal wave of combinatorial firepower would swamp all other approaches to drug development and establish itself as the dominant technology for some time to come. The reality was somewhat less grandiose. Clearly, the idea of creating hundreds of thousands to millions of compounds quickly and being able to screen them cheaply is alluring. But major limitations of the technology were immediately clear. We summarize below some of these challenges and efforts to surmount them. Due to space limitations, the discussion below cannot be comprehensive. Instead, we highlight innovations and advances that are most relevant to the development of DEL technology and that might yet play a role in future combinatorial chemistry efforts. We apologize to authors that contributed significantly to this effort but are not mentioned here.

A seminal issue was the structural diversity of the libraries, which continues to be an important consideration for DEL design today. In what now might be called "classical high-throughput screening (HTS)," compound collections that covered chemical space broadly were screened in some type of functional assay for agonists or antagonists of a target protein. The goal was to identify a scaffold suitable as a starting point for a drug development program. The expectation was that once such a scaffold was identified, the medicinal chemists would create hundreds of derivatives of that core molecule to provide a true lead. In stark contrast, the combinatorial chemistry libraries created in the early '90s were limited to peptides, which have many drawbacks as drug candidates. As will be discussed below, ribosome-assembled peptide libraries have come a long way since their inception and powerful methods exist that allow a significant expansion of the accessible chemical space. But there remains a major advantage to tackling intracellular targets with cell permeable, more drug-like small molecules.

In principle, libraries created via a synthetic split and pool approach could be constructed from almost any type of building block. The catch is that the split and pool process delivers all of the beads in a single tube, making it impossible to track what compound is on what bead. Thus, the structure of screening hits must be determined de novo, demanding a sensitive and general analytical technique to do so. Edman sequencing or tandem mass spectrometry (after release of the peptide from "hit beads") works well for this application [9, 10]. Unfortunately, no such general solution is available for structural characterization of most non-peptidic structures. There is not nearly enough compound on the resin to support NMR analysis, which is relatively insensitive, and no other technique is capable of routinely solving complex structures.

Thus, we first discuss the landscape of methods available to chemists interested in small-molecule drug development in the early days of combinatorial chemistry prior to the advent of DECL technology.

3 High-Throughput Parallel and Combinatorial Synthesis (Without Encoding)

Project teams seeking bioavailable small-molecule drugs generally begin the discovery process with a protein target in mind, and afterwards seek hit molecules to begin medicinal chemistry optimization. Hit molecules may be acquired from numerous sources including mining of public information (journals and patents), in silico screening of cheminformatics databases, high-throughput screening (HTS), fragment screening, and DNA-encoded chemical library (DECL) selections [11, 12]. Each of these methods has benefits and shortcomings (Table 1).

Notably, HTS has played a key role in hit discovery for several decades, and two key drawbacks are worth discussing. First, it's extremely expensive to build an HTS compound collection, particularly if you wish to stock the library with proprietary compounds. This challenge is a key aspect of why the pharmaceutical industry has had a historic interest in combinatorial and high-throughput parallel chemistry. It also partly explains the interest that both academic and industrial labs have in DECL technology [13]. Only the largest pharmaceutical companies are well positioned to create the best HTS compound collections, possibly putting all others at a competitive disadvantage [14]. Second, once an HTS compound collection is created, it's extremely expensive to alter its contents. In other words, the HTS compound collection is largely static. This can be problematic as the pharmaceutical industry seeks to investigate novel chemical space outside of what currently exists in their compound collections (for example; macrocycles, molecular glues, bifunctional degraders, covalent inhibitors, and so on) [15].

We would also like to point out an important difference between how HTS and DECL libraries are interrogated. While the terminology is not fully agreed upon by all practitioners (the terms "screen" and "selection" are often used interchangeably), an HTS is a "screen," where each compound in the library is individually assessed. In contrast, a DECL undergoes a "selection," where all the library members are contained in a single mixture, which is incubated with the target. Obviously, the latter is more cost and time efficient, although generally limited to only detection of direct binding (biochemical assays are generally not possible).

Table 1 Small-molecule hit finding methods employed in the pharmaceutical industry

	Pros	Cons
Mining public information	• Rapid jumpstarting of project • Inexpensive • Possibly potent & drug-like chemical matter	• Lack of competitive advantage & risk of being "scooped"
In silico screening	• Inexpensive & fast	• Requires pre-existing ligands and structural information • Unlikely to discover hits with novel binding modes
High-throughput screening (HTS)	• Novel chemical matter and/or binding modes • Compatible with a wide range of screens including biochemical, cellular, and competitive binding	• High cost • Only as good as the compound collection (hundreds of thousands to millions of compounds) • Compound collection is difficult to acquire and static • Many assays are sensitive to trace impurities • Assay development is time consuming
Fragment screening	• Less expensive than HTS • Small compound collection (thousands) • Focus on finding hits with very low molecular weights • Direct binding assays allow for discovery of novel binding sites	• Discovered hits are weakly potent and require expensive and time-consuming medicinal chemistry optimization
DNA-encoded libraries	• Less expensive than HTS • Compound collection can rapidly evolve • Little or no assay development • Interrogated by a "selection" instead of a "screen"	• Only compatible with a direct binding selection • Hit confirmation is expensive
One-bead-one-compound DNA-encoded libraries	• Compound collection can rapidly evolve • Potential to conduct a wide range of screens including cellular functional assays	• Assay development is time consuming • Hit confirmation is expensive • Unproven technology

4 What Is a Hit Molecule?

As the purpose of making large collections of compounds and screening them is to find hits, it would make sense to briefly discuss what a hit molecule actually is [16]. The pharmaceutical industry generally thinks of a hit as a chemical starting point for a drug discovery project. The criteria for a molecule to be considered a hit vary widely and are (or should be) defined prior to any screening. However, there are a few rough guidelines, particularly when pursuing a bioavailable small-molecule drug. The molecule should be measurably potent (usually $<10\ \mu M$ IC_{50} or K_D) in a primary assay of choice (usually biochemical or direct binding). Of course, the more potent, the better. Additionally, the physical properties of the potential hit should meet (or come close to) Lipinski's rule of 5 [17]. The more challenging the target (i.e., the less druggable), the less strict these criteria become. On the other hand, a project team working on a readily druggable target may require cellular activity, observed selectivity against a challenging off-target, and/or high potency and ligand efficiency.

After a hit molecule is found, it must be optimized by medicinal chemists. A waypoint between a "hit" and a drug is a "lead." Again, criteria for a lead molecule vary for different projects, but it generally implies optimization of potency and physical properties such that the compound is appropriate for in vivo testing. There is much attrition when optimizing hits into leads, and the effort is largely trial and error. And so, project teams desire not just one, but a number of structurally distinct hit molecules to begin with. And so, we can appreciate the importance of a robust hit finding platform. Project teams that discover numerous structurally diverse hits for their target, enjoy a competitive advantage over those that don't. Obviously, larger and more compound collections provide a better chance to find numerous hit molecules.

The medicinal chemical optimization process often proceeds rationally, using known guidelines for bioavailability (Lipinski's rule of 5) and structure-activity relationships (SAR). In brief this means that a single (or handful) of molecules are synthesized, each containing a single or small alteration to its chemical structure. The potency of each new analog is then measured, and in this manner a relationship between chemical structure and activity is slowly (and unfortunately expensively) deconvoluted. To investigate SAR, each batch of new compounds to be synthesized is carefully chosen (altering one chemical variable at a time), and thus each molecule generally requires custom synthesis. The pharmaceutical industry has a long running interest in the potential of combinatorial or high-throughput chemistry to decrease the cost of the hit-to-lead process.

5 Relationship Between Assay and Compound Purity Requirements

A key, but not often discussed, aspect of building a compound collection is the intended assay and the compound purity thus required. HTS boasts access to a very wide range of biochemical, competitive binding, and cellular functional assays. However, many of these assays are sensitive to trace impurities. Indeed, perhaps all HTS assays employ extremely low concentrations (nM) of target protein, making them susceptible to trace (but highly potent) reagents and by-products. The result of potent trace impurities can be catastrophic to some screens and the hopes of the project team, resulting in a large amount of wasted effort [18]. Most notorious is the effect of trace metals on particular HTS assays, but similar results arise from

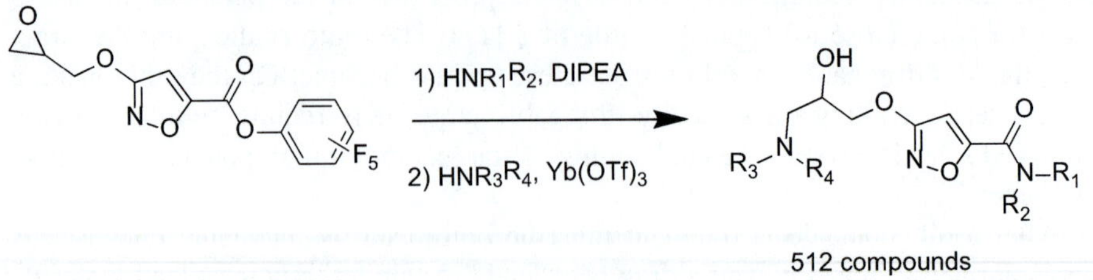

Fig. 3 Combinatorial synthesis of a library mixture. Employing split and pool synthesis, a mixture of 512 desired products was generated. Final products were not purified. The mixture was then interrogated by ASMS, and potential hit molecules identified by mass spectroscopy

degradation of compounds to give reactive intermediates. Unfortunately, with few exceptions, none of these disappointing results are published.

Due to the above impurity issues, the pharmaceutical industry takes a certain approach when it comes to building compound collections; that they must maintain the highest compound purity possible. Unfortunately, HTS compound collections are known to degrade over time, with roughly half the molecules possessing $<60\%$ purity after repeated freeze thaw cycles [19]. Additionally, compounds placed into large collections are derived from sources and chemistries too numerous to count, and thus every compound in the collection may contain a unique and difficult to identify contaminant.

Not all assays employ a low concentration of target and a high concentration of small molecule. Flipping this relationship around (a high concentration of protein target and a low concentration of small molecule) alters the requirements for compound purity and is a major factor in the success of DECLs [13]. However, such assays can also be used to interrogate HTS compound collections, as is illustrated by the direct binding assay, affinity size exclusion mass spectrometry (ASMS) [20]. In this case, hundreds of molecules taken from the HTS compound collection are purposefully COMBINED prior to the assay. The molecules are incubated with target protein, filtered by size exclusion (protein and bound ligands are eluted, while unbound ligands are retained), and the bound ligands then identified by mass spectrometry. Because the concentration of target protein used in the assay is far greater than that of any single compound being assayed, the presence of a trace impurity, even if highly potent, will not interfere with the assay [21]. Since ASMS is compatible with mixtures of compounds, it certainly makes sense to combine the assay with combinatorial chemistry. Indeed, small libraries of hundreds or thousands of compounds have been produced as mixtures (Fig. 3) and then assayed via ASMS [21]. Presumably, libraries of up to a million novel compounds have been screened against a single target [22].

Unfortunately, ASMS does have some shortcomings, which perhaps has prevented more widespread use of the technology as a screening platform. First, the greatest limitation of HTS is the static nature of the compound collection. And so, the use of ASMS to interrogate an existing HTS compound collection fails to tackle the most pressing issue for HTS, access to new chemical matter. Alternatively, as demonstrated in Fig. 3, it is possible to combine an ASMS screen with combinatorial chemistry. However, it's unclear how practical this approach is, in particular the relationship between product yields and detection, and also the total number of molecules that can be readily synthesized and screened. As reported by Coburn et al. [22], over one million compounds were synthesized and screened by ASMS, which is equivalent to that of a large HTS compound collection. However, we must look closely at the library constructed in Fig. 3 and realize that combinatorial chemistry (and usually high-throughput chemistry) generates large numbers of structurally similar compounds (see Sect. 6 for further discussion).

6 Target Agnostic Libraries

Pharmaceutical HTS compound collections are usually built and continuously supplemented with the products of medicinal chemistry optimization. As mentioned above, hits are progressed to leads through the rational design and custom synthesis of batches of discreet compounds. Each new iterative cycle of compounds is then tested for activity to generate increasing SAR knowledge for the target of choice. Each molecule investigated by the medicinal chemistry team is synthesized in a relatively large scale (preferably >20 mg) and high purity (for instance, >90% as judged by analytical LC). Thus, every molecule generated for the purpose of investigating SAR can also be submitted to the HTS compound collection and be screened against future targets for years (decades) to come. The compound collections generated from decades of medicinal chemical optimization projects naturally consist of drug-like molecules and tend to be structurally diverse, as they are derived from an extremely large number of different projects, scaffolds, and chemistries.

As previously mentioned, the building of a new HTS compound collection via custom compound synthesis is extremely expensive, and so the pharmaceutical industry has always desired methods to reduce this cost. Large companies wish to supplement their existing collections, and small companies wish to generate new collections with a competitive advantage. A primary technique to lower the cost of compound synthesis is high-throughput parallel synthesis. While such synthesis can be conducted in a variety of different ways, a common workflow is the synthesis of a rigid scaffold which can then be appended with a variety of different building blocks (Fig. 4) [23]. As was once common, to further reduce costs, synthetic methods which didn't require column purification were desired. And so, as illustrated in Fig. 4, purification was instead achieved by acid/base dissolution or washing, with purities of observed final products >90%. Using a related logic, a library of >1,600 benzodiazepines was synthesized employing solid-phase resins [24, 25]. Again, the reaction conditions were optimized such that presumably reasonable purities could be achieved without requiring column purification. The benzodiazepine library was screened against a number of future targets, including using biochemical assays.

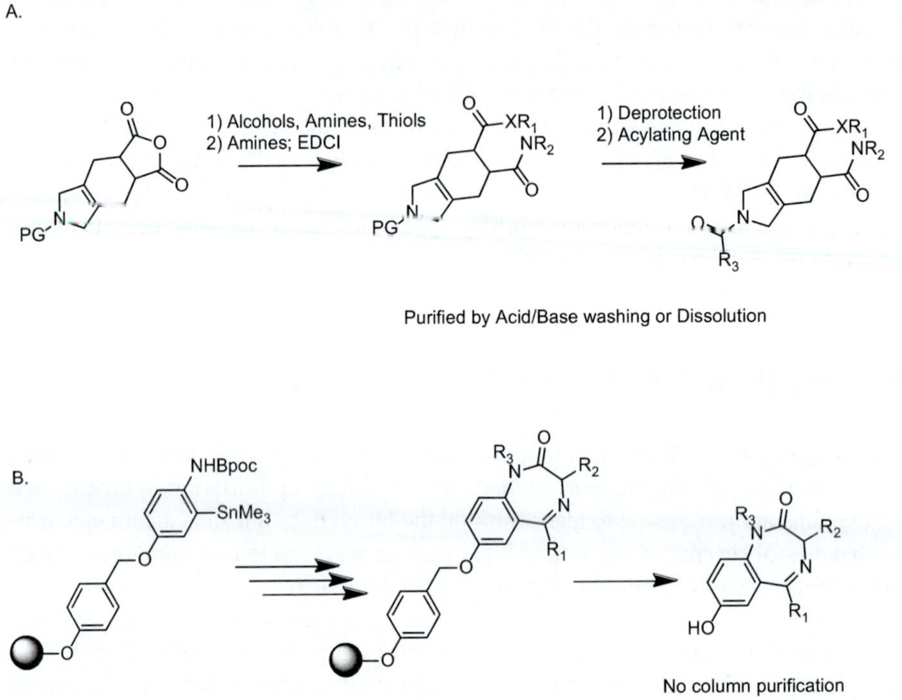

Fig. 4 High-throughput parallel synthesis. The compounds shown were synthesized in parallel (not as mixtures). However, final products were not purified by column chromatography prior to being employed in assays. (**a**) Solution-phase synthesis. (**b**) Solid-phase synthesis

High-throughput parallel synthesis as mentioned above was published in the 1990s and has generally fallen out of favor in the pharmaceutical industry for two main reasons. First, the molecules were not individually column purified and characterized. Yet, these libraries were still sometimes used to supplement existing HTS compound collections and saw use in numerous biochemical and cellular assays. As already mentioned, HTS assays are often sensitive to trace impurities. Of course, with proper controls, it may be possible to prudently use such impure libraries in biochemical assays, such as demonstrated by Weber et al. [26]. In this example successive libraries were produced using a 4-component Ugi reaction without column purification of final products. Key to this workflow, the library was produced for a specific biochemical assay (a thrombin chromogenic assay), and control products were thoroughly tested for false positives and other issues. Of course, this workflow implies that library products cannot simply be added to the HTS compound collection. On the other hand, as previously mentioned, direct binding assays which employ a relatively high concentration of target protein to ligand are generally amenable to interrogation of less pure libraries. One example was reported by Houghten where a solid-phase library of 248 short peptides were generated, cleaved from resin (and directly used without further purification), the peptides absorbed to wells in microtiter plates, and then investigated for antibody binding (ELISA) [6].

The current approach to high-throughput parallel synthesis is to purify each compound produced, such that these compounds can be added to HTS compound collections and screened against any future target. Towards that goal, effort has been focused on generating automated synthetic and high-throughput purification laboratories [27, 28]. In these instrument intensive chemistry labs, the goal is to reduce the cost per compound synthesized, while maintaining a high level of compound purity and characterization. Unfortunately, such efforts are still relatively expensive and lead us to the second drawback of high-throughput parallel chemistry addressed below.

It's challenging (perhaps impossible) to quantify the difference in the value of compounds produced via combinatorial methods versus an equal sized well-manicured HTS compound collection. However, we postulate that the former will be less productive than the latter in most cases. For instance, DECL routinely conducts selections employing billions of combinatorically generated molecules and may be considered roughly equally productive compared to HTS [13]. And so, perhaps orders of magnitude more combinatorically generated molecules are required to be equally productive as a compound collection where the molecules are derived from varied sources and chemistries. Another way of looking at this is a scaffold-poor combinatorial library is unlikely to provide a hit that will eventually result in the development of a potent lead unless one gets very lucky, and one's protein target happens to be particularly well suited to one of the few scaffolds in the library.

Burke and Schreiber proposed a method to ameliorate the above shortcoming by employing smartly designed chemical reactions to convert a single common substrate into products with varied skeletal cores (Fig. 5) [29]. The general approach has been termed diversity oriented synthesis (DOS). In this manner, unlike earlier iterations of parallel chemistry, starting with a single scaffold doesn't result in every product being structurally similar (as judged by the eye of a medicinal chemist). This approach can be used to build a novel HTS compound collection and has been employed to discover novel anti-malarial hit molecules following a phenotypic HTS of 100,000 compounds [30]. The relative productivity of screening molecules acquired from historic medicinal chemistry SAR, traditional parallel chemistry or combinatorial chemistry, or DOS, remains the subject of debate.

7 Focused Libraries

The discussion above has primarily focused on the generation and screening of target agnostic libraries. In these cases, libraries of compounds are generated, "placed on a shelf," and then interrogated as needed in years to come. The main goal in generating such libraries is to produce structurally diverse collections of compounds possessing lead- and drug-like physical properties (sometimes referred to as "prospecting

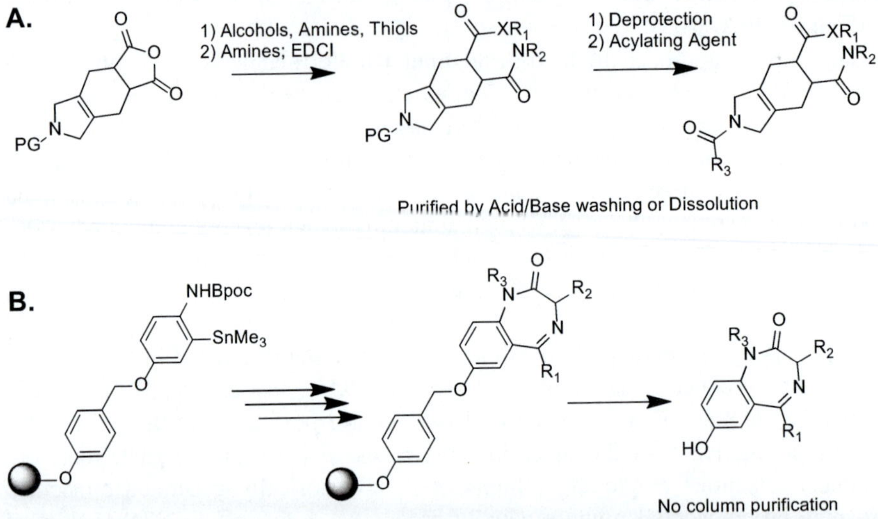

Fig. 5 An example of a strategy to generate skeletal diversity from a common precursor. The furan precursor was subjected to various conditions to produce different scaffolds shown as products. The type of chemistry employed was encoded using haloaromatic GC tags. *BB* Building block. *Ar* aromatic unit

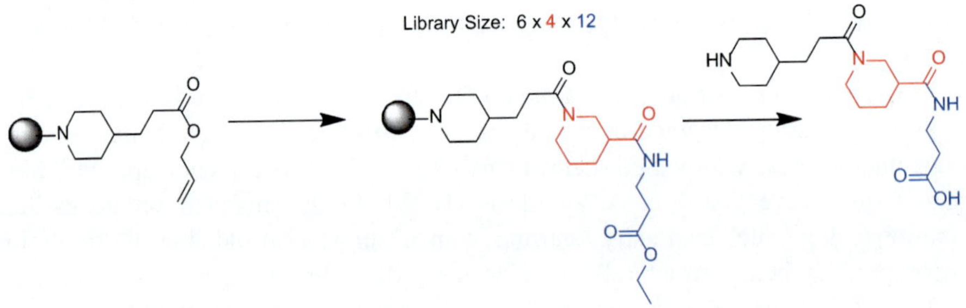

Fig. 6 Use of high-throughput parallel chemistry to generate a focused library. Optimization of the potency of a parent molecule via the synthesis of a small library of analogs. The library was comprised of three cycles of chemistry (shown in black, red, and blue). The parent molecule is shown in the figure. Some analogs were assayed without employing column purification

libraries") [31]. An alternative is to synthesize a "focused" library. The purpose of a focused library is to support the hit-to-lead medicinal chemistry effort. The challenges faced when synthesizing a focused library as part of hit optimization are quite different from those when building target agnostic libraries. Typically, a hit molecule lacks potency, and the purpose of the focused library is to optimize the hit more quickly as compared to traditional medicinal chemistry SAR efforts. One example is provided by Hoekstra et al. where >200 analogs of an initial hit molecule (a fibrinogen receptor antagonist) were synthesized on solid phase (Fig. 6), tested for in vitro activity (without column purification of the final products), and an analog with 25-fold greater potency discovered [32].

Focused libraries are still commonly employed to enhance hit optimization. However, the current trend is to column purify each product (usually preparative LC-MS), synthesize relatively small libraries (10–50 products), and design and build the libraries in an iterative manner (i.e., use SAR to drive rational library design) when possible. Indeed, the difference between traditional medicinal chemistry and the production of small libraries is sometimes not obvious. One of numerous such exemplars is reported by Cavallaro et al., where a moderately potent hit (an antagonist of CC Chemokine Receptor 1, found by HTS screening of a proprietary compound collection) is optimized by making several relatively small libraries of compounds [33]. The greatest challenge in such library construction is the time required to design and test library synthetic pathways. Traditional medicinal chemistry optimization can proceed rapidly compared to library validation, and it can happen that the direction of the SAR advances more quickly than library design and synthesis. In such cases, library production may lag SAR knowledge, and resulting library products may end up of little value to the project team. Thus, current focus is on the nimble production of very small libraries in close cooperation with developing SAR. Alternatively, some hit molecules discovered by HTS efforts are not synthetically amenable to library production. The report by Cavallaro et al. [33] illustrates a common theme of focused library production, a reliance on simple and robust reactions including acylation, sulfonylation, and carbamoylation of an accessible advanced intermediate.

Lastly, we note that the generation and screening of focused DECLs has been reported [34–36]. The interest in focused DECLs does appear larger in academic rather than industrial labs. This is likely due to academic labs being able to pick and choose their targets, having more time to synthesize and screen the focused DECL, and that generating a focused DECL generally requires fewer building blocks, and thus has lower labor and reagent costs.

8 Encoding Techniques

As detailed above, parallel synthesis approaches to compound generation are expensive and labor-intensive. Therefore, it is not surprising that many chemists were reluctant to abandon the scale and efficiency of split and pool synthesis. They explored two general approaches to expand the chemical space of such libraries. The first was to develop encoding techniques, while the second was to invent new types of oligomers that could be structurally characterized de novo using sensitive analytical techniques such as MS/MS. Almost all this work involved the use of OBOC libraries or related techniques.

As mentioned at the beginning of this chapter, Brenner and Lerner proposed in 1992 the idea that the sequence of building blocks and reactions used to create a combinatorial library by split and pool synthesis could be recorded by the sequential addition of DNA oligonucleotides to the molecules [1]. A year later, Gallop and colleagues at Affymax reported the first experimental demonstration of a DEL [2]. This was a library of peptides (containing both D and L amino acids, which would not have been possible with phage display) constructed by solid-phase synthesis on tiny (10 μm in diameter) amine-functionalized polystyrene beads. The beads were prepped by first reacting them with a mixture of an N-protected amino acid and an omega-hydroxy carboxylic acid. The amino and hydroxyl groups served as the starting points for the library peptides and a single-stranded encoding molecule, respectively. First, the invariant segments of a single-stranded oligonucleotide were synthesized using multiple cycles of phosphoramidite chemistry. Split and pool synthesis was then employed to create both the peptide chain and the encoding single-stranded oligonucleotide sequences, with both an amino acid and a deoxynucleoside addition being done at each split, taking advantage of the chemically orthogonal protecting groups on each unit. The synthesis was finished by construction of the remainder of the invariant section of the encoding tag followed by deprotection of the peptide. Interestingly, the authors were forced to employ 7-deaza-2′-deoxyadenosine in place of the canonical adenosine unit since this analog is more resistant to depurination during the trifluoroacetic acid (TFA)-mediated deprotection of the peptide side chains. They also did not use guanine nucleotides. This type of library is called a one-bead two-compound (OBTC) library, since two different compounds are created on the same bead. The OBTC concept is key for the creation of encoded libraries and has been used many times subsequently (vide infra).

The library was screened by incubating the beads with a fluorescently labeled monoclonal antibody. After thorough washing, beads displaying peptides that retained the antibody were isolated using a flow cytometer (Fig. 7). This method is a significant improvement over manual bead picking and a variant of it is still used today to screen bead-based DELs [37–39].

It is ironic that this first encoded library of any kind used DNA as the encoding material. Yet 17 more years would go by before the first report of a truly practical DNA-recorded synthesis and screening platform appeared! [40] This is because deep sequencing technology was unavailable in 1992-3. Individual hit beads had to be collected in separate tubes so that the encoding tags could be PCR amplified and sequenced using the Sanger method. There was no way to deal with mixtures. Given this limitation, it was not obvious at the time that DNA recorded synthesis was superior to other encoding schemes. After all, DNA is relatively fragile chemically. Its use as a recording tag precluded employing many types of reagents that are used commonly in small-molecule synthesis, such as strong oxidizing or reducing agents, acids, and bases.

One of the earliest alternative schemes was reported by Still and colleagues in 1993 [41]. They created a collection of 18 tags with the general structure shown in Fig. 8. These tags could be attached to amine-displaying resin through their carboxylic acid group and released from beads when desired through photolysis. Halogenated hydrocarbons were chosen because they were readily separable by gas chromatography and could be detected very sensitively using electron capture

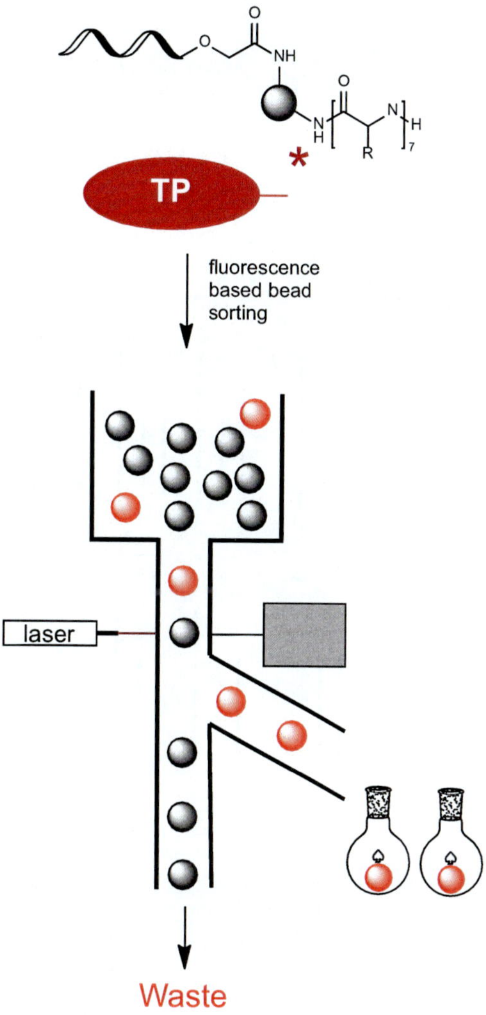

Fig. 7 Schematic view of FACS-based screening of an OBOC DEL. Beads displaying a peptide and an encoding DNA were incubated with a fluorescently labeled target protein. After washing, the small (10 μm) beads were passed through a FACS gated to collect those that displayed a high level of fluorescence. The encoding tags on these beads were amplified and Sanger sequenced

(EC) technology. Indeed, the material from a single 50 μm bead could easily be characterized.

They employed this technology to create an encoded library of 117,649 peptides with the general structure NH2-XXXXXXEEDLGGG using seven different amino acids at the six variable positions. The glycines were included to provide a linker separating the peptides from the polystyrene matrix and the other invariant residues were present to ensure that the library contained a peptide sequence known to bind an anti-Myc antibody that was employed as a model target.

The encoding strategy relied on a three-bit digital barcode. They chose three of the 18 unique tags to encode each position. For each amino acid, some combination of the tags was added to a small percentage of the molecules on the resin. For example, in one split, they might add tags T1 and T2, but not T3. This bead would

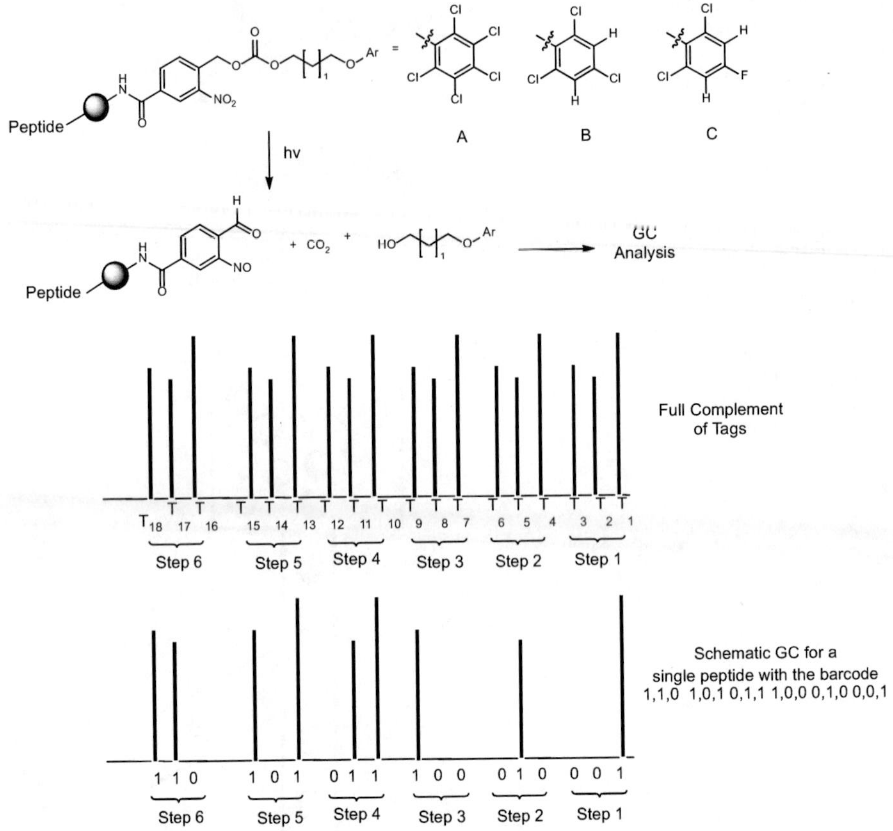

Fig. 8 A haloalkane encoding system developed by Still and colleagues. At each step in the synthesis of a combinatorial library, some combination of three haloaromatic tags are connected to a small percentage of the amine groups on the resin. The presence or absence of each of the three tags provides a digital barcode (1,0,0; 1,0,1; etc.) that uniquely encodes up to eight (2^3) units at each position (seven in practice, since the code 0,0,0 is not employed). Each tag is appended to the bead via a photocleavable linker. When a bead is chosen as a screening hit, it is photolyzed and the released tags are analyzed by gas chromatography to determine the "fingerprint" of encoding tags on that bead

thus acquire the barcode 1,1,0. A bead modified with tag T2, but not T1 or T3, would have the barcode (0,1,0). This provides eight (2^3) encoding units for each position. The tags used for encoding steps 1–6 in the synthesis had different molecular masses, allowing them to be easily separated by GC. Since the encoding tags were employed at $<1\%$ stoichiometry, the number of truncated molecules was small.

The library was screened against the anti-Myc antibody. Beads that retained the target were visualized by staining with a labeled secondary antibody. The stained beads were picked manually, placed in separate tubes, then photolyzed. The GC/EC trace then revealed the predicted structure of the peptide ligand (Fig. 8).

Several aspects of this clever encoding strategy were unique. It did not rely on the synthesis of a polymeric encoding tag that would have to be sequenced. It also seemed unlikely that the halocarbons would themselves function as protein ligands, and so the presence of small amounts of these compounds on the bead surface would likely not provide false positives. Finally, and most importantly, these tags are inert to most reactions used routinely by synthetic chemists. Certainly, they are far more robust than single-stranded oligonucleotides. While the Still study employed this technology to make a peptide library, it was clear that this was not a limitation of the approach.

Indeed, in a beautiful 2004 paper, Schreiber and colleagues used this encoding technology to create an encoded library of 1,260 small molecules by split and pool solid-phase synthesis [29]. While the size of the library was small and no screening results were reported, this study was noteworthy in two respects. First, it clearly demonstrated the ability of Still's encoding tags to stand up to various reagents and conditions that would have shredded DNA. Second, these authors used the tags to encode the type of transformation that was carried out, not simply the identity of a building block. This was important because it allowed them to address a critical issue in library design mentioned above: skeletal diversity. They encoded a set of rearrangement reactions that transformed a relatively simple furan precursor into a number of more complex systems (Fig. 5). In addition, stereochemical diversity was also built into this scheme. Interest in chemical strategies to create more skeletally diverse combinatorial libraries remains high [42, 43].

Lam and colleagues reported another clever encoding strategy in 2002 [44]. They also employed the OBTC concept, but in this case with a twist. By this time, TentaGel (TG) resin had become the solid support of choice for the creation of OBOC libraries. TentaGel beads are comprised of a hydrophobic, polystyrene-based core onto which is grafted a thick layer of an amine-terminated polyethylene glycol (PEG). This provides a surface that resists non-specific protein binding, which was a significant problem using simple polystyrene beads or other hydrophobic matrices for protein binding assays.

The authors took advantage of the very different hydrophobicities of the exterior and interior of TentaGel resin to develop conditions where the surface amines could be Fmoc-protected without significant reaction of the interior amine groups. This allows modification of the interior groups with an encoding unit, for example a Boc-protected amino acid. The surface amines can then be deprotected and operated on in some other way. The process is repeated until the OBTC library is created with surface-exposed library molecules and peptide-encoding tags in the protein inaccessible interior of the beads.

Lim and colleagues later reported a similar strategy except they used peptoids (oligomers of N-terminal glycines [45]) as the encoding unit to allow the creation of an innovative library of 1,458 non-peptidic helix mimetics (Fig. 9) [46]. They were able to identify good ligands for MCL-1 and α-Synuclein. Indeed, they demonstrated that the helix mimetic shown in Fig. 9 was a sub-μM ligand for the aggregation-prone A53T α-Synuclein mutant and that it inhibited aggregation, presumably by acting as a molecular chaperone.

The peptide/peptoid-based encoding strategy has the advantage of "hiding" the encoding tag from the target protein by sequestration in the interior of the bead. It

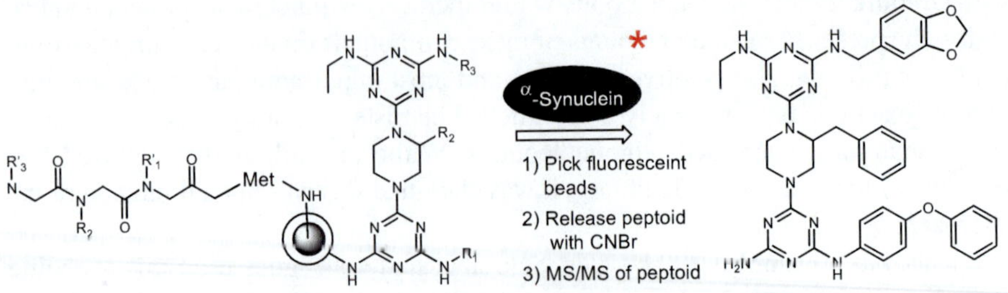

Fig. 9 A one-bead two-compound library of peptide-encoded α-helix mimetics was successfully screened against a-Synuclein, resulting in the isolation of a molecular chaperone (see text for details)

also is highly scalable since hundreds of amino acids and amines are commercially available to use as encoding tags.

9 Split and Pool Solid-Phase Synthesis of Libraries of Non-peptidic Oligomers

Beginning in the late 1990s and early 2000s, several groups began to explore the idea that the split and pool method could be used to create libraries of molecules with better drug-like properties than peptides without having to resort to encoding strategies. Some of this attention was focused on peptoids, which are oligomers of N-substituted glycines [47]. Like peptides, peptoids can be sequenced using Edman degradation, tandem mass spectrometry, or a combination of both with high sensitivity [9, 10].

Peptoids were an attractive first step away from peptides for several reasons. First, they are essentially immune to peptidases and proteases [48]. Second, Zuckermann and colleagues had reported a convenient "sub-monomer" solid-phase synthesis for these molecules [46]. It involves acylation of a resin-displayed amine with activated bromoacetic acid, followed by nucleophilic displacement of the bromide with an amine, allowing the latter to be employed as the diversity element in a split and pool synthesis (Fig. 10) These two reactions, which are essentially quantitative, are repeated until the desired oligomer is created. The excellent efficiency of the sub-monomer method and the ready availability of large numbers of cheap building blocks are perfect for combinatorial chemistry. Third, there was reason to believe that peptoids might be far more cell permeable than standard peptides. Studies of

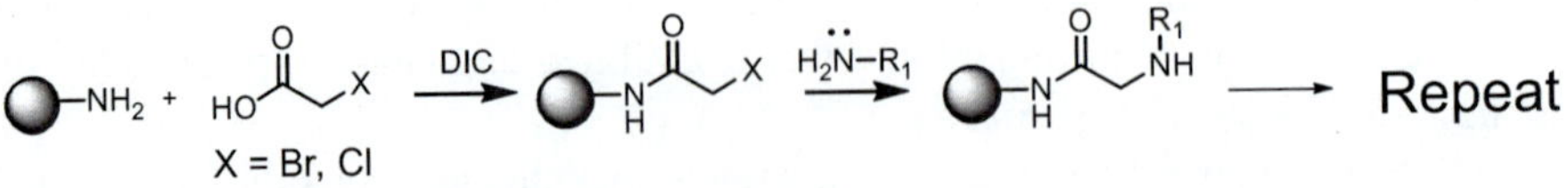

Fig. 10 The "sub-monomer" synthesis of peptoids

certain natural products such as cyclosporin, as well as model peptides, had suggested that replacement of the highly polar N-H amide bond of a peptide with a methyl group greatly increased the cell permeability of the molecule (reviewed in [49–51]). This indicated that peptoid libraries might be a better choice for the discovery of ligands to intracellular protein targets. Indeed, the Kodadek laboratory, using a novel high-throughput cell assay, validated this idea and showed that peptoids of 6–8 residues are indeed highly cell permeable unless they carry negatively charged units in their side chains [52].

The first combinatorial peptoid libraries were reported by Zuckermann and colleagues [53]. They employed a modest library of peptoid trimers in a screen in which mixtures of many hundreds of peptoids were released from the beads and assayed for binding to α-adrenergic receptor using a radioligand displacement assay. The identities of the screening hits were eventually determined using an iterative deconvolution strategy similar to that reported by Houghten and colleagues [6]. The results were highly encouraging. The best hit, CHIR 2279, had a high affinity for the receptor (Ki = 5 nM) and several other ligands with affinities in the nM range were described. CHIR 2279 proved to be highly active in vivo [54].

These reports suggested that peptoid library screening might provide a rapid, efficient, and inexpensive method for the discovery of bioactive probe molecules and perhaps even drug leads. Unfortunately, the high affinity of the CHIR 2279 ligand for the α-adrenergic receptor proved to be the exception rather than the rule for peptoid ligands. In addition, the methods for screening OBOC libraries left much to be desired as one of us (T.K.) learned when we began working with OBOC peptoid libraries in circa 2002. This was in response to a need for probe molecules to support various biological projects that were focused on transcription factors and other proteins considered "undruggable" using traditional Lipinski-compliant molecules [55]. First, we learned that if one wishes to screen on resin, it is critical to use TentaGel beads. Almost any other resin is either too fragile to survive the split and pool process intact or exhibits an unacceptable level of non-specific binding to many protein targets [9]. Unfortunately, for unknown reasons, this resin exhibits an annoyingly high level of autofluorescence that complicates clearly distinguishing screening hits from background. This autofluorescence is less severe at longer wavelengths (>600 nm) and the use of dyes such as Texas Red to label the target protein provided acceptable, albeit far from perfect, contrast between beads that display protein ligands and those that do not under a fluorescence microscope. Several strategies were described by various groups using OBOC libraries to overcome this problem. The Lam group utilized their "two-layer" methodology described above to incorporate a nitrobenzene moiety selectively in the interior of the beads [56]. This acted as a quencher of much of the bead autofluorescence without quenching the fluorescence of a labeled protein captured on the bead surface. Our group (T.K.) turned to the use of quantum dots (QDs) as labeling reagents. QDs exhibit massive Stokes shifts, far in excess of those of typical organic dyes, and are also exceptionally bright. Thus, beads that retained a target protein could be stained with a QD-conjugated antibody and irradiated with UV light, resulting in intense QD fluorescence on the surface of the bead at a wavelength where TentaGel autofluorescence was negligible [57]. Other reports described methods to separate hit beads from the bulk of the library using non-fluorescent methods such as magnetic recovery or colorimetric staining [58].

OBOC DECLs that utilize the Praecis/GSK encoding strategy [40] are now in use in a few laboratories and companies [38, 38, 59–62]. Fortunately, the 10 μm TG beads used to create modern OBOC DECLs have a much lower level of this autofluorescence, but it is not zero. Moreover, some of the molecules in a library can be intrinsically fluorescent. Therefore, strategies to mitigate this background remain relevant.

Another major problem with on-resin OBOC screens is the very high level of false positives. To illustrate the magnitude of this problem, it is instructive to consider a report from Pei and colleagues of a screen of an OBOC library of bicyclic peptides against Tumor Necrosis Factor-α (TNF-α) [58]. They began with a library of about 100,000 beads. The initial target was TNF-alpha affixed to a magnetic particle. ≈ 400 macrocyclic-displaying TG beads that retained this conjugate were separated from the bulk of the library using a magnet. A series of stripping and rebinding experiments followed where protein binding to the beads was marked using different techniques (dye staining and fluorescence). The compounds resident on the 44 beads that survived this protocol were then analyzed for binding to TNF-α in solution, and only six did so to any significant extent. This remarkably high false positive rate of 98.5% (6/400) is sadly not atypical and was also encountered in screens of OBOC peptoid libraries as well.

An extremely important technical advance was reported by Doran et al. in 2014 that ameliorated this problem greatly [63]. They screened multiple copies of an OBOC library against a protein target. For example, using 500,000 beads of a library with a theoretical diversity of 100,000 unique molecules would provide approximately 5 copies of the library. Upon analysis of the hits obtained, they demonstrated that compounds identified on multiple different beads were almost always bona fide ligands, whereas those found on only a single bead were usually false positives that failed to bind the protein well in solution. They discovered the high false positive rate was due, in large part, to the inherent heterogeneity of the number of reactive sites on individual TG beads in any given batch of resin, which displays a Poisson distribution of amine density on individual beads, likely a result of the grafting method used to make the microspheres. This leads to widely varying densities of compounds displayed on the surface from bead to bead. They speculated that a small percentage of the population at the dense end of the distribution acted as a sort of "kelp forest" that was able to trap any protein that entered this extremely concentrated microdomain even if the compound had a poor intrinsic affinity for the labeled target protein. Indeed, this was the basis of the "focus only on redundant hits" strategy. The idea was that the kelp forest beads represent only a few percent of the population, so in a large library it would be unlikely for the same compound to randomly appear in the hit pool on more than one bead unless it is truly a good ligand for the target. While this strategy necessitates screening many more beads than previous protocols, it resulted in enormous savings in time and reagents in the post-screening validation phase of a screening effort.

Fig. 11 A two-color screen for ligands to integral membrane proteins. An OBOC library is incubated with cells that do or do not express the protein target (red ovals) and are labeled with red and green dyes, respectively. After washing, the beads are examined under a dissecting fluorescence microscope and beads that retain only red cells are picked. Typical micrographs of beads that bind no cells, both colored cells and only red cells are shown

Again, this study is mentioned here because it remains relevant to the screening of modern OBOC DELs. Unlike the Gallop study mentioned above, FACS-based screens of OBOC DELs constructed on 10 μm beads make no attempt to separate individual hit beads into different tubes [38, 59]. Rather they are collected as a batch, which is far more efficient. The entire population of encoding tags on these many beads are then amplified and deep sequenced. In order to make use of the redundant library strategy in such screens, Mendes et al. included a bead-specific barcode into the encoding tag by carrying out two rounds of split and pool oligonucleotide addition after all of the chemistry for the DEL synthesis had been completed [38]. This readily allowed compounds that had appeared in the hit pool on multiple beads to be distinguished from those that appeared on only a single bead by inspection of the sequence files.

Another innovation developed in the context of peptoid OBOC libraries that addressed the false positive problem, but also provided other important advantages, was the development of two-color screening (Fig. 11). This work built on pioneering studies of Lam and colleagues, who showed that intact cells could be screened against OBOC peptide libraries to identify ligands for surface-exposed proteins [64, 65]. Kodadek and co-workers carried out a cell-binding screen using a mixture of cells that were identical except that one expressed a high level of the Vascular Endothelial Growth Factor Receptor 2 (VEGFR2) and the other did not [66]. The two cell types were stained red and green, respectively, with quantum dots that were internalized via endocytosis. The mixed cells were exposed to a library of $\approx$ 250,000 hexameric peptoids. After washing, the beads were examined under a low power fluorescence microscope to identify those that had retained only the red-labeled cells, but not the green-labeled cells (Fig. 11). These were presumably highly selective ligands for VEGFR2. Only six such beads were found in the population, whereas >2,000 beads bound both cell types. This ratio attested to the stringency of the two-color screen, which demanded a potential VEGFR2 ligand ignores any other protein on the cell surface. The best of these hits proved to be a 3 μM ligand for the extracellular domain of VEGFR2. Since this receptor is a native dimer, fusing two copies of the ligand together with an appropriate linker gave a low nM ligand with extremely high selectivity for VEGFR2. This dimer proved to be a potent inhibitor of VEGFR2-mediated angiogenesis in vivo [67, 68].

Udugamasooriya and co-workers used this two-color screening technology to identify peptoid ligands that bind selectively to lung cancer cells over normal lung epithelial cells [69] as well as peptoids that bind selectively to cancer stem cells [70]. While the screens were unbiased, these workers were later able to track down the molecular targets of their cell-discriminating peptoids [71]. This elegant work suggests that two-color OBOC screening against cells will serve as a powerful tool to identify highly selective reagents for the engagement of cell-surface targets.

The two-color screening technology has continued to prove extremely useful in the OBOC DECL age. Whereas the type of cell-binding screens described above have not yet been reported in this context, FACS-based two color-screens using differentially tagged proteins, serum samples [38], and RNAs [60] have shown the utility of this approach for the isolation of highly selective ligands from OBOC DECLs.

With the methodological improvements discussed above in place, the identification of peptoid ligands for proteins finally began to become more routine. Unfortunately, with this progress came the realization that this class of molecules rarely binds proteins with high affinity. Whereas reports of primary screening hits with affinities in the mid- to low-μM range were relatively common, few, if any, peptoid ligands with affinities rivaling that of the α-adrenergic receptor ligand were published. Moreover, in unpublished work, the Kodadek laboratory found that it was difficult to drastically improve the affinity of primary peptoid screening hits for their target through systematic substitution of the N-linked side chains. This was thought to be a result of the extreme "floppiness" of peptoids. There is little barrier to rotation around any of the bonds in these oligomers other than the amide linkage and even this bond exists as a nearly equal mixture of cis and trans isomers, with rare exceptions [72, 73]. Peptoids with multiple α-chiral side chains spaced a certain distance apart fold into helices [74–76] but these molecules, while of fundamental interest as foldamers [77], are too large to be attractive candidates for probe molecule development, especially against intracellular targets. Libraries of macrocyclic peptoids [78] have been explored to a limited extent as protein ligands. But in one comparative study, they were found to be no better than linear peptoids as a source of streptavidin ligands unless the ring size became fairly small (16 atoms) and even then, only low micromolar hits were identified [79]. This may reflect the fact that the tertiary amide bond populates both the cis and trans conformations, unlike the secondary amide of most peptides, which is predominantly trans, resulting in macrocyclic peptoids being more flexible than macrocyclic peptides. However, this study only employed a single protein target so general conclusions cannot be made. Several studies have reported the synthesis of macrocyclic peptide-peptoid hybrids [80] and useful ligands from such libraries have been reported. There are indications that peptide-peptoid hybrid macrocycles may exhibit better drug-like properties than peptides and exhibit more conformational constraint than peptoids. These molecules thus represent a potentially promising source of protein ligands [81].

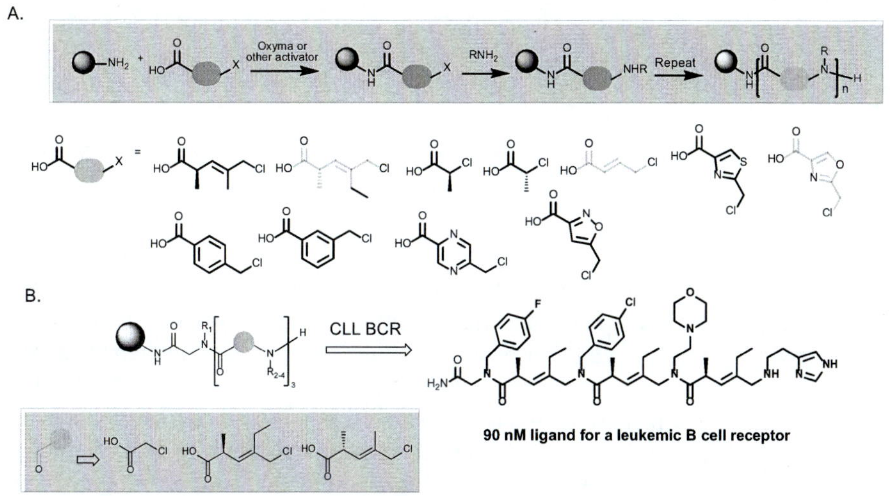

Fig. 12 Peptoid-inspired conformationally constrained oligomers (PICCOs) and chiral oligomers of pentenoic acids (COPAs), two classes of conformationally constrained oligomers. (**a**) Synthesis of PICCOs (box) and the structures of various chloroacids employed to make PICCO libraries. (**b**) General structure of a library employed in a screen for ligands against an antigen-specific B cell receptor. The structure of the best hit obtained in the screen is shown

Given the limitations of peptoids a few laboratories started experimenting with the creation of large libraries of more conformationally constrained oligomers. Notable amongst these efforts was the development of the PICCO (peptoid-inspired conformationally constrained oligomers) class of molecules [82]. PICCOs are synthesized using peptoid-like chemistry, except that more complex chloroacids are employed in place of 2-bromo- or 2-chloro-acetic acid (Fig. 12a). Various "stiffening" moieties are inserted between the COOH and CH_2Cl units of the chloroacids, providing both much greater main chain chemical diversity and some degree of conformational restriction. Direct comparisons of some of these PICCO molecules with peptoids demonstrated the superiority of the former. An excellent example is a study by Sarkar et al. [83] in which they screened an OBOC library of tetramers, with the general structure shown in Fig. 12B, against a monoclonal antibody representing a soluble analog of a pathogenic B cell receptor. At three of the "main chain" positions in the library, either a chloroacetic acid or one of two chiral pentenoic acid units was used as a variable element. The library contained over one million molecules. The chiral pentenoic acids had been shown previously to enforce significant conformational restriction on the oligomers through allylic 1,3 strain effects. 16 validated ligands were identified in the library and all of them had a pentenoic acid unit at each of the three-C-terminal positions. No peptoid ligands were isolated, arguing that none of the peptoids had sufficient affinity for this difficult target to be scored as a hit. The highest affinity ligand for this protein–protein interaction target exhibited a K_D of 90 nM. Changing the stereochemistry at any of the chiral centers abolished binding.

The pentenoic acid oligomers were attractive molecules for OBOC library studies because, like peptoids, their structures could be determined de novo by tandem MS. [84] Unfortunately, this did not prove to be true for PICCOs in general. When mixed oligomers were made using various chloroacids shown in Fig. 12, it proved difficult or impossible to deduce their structure by tandem MS. The cleavage efficiency of the amide bond varied widely depending on the nature of the chloroacid-derived unit and readily interpretable fragmentation patterns were not produced routinely. This made it impossible to take full advantage of the potential molecular diversity of PICCOs and restricted the library design to molecules with homogenous or nearly homogenous "main chains." Clearly, overcoming this limitation would require the adoption of an encoding strategy. As mentioned above, while some non-DNA options were available, these would have required using the tedious manual bead picking methods since the 10 μm TG beads that pass readily through a FACS do not have enough material on them to detect by MS. Therefore, a commitment was made to adopt a DNA-encoding strategy. This, in turn, has now resulted in the development of OBOC DELs of PICCOs (both linear and macrocyclic) [38, 39, 59, 85].

10 Summary of Pre-DEL Combinatorial Chemistry Efforts

In summary, 30 years of OBOC library studies have resulted in a number of interesting advances in both novel oligomer chemistry and the development of sophisticated screening strategies. However, it is fair to say that the real-world impact of these efforts has been modest. Few probe molecules have been produced that are used routinely in chemical biology and almost no true drug leads. There are several reasons for this. For quite some time, the field was dominated by peptide chemistry. As mentioned above, peptides have several drawbacks, particularly for engaging intracellular proteins. In addition, synthetic peptide library chemistry evolved in parallel with genetically encoded peptide libraries. While the former had the advantage of being to use a wider range of building blocks, the much larger size of phage and ribosome display libraries combined with the relative ease of screening them provided the genetically encoded libraries an enormous advantage. Because of analytical limitations, it was difficult to move away from peptides or close relatives such as peptoids on the OBOC platform. Still's halocarbon and Lam's peptide-encoding strategies were helpful but never widely adopted. Thus, the chemical space explored by most OBOC libraries was far from optimal from the point of view of the larger chemical biology community.

But, as reviewed above, it is easy to see in retrospect that this effort was critical in laying the groundwork for the modern era of combinatorial chemistry, which will surely be dominated by DELs for some time to come. Major advances included: (1) the development of the split and pool synthesis strategy, (2) the concept of library encoding, (3) the development of conformationally constrained oligomers, including macrocycles, and (4) the establishment of methods for two-color screening of both soluble and membrane-integral proteins, to name a few. These will all see heavy use in the DEL era. Moreover, it is our opinion that OBOC DELs, in particular, will see increased use. DNA-encoding technology has finally expanded the chemical space that can be explored conveniently using this platform. Even more importantly, OBOC libraries are fundamentally different from DELs created by solution-phase split and pool synthesis in that the beads can be separated, whereas most DELS are produced as an intractable mixture of molecules. This provides opportunities to employ OBOC DELs in functional and phenotypic screens [62, 86] that would be impossible to carry out with other types of DELs. We believe the future is very bright for this technology.

11 The Maturation of Genetically Encoded Libraries

While the development of synthetic combinatorial libraries after 1991 proceeded in fits and starts as major methodological problems were addressed and overcome, it is fair to say that the further development of genetically encoded combinatorial libraries was much smoother. Indeed, the most recent versions of these methodologies are, without question, the most powerful ligand discovery platforms ever developed.

Phage display and other genetically encoded libraries (vide infra) have a huge advantage over synthetic libraries in that serial screens with the amplification of hits in between rounds can be done easily. For example, in a phage display screen, the primary library is incubated with a target protein, and the target is somehow separated from unbound phage, for example using a simple "pull-down" approach via a tag attached to the target protein. The bound phage are eluted and used to infect E. coli, which amplifies them greatly. The process is then repeated several times (Fig. 1). In the old days, prior to next generation sequencing, a sampling of individual phage colonies on a plate would be Sanger sequenced after later rounds of panning to determine if the population had begun to narrow to a few specific peptide-encoding sequences. This is now much more straightforward since the entire population can be analyzed comprehensively after each round of panning by deep sequencing. It is straightforward to vary conditions in each round of selection. For example, when the bound phages are isolated by, for example, pulling down a GST-target protein fusion on glutathione agarose resin, a round that excludes the target protein can be done to remove GST or matrix binders from the population. Alternatively, one can decrease the concentration of the target or increase the concentration of untagged competitor proteins in the screen to demand increased affinity or selectivity.

Therefore, the challenge with genetically encoded libraries has never been establishing effective screening methods. Instead, it is to extend the chemical space one can cover beyond linear peptides containing the 20 canonical amino acids. We review below a few major advances towards this goal.

Cysteine residues are easily alkylated selectively in proteins given the exceptional nucleophilicity of the thiol group. Winter, Heinis, and colleagues took advantage of

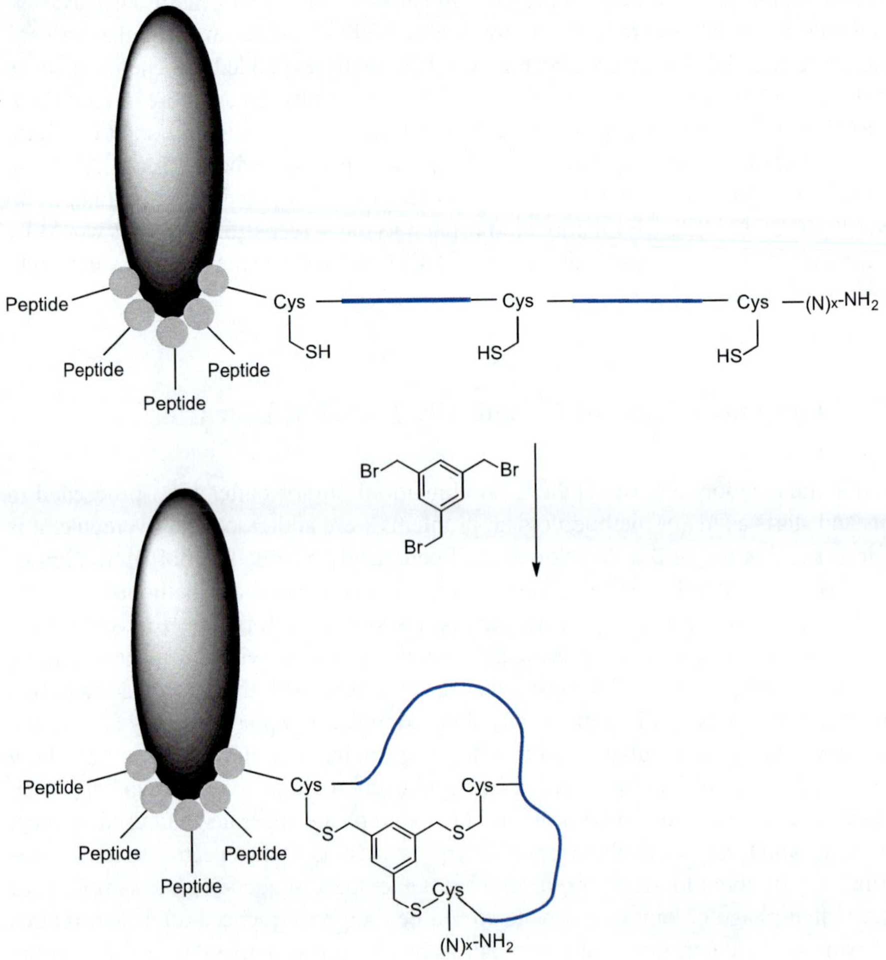

Fig. 13 Synthesis of a library of bicyclic peptides on the surface of phage. See text for details. The blue line represents the peptide chain, comprised of variable residues

this to produce phage display libraries that display conformationally constrained bicyclic peptide libraries on their surface [87, 88]. They encoded a library to contain three invariant cysteines. After the phage population was produced, it was treated with tris-(bromomethyl) benzene (TBMB) as shown in Fig. 13. This results in a bicyclic peptide with much reduced conformational flexibility relative to the starting linear peptide and much greater stability to proteases and peptidases. The biotechnology company Bicycle and the Heinis laboratory have reported a number of successful screens using these bicyclic libraries [89–94], which have resulted in the discovery of many high affinity protein ligands that demonstrate excellent selectivity. By varying the position of the cysteines and the nature of the alkylating agent, one can generate further scaffold diversity with respect to the bicycle [95].

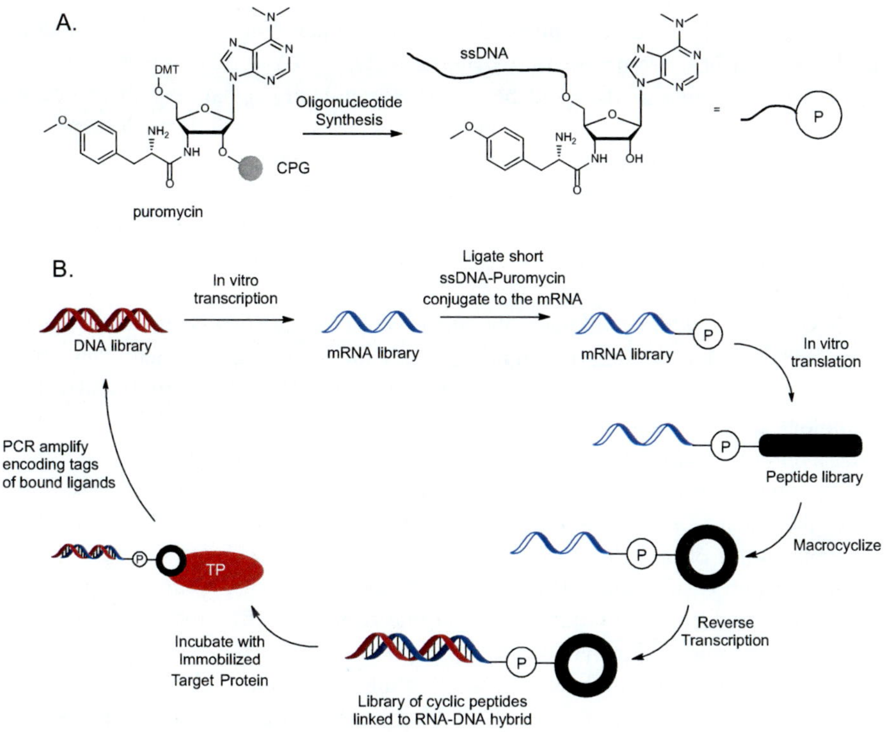

Fig. 14 mRNA display. A. Schematic depiction of the construction of a puromycin-oligonucleotide conjugate. B. Representation of the workflow of the mRNA display library screening workflow for the case of macrocyclic peptides

The requirement that phage display libraries must be made in E. coli limits their size to $\approx 10^7$–10^9 peptides and their composition to the 20 canonical amino acids. Therefore, a major advance in the field was the development of mRNA display [96, 97], in which peptide libraries are constructed in a cell-free fashion by in vitro transcription and translation. The critical initial innovation that made this possible was the use of the antibiotic puromycin to create a stable physical linkage between the peptide and its encoding RNA. Puromycin is a mimic of the 3′ end of a tyrosine aminoacyl tRNA (Fig. 14). Puromycin can enter the A site of the ribosome and terminate translation, forcing the ribosome to release the peptide product. When puromycin is attached covalently to the 3′ end of the encoding RNA, it thus serves a stable covalent bridge between the peptide and the encoding RNA. mRNA displays allowed peptide libraries of 10^{12}–10^{14} molecules to be created. However, it remained difficult to employ unnatural amino acids.

This technology was made far more powerful by two further advances. The first was the use of purified components for the in vitro translation reaction, dubbed the "PURE" (protein synthesis using recombinant elements) system. The use of an entirely defined translation mix, rather than a crude extract, made it possible to leave out specific components, for example a canonical amino acid, and replace it with something else, such as a non-canonical amino acid that is nonetheless recognized by the tRNA synthetase [98]. Alternatively, tRNA synthetases pre-charged with unnatural amino acids could be added, significantly expanding the scope of substitutions possible [99]. Finally, and perhaps most importantly with respect to expanding the chemical space of genetically encoded libraries, was the development by Suga and colleagues of a ribozyme capable of linking activated amino acids onto full-length, in vitro transcribed tRNAs. This remarkable ribozyme, appropriately named "Flexizyme" does not recognize the side chain of the amino acid and thus displays extraordinary substrate scope. [100–102]

These advances have allowed the creation of exceptionally diverse peptide libraries that include unnatural amino acids, N-methylated amino acids and even hydroxy acids (for the synthesis of depsipeptides) [103, 104]. Very importantly, this technology also allows the facile synthesis of macrocyclic peptides. This is achieved using a chloroacetyl-containing building block that is inserted at the N-terminus of the peptide (via charging it onto a tRNA). The reactive chloride is displaced at the end of the translation process by an efficient, proximity-mediated S_N2 reaction with a cysteine sulfur placed elsewhere in the peptide [105, 106].

More recently, Suga and colleagues have further increased the chemical diversity accessible in this format by demonstrating that residues in the library can be post-translationally modified enzymatically [107].

As mentioned above, the power of modern mRNA display technology for ligand discovery is unmatched [108]. This technology has been industrialized and is now employed by several biotechnology and large pharmaceutical companies. Large macrocyclic peptides of 12–20 residues with nM affinities for even difficult protein–protein interaction targets are routinely pulled out of screens using mRNA display libraries. This outstanding track record is, in part, due to the number of molecules in the library. They are comprised of up to 10^{14} peptides and "mutants" undoubtedly appear due to PCR errors during amplification. The relatively large size of these molecules likely also contributes to their efficacy. Their "wingspan" is sufficient to make multiple contacts on the surface of a target and fill more than one pocket, in contrast to a traditional drug-like small molecule. Indeed, these macrocyclic peptides rival antibodies in terms of their affinity and selectivity. However, they have significant advantages over antibodies for certain applications. Since they are 50–100 times smaller than an IgG, they are likely to have much better tissue penetration and their circulating half-life is much lower. This makes them perfect for acting as vehicles to deliver radiotherapeutics and imaging agents to molecular targets located on the surface of tumors, for example. The fact that macrocyclic peptides can be made synthetically also makes it far easier to make drug or imaging conjugates.

Of course, such large peptides are generally not cell permeable and so are not a good starting point for drug development for intracellular targets. Nonetheless, pharma companies routinely run difficult intracellular targets through mRNA display screens for several reasons. One is to gauge ligandability using this highly reliable ligand discovery technology. Furthermore, it is sometimes easier to crystallize the target-macrocyclic peptide complex than the target itself due to induced conformational constraint. Finally, the macrocyclic peptide can be labeled and used in a simple, high-throughput displacement assay using a collection of non-peptidic molecules. This obviates the need to develop a more elaborate functional assay for the target protein [109].

It is worth mentioning that many laboratories are focused on establishing methods to introduce macrocyclic peptides (as well as other types of cell impermeable "cargo") into cells via mechanisms other than passive diffusion [110]. Success in this endeavor would open up new frontiers in drugging difficult intracellular targets.

12 Target Agnostic Solution-Phase DECLs

In this section we attempt to bridge the gap between the history of combinatorial chemistry, presented in the previous sections, and the real-world solutions provided by DNA-encoded chemical libraries (DECLs). The most frequent application of DECLs in industry today is towards target agnostic libraries. The basic methods for rapid DECL synthesis and selections, still in use today, were reported by Clark et al. in 2009 [40]. The report describes in detail the synthesis of two small-molecule DECLs employing split and pool combinatorial synthesis. The barcodes of the DECLs were constructed in an iterative fashion, employing dsDNA precursors, and remains the industry standard. The quality control (QC) reported in the supplemental data of Clark et al. [40] remains relevant and includes (1) workflows for building block validation, (2) production and analysis of a test library, and (3) methods of analysis of tag ligation and chemical conversions throughout library production. Clark et al. [40] also provided detailed protocols for the DECL affinity selection, including barcode amplification in preparation of DNA sequencing. Clark et al. [40] were an early adopter of high-throughput DNA sequencing to interrogate library composition (along with work by Buller et al. [111, 112]). Such sequencing allowed the observation of a wide range of enriched chemical structures and did not require convergence to a small number of the strongest binders. This advancement was particularly important for small-molecule drug discovery, since the physical properties of the hit molecules (and chemical novelty) are often more important than their dissociation constants. Lastly, some basic aspects of data analysis, including visualization of enrichment with a 3-D cube, were exemplified by Clark et al. [40] It's important to note that the work of Clark et al. [40] built upon the work of many others including (1) the general concept of DNA encoding of small molecules [1], (2) early iterations of DNA-encoded peptide libraries [2, 113], (3) early iterations of DNA-encoded small molecules [114–116], (4) the general concept of DNA tag ligation for constructing a barcode [117], (5) the first reported deep sequencing of a DEL screen [112], and (6) reports of templated DELs. [118]

The report by Clark et al. [40] exemplified both the potential and shortcomings of DECL technology. We briefly list some of these shortcomings and leave them to be addressed in the forthcoming chapters. Firstly, the DECL selection method was limited to purified and soluble proteins and appears incompatible with biochemical or cellular functional screens. Second, the limitations of DNA-compatible chemical reactions, and the chemical space accessed by combinatorial chemistry methods in general, have consistently raised concerns regarding the lead-like nature (and overall value) of DECL chemical space. And lastly, the inability to purify away impurities during library synthesis, or quantify product yields, made it impossible to know exactly what was in the DECL.

13 Focused DECLs

In addition to target agnostic libraries, the pharmaceutical industry has also sought to lessen the cost of hit optimization. An excellent example of the technology addressing this goal was reported by Wichert et al. [35], using a so-called dual-display DECL. In this particular instance, the DECL selection was used to interrogate hundreds of fragment pairs against carbonic anhydrase IX (CAIX), an extracellular target for tumor targeting applications. One of the fragments was an acetazolamide (AAZ) derivative, a known binder to CAIX with a direct binding of 16.7 nM as measured by SPR. The goal of the DECL focused selection was to then find a second fragment to pair with AAZ to give a probe with increased affinity. In the end, Wichert et al. [35] were able to discover a new compound with ~40-fold increased potency. Perhaps most importantly, this newly discovered compound-dye conjugate shows vastly improved residence time (at the tumor) in mice, as compared to the parent compound AAZ (Fig. 15).

While the above example successfully demonstrates the use of a focused DECL, it also suggests some potential shortcomings. Probably most important, the discovered molecule (Fig. 15) while ~40-fold more potent, is also more than twofold larger in terms of molecular weight. While this may not be an important issue when targeting an extracellular protein, for programs seeking a bioavailable drug, this is

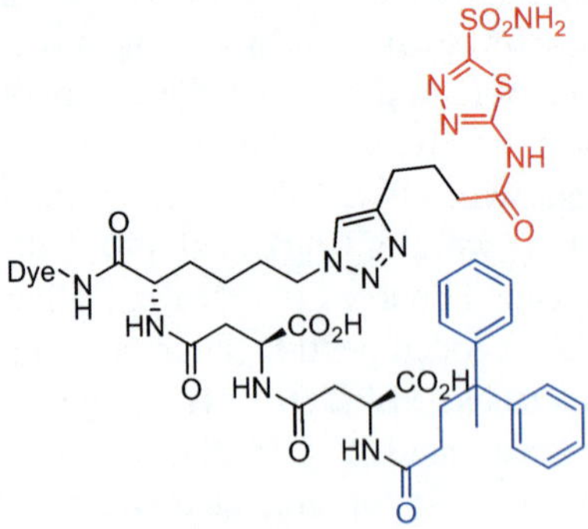

Fig. 15 Results of a DECL focused selection. A dual-display DECL was produced with the goal of optimizing the potency of acetazolamide (AAZ, shown in red) towards carbonic anhydrase IX. The selection suggested pairing AZZ with the newly discovered fragment shown in blue. The combined molecule shows an ~10-fold increase in binding affinity over the parent molecule AAZ.

very concerning. Also, the use of focused DECLs in the pharmaceutical industry depends upon its value relative to more traditional rational design approaches. To date, we lack clear examples of DECL outperforming (or supplementing) traditional medicinal chemistry compound optimization approaches within an industrial setting. Instead, the use of focused DECLs appears to remain largely in academic labs [34, 36].

14 Conclusion

The path of combinatorial chemistry has followed the desires and realities of the pharmaceutical industry. Industry desired a way to synthesize molecules more cheaply than one compound at a time. Combinatorial chemistry offered relatively inexpensive mixtures. However, the compounds in these mixtures lacked structural diversity and drug-likeness, and it was challenging to deconvolute the source of their biochemical activity. An alternative to mixtures was high-throughput parallel synthesis, yet these molecules also suffered from relatively poor chemical diversity (as products were still synthesized from a small number of common precursors). Expensive column purification could be avoided, but such impure products ran a higher risk of false positives in desired biochemical and cellular assays. Or, the compounds could be column purified, but then the synthesis costs remained relatively high. Returning to the concept of combinatorial mixtures, the idea of tagging molecules, e.g. with peptides or halogenated phenyls, to aid in deconvolution took hold. However, the number of tags that could be acquired and interrogated was relatively small. And so, the resulting libraries were generally numerically small, and again suffered from low chemical diversity. These drawbacks resulted in low hit finding productivity. At this point in history, around the year 2000, the drug discovery community had largely lost patience with combinatorial chemistry and, erroneously, concluded that compound mixtures were incompatible with screening. In this environment of disappointment and underwhelming results, DNA-encoded chemical libraries arose. The use of DNA tag precursors allowed the use of large split sizes during library synthesis, resulting in increased chemical diversity (due to the ability to input and encode large numbers of chemically diverse building blocks). And the commercial availability of high-throughput DNA sequencing (around the year 2005) permitted the inexpensive interrogation of numerically large library mixtures before and after affinity selections. Thus, DNA barcoding alleviated the main problems with earlier combinatorial libraries; a lack numeric size and chemical diversity. DECLs were big enough, and diverse enough, to routinely find hit molecules. Still, the hangover of poor results from earlier combinatorial efforts took many years to diminish, and remnants of distrust still remain. Nonetheless, today DECL is widely employed in both academia and industry, and the technology is steadily improving, as more researchers seek to further exploit its benefits.

Declaration of Conflicting Interests

This is a financial conflict of interest and funding statement for the chapter.

Alexander Lee Satz is employed by Wuxi Aaptec, his authorship of this chapter was completed within the scope of his employment with Wuxi Aaptec. Thomas Kodadek is a founder of, an advisor for, and major shareholder in Deluge Biotechnologies and Triana Biosciences.

Funding The authors received no financial support for the research, authorship, and/or publication of this article.

This chapter is a review and no humans or animals were involved in the preparation of the chapter.

References

1. Brenner S, Lerner RA (1992) Encoded combinatorial chemistry. Proc Natl Acad Sci 89(12): 5381–5383. https://doi.org/10.1073/pnas.89.12.5381
2. Needels MC, Jones DG, Tate EH, Heinkel GL, Kochersperger LM, Dower WJ, Barrett RW, Gallop MA (1993) Generation and screening of an oligonucleotide-encoded synthetic peptide library. Proc Natl Acad Sci U S A 90:10700–10704
3. Scott JK, Smith GP (1990) Searching for peptide ligands with an epitope library. Science 249: 386–390
4. Devlin JJ, Panganiban LC, Devlin PE (1990) Random peptide libraries: a source of specific protein binding molecules. Science 249:404–406
5. Cwirla SE, Peters EA, Barrett RW, Dower WJ (1990) Peptides on phage: a vast library of peptides for identifying ligands. Proc Natl Acad Sci U S A 87:6378–6382
6. Houghten RA, Pinilla C, Blondelle SE, Appel JR, Dooley CT, Cuervo JH (1991) Generation and use of synthetic peptide combinatorial libraries for basic research and drug discovery. Nature 354(6348):84–86. https://doi.org/10.1038/354084a0
7. Lam KS, Salmon SE, Hersh EM, Hruby VJ, Kazmierski WM, Knapp RJ (1991) A new type of synthetic peptide library for identifying ligand-binding activity. Nature 354(6348):82–84. https://doi.org/10.1038/354082a0
8. Furka A, Sebestyén F, Asgedom M, Dibó G (1991) General method for rapid synthesis of multicomponent peptide mixtures. Int J Pept Protein Res 37(6):487–493. https://doi.org/10.1111/j.1399-3011.1991.tb00765.x
9. Alluri PG, Reddy MM, Bachhawat-Sikder K, Olivos HJ, Kodadek T (2003) Isolation of protein ligands from large peptoid libraries. J Am Chem Soc 125(46):13995–14004. https://doi.org/10.1021/ja036417x
10. Paulick MG, Hart KM, Brinner KM, Tjandra M, Charych DH, Zuckermann RN (2006) Cleavable hydrophilic linker for one-bead-one-compound sequencing of oligomer libraries by tandem mass spectrometry. J Comb Chem 8(3):417–426
11. Dragovich PS, Haap W, Mulvihill MM, Plancher J-M, Stepan AF (2022) Small-molecule Lead-finding trends across the Roche and Genentech research organizations. J Med Chem 65(4):3606–3615. https://doi.org/10.1021/acs.jmedchem.1c02106
12. Leveridge M, Chung C-W, Gross JW, Phelps CB, Green D (2018) Integration of Lead discovery tactics and the evolution of the lead discovery toolbox. SLAS Discov 23(9): 881–897. https://doi.org/10.1177/2472555218778503
13. Satz AL, Brunschweiger A, Flanagan ME, Gloger A, Hansen NJV, Kuai L, Kunig VBK, Lu X, Madsen D, Marcaurelle LA, Mulrooney C, O'Donovan G, Sakata S, Scheuermann J (2022) DNA-encoded chemical libraries. Nat Rev Methods Primers 2(1):3. https://doi.org/10.1038/s43586-021-00084-5

14. van Vlijmen H, Ortholand J-Y, Li VM-J, de Vlieger JSB (2021) The European Lead factory: an updated HTS compound library for innovative drug discovery. Drug Discov Today 26(10): 2406–2413. https://doi.org/10.1016/j.drudis.2021.04.019
15. DeGoey DA, Chen H-J, Cox PB, Wendt MD (2018) Beyond the rule of 5: lessons Learned from AbbVie's drugs and compound collection. J Med Chem 61(7):2636–2651. https://doi.org/10.1021/acs.jmedchem.7b00717
16. Hughes JP, Rees S, Kalindjian SB, Philpott KL (2011) Principles of early drug discovery. Br J Pharmacol 162(6):1239–1249. https://doi.org/10.1111/j.1476-5381.2010.01127.x
17. Lipinski CA (2004) Lead- and drug-like compounds: the rule-of-five revolution. Drug Discov Today Technol 1(4):337–341. https://doi.org/10.1016/j.ddtec.2004.11.007
18. Hermann JC, Chen Y, Wartchow C, Menke J, Gao L, Gleason SK, Haynes N-E, Scott N, Petersen A, Gabriel S, Vu B, George KM, Narayanan A, Li SH, Qian H, Beatini N, Niu L, Gan Q-F (2013) Metal impurities cause false positives in high-throughput screening campaigns. ACS Med Chem Lett 4(2):197–200. https://doi.org/10.1021/ml3003296
19. Bowes S, Sun D, Kaffashan A, Zeng C, Chuaqui C, Hronowski X, Buko A, Zhang X, Josiah S (2006) Quality assessment and analysis of Biogen Idec compound library. J Biomol Screen 11(7):828–835. https://doi.org/10.1177/1087057106290993
20. Comess KM, Schurdak ME, Voorbach MJ, Coen M, Trumbull JD, Yang H, Gao L, Tang H, Cheng X, Lerner CG, McCall JO, Burns DJ, Beutel BA (2006) An Ultraefficient affinity-based high-throughout screening process: application to bacterial cell wall biosynthesis enzyme MurF. J Biomol Screen 11(7):743–754. https://doi.org/10.1177/1087057106289971
21. Annis DA, Athanasopoulos J, Curran PJ, Felsch JS, Kalghatgi K, Lee WH, Nash HM, Orminati J-PA, Rosner KE, Shipps GW, Thaddupathy GRA, Tyler AN, Vilenchik L, Wagner CR, Wintner EA (2004) An affinity selection–mass spectrometry method for the identification of small molecule ligands from self-encoded combinatorial libraries: discovery of a novel antagonist of E. Coli dihydrofolate reductase. Int J Mass Spectrom 238(2):77–83. https://doi.org/10.1016/j.ijms.2003.11.022
22. Coburn CA, Stachel SJ, Li Y-M, Rush DM, Steele TG, Chen-Dodson E, Holloway MK, Xu M, Huang Q, Lai M-T, DiMuzio J, Crouthamel M-C, Shi X-P, Sardana V, Chen Z, Munshi S, Kuo L, Makara GM, Annis DA, Tadikonda PK, Nash HM, Vacca JP, Wang T (2004) Identification of a small molecule nonpeptide active site beta-secretase inhibitor that displays a nontraditional binding mode for aspartyl proteases. J Med Chem 47(25): 6117–6119. https://doi.org/10.1021/jm049388p
23. Boger DL, Tarby CM, Myers PL, Caporale LH (1996) Generalized dipeptidomimetic template: solution phase parallel synthesis of combinatorial libraries. J Am Chem Soc 118(8): 2109–2110. https://doi.org/10.1021/ja9539711
24. Thompson LA, Ellman JA (1996) Synthesis and applications of small molecule libraries. Chem Rev 96(1):555–600. https://doi.org/10.1021/cr9402081
25. Plunkett MJ, Ellman JA (1995) Solid-phase synthesis of structurally diverse 1,4-benzodiazepine derivatives using the Stille coupling reaction. J Am Chem Soc 117(11): 3306–3307. https://doi.org/10.1021/ja00116a049
26. Weber L, Wallbaum S, Broger C, Gubernator K (1995) Optimization of the biological activity of combinatorial compound libraries by a genetic algorithm. Angew Chem Int Ed Engl 34(20): 2280–2282. https://doi.org/10.1002/anie.199522801
27. Farrant E (2020) Automation of synthesis in medicinal chemistry: progress and challenges. ACS Med Chem Lett 11(8):1506–1513. https://doi.org/10.1021/acsmedchemlett.0c00292
28. Tu NP, Searle PA, Sarris K (2016) An automated microwave-assisted synthesis purification system for rapid generation of compound libraries. J Lab Autom 21(3):459–469. https://doi.org/10.1177/2211068215590580
29. Burke MD, Berger EM, Schreiber SL (2004) A synthesis strategy yielding skeletally diverse small molecules combinatorially. J Am Chem Soc 126(43):14095–14104. https://doi.org/10.1021/ja0457415

30. Kato N, Comer E, Sakata-Kato T, Sharma A, Sharma M, Maetani M, Bastien J, Brancucci NM, Bittker JA, Corey V, Clarke D, Derbyshire ER, Dornan GL, Duffy S, Eckley S, Itoe MA, Koolen KMJ, Lewis TA, Lui PS, Lukens AK, Lund E, March S, Meibalan E, Meier BC, McPhail JA, Mitasev B, Moss EL, Sayes M, Van Gessel Y, Wawer MJ, Yoshinaga T, Zeeman A-M, Avery VM, Bhatia SN, Burke JE, Catteruccia F, Clardy JC, Clemons PA, Dechering KJ, Duvall JR, Foley MA, Gusovsky F, Kocken CHM, Marti M, Morningstar ML, Munoz B, Neafsey DE, Sharma A, Winzeler EA, Wirth DF, Scherer CA, Schreiber SL (2016) Diversity-oriented synthesis yields novel multistage antimalarial inhibitors. Nature 538(7625):344–349. https://doi.org/10.1038/nature19804
31. Spaller MR, Burger MT, Fardis M, Bartlett PA (1997) Synthetic strategies in combinatorial chemistry. Curr Opin Chem Biol 1(1):47–53. https://doi.org/10.1016/s1367-5931(97)80107-x
32. Hoekstra WJ, Maryanoff BE, Andrade-Gordon P, Cohen JH, Costanzo MJ, Damiano BP, Haertlein BJ, Harris BD, Kauffman JA, Keane PM, McComsey DF, Villani FJ, Yabut SC (1996) Solid-phase parallel synthesis applied to Lead optimization: discovery of potent analogues of the GPIIb/IIIa antagonist RWJ-50042. Bioorg Med Chem Lett 6(20): 2371–2376. https://doi.org/10.1016/0960-894X(96)00438-6
33. Cavallaro CL, Briceno S, Chen J, Cvijic ME, Davies P, Hynes J, Liu R-Q, Mandlekar S, Rose AV, Tebben AJ, Van Kirk K, Watson A, Wu H, Yang G, Carter PH (2012) Discovery and lead optimization of a novel series of CC chemokine receptor 1 (CCR1)-selective piperidine antagonists via parallel synthesis. J Med Chem 55(22):9643–9653. https://doi.org/10.1021/jm300896d
34. TEAD–YAP interaction inhibitors and MDM2 binders from DNA-encoded indole-focused Ugi peptidomimetics – Kunig – 2020 – Angewandte Chemie International Edition – Wiley Online Library. https://onlinelibrary.wiley.com/doi/full/10.1002/anie.202006280. Accessed 5 May 2022
35. Wichert M, Krall N, Decurtins W, Franzini RM, Pretto F, Schneider P, Neri D, Scheuermann J (2015) Dual-display of small molecules enables the discovery of ligand pairs and facilitates affinity maturation. Nat Chem 7(3):241–249. https://doi.org/10.1038/nchem.2158
36. Yuen LH, Dana S, Liu Y, Bloom SI, Thorsell A-G, Neri D, Donato AJ, Kireev D, Schüler H, Franzini RM (2019) A focused DNA-encoded chemical library for the discovery of inhibitors of NAD+-dependent enzymes. J Am Chem Soc 141(13):5169–5181. https://doi.org/10.1021/jacs.8b08039
37. Erharuyi O, Simanski S, McEnaney PJ, Kodadek T (2018) Screening one bead one compound libraries against serum using a flow cytometer: determination of the minimum antibody concentration required for ligand discovery. Bioorg Med Chem Lett 28(16):2773–2778. https://doi.org/10.1016/j.bmcl.2018.01.033
38. Mendes K, Malone ML, Ndungu JM, Suponitsky-Kroyter I, Cavett VJ, McEnaney PJ, MacConnell AB, Doran TM, Ronacher K, Stanley K, Utset O, Walzl G, Paegel BM, Kodadek T (2017) High-throughput identification of DNA-encoded IgG ligands that distinguish active and latent mycobacterium tuberculosis infections. ACS Chem Biol 19:234–243
39. Koesema E, Roy A, Paciaroni NG, Coito C, Tokmina-Roszyk M, Kodadek T (2022) Synthesis and screening of a DNA-encoded library of non-peptidic macrocycles. Angew Chem Int Ed Engl:e202116999. https://doi.org/10.1002/anie.202116999
40. Clark MA, Acharya RA, Arico-Muendel CC, Belyanskaya SL, Benjamin DR, Carlson NR, Centrella PA, Chiu CH, Creaser SP, Cuozzo JW, Davie CP, Ding Y, Franklin GJ, Franzen KD, Gefter ML, Hale SP, Hansen NJ, Israel DI, Jiang J, Kavarana MJ, Kelley MS, Kollmann CS, Li F, Lind K, Mataruse S, Medeiros PF, Messer JA, Myers P, O'Keefe H, Oliff MC, Rise CE, Satz AL, Skinner SR, Svendsen JL, Tang L, van Vloten K, Wagner RW, Yao G, Zhao B, Morgan B, Design A (2009) Synthesis and selection of DNA-encoded small-molecule libraries. Nat Chem Biol 5(9):647–654. nchembio.211 [pii]. https://doi.org/10.1038/nchembio.211
41. Ohlmeyer MH, Swanson RN, Dillard LW, Reader JC, Asouline G, Kobayashi R, Wigler M, Still WC (1993) Complex synthetic chemical libraries indexed with molecular tags. Proc Natl Acad Sci U S A 90:10922–10926

42. Dickson P, Kodadek T (2019) Chemical composition of DNA-encoded libraries, past present and future. Org Biomol Chem 17(19):4676–4688. https://doi.org/10.1039/c9ob00581a
43. Paciaroni NG, Ndungu JM, Kodadek T (2020) An aldehyde explosion strategy for the synthesis of structurally complex DNA-encoded libraries. Chem Commun 56:4656–4659
44. Liu R, Marik J, Lam KS (2002) A novel peptide-based encoding system for "One-Bead One-Compound" Peptidomimetic and small molecule combinatorial libraries. J Am Chem Soc 124:7678–7680
45. Zuckermann RN, Kerr JM, Kent SBH, Moos WH (1992) Efficient method for the preparation of peptoids (oligo(N-substituted Glycines)) by sub-monomer solid-phase synthesis. J Am Chem Soc 114:10646–10647
46. Moon H, Lim H-S (2015) Synthesis and screening of small-molecule α-helix mimetic libraries targeting protein-protein interactions. Curr Opin Chem Biol 24:38–47. https://doi.org/10.1016/j.cbpa.2014.10.023
47. Simon RJ, Kania RS, Zuckermann RN, Huebner VD, Jewell DA, Banville S, Ng S, Wang L, Rosenberg S, Marlowe CK et al (1992) Peptoids: a modular approach to drug discovery. Proc Natl Acad Sci U S A 89(20):9367–9371
48. Miller SM, Simon RJ, Ng S, Zuckermann RN, Kerr JM, Moos WH (1994) Proteolytic studies of homologous peptide and N-substituted glycine peptoid oligomers. Bioorg Med Chem Lett 4(22):2657–2662
49. Pye CR, Hewitt WM, Schwochert J, Haddad TD, Townsend CE, Etienne L, Lao Y, Limberakis C, Furukawa A, Mathiowetz AM, Price DA, Liras S, Lokey RS (2017) Nonclassical size dependence of permeation defines bounds for passive adsorption of large drug molecules. J Med Chem 60(5):1665–1672. https://doi.org/10.1021/acs.jmedchem.6b01483
50. Matsson P, Doak BC, Over B, Kihlberg J (2016) Cell permeability beyond the rule of 5. Adv Drug Deliv Rev 101:42–61. https://doi.org/10.1016/j.addr.2016.03.013
51. Whitty A, Zhong M, Viarengo L, Beglov D, Hall DR, Vajda S (2016) Quantifying the chameleonic properties of macrocycles and other high-molecular-weight drugs. Drug Discov Today 21(5):712–717. https://doi.org/10.1016/j.drudis.2016.02.005
52. Yu P, Liu B, Kodadek T (2005) A high-throughput assay for assessing the cell permeability of combinatorial libraries. Nat Biotechnol 23(6):746–751. nbt1099 [pii]
53. Figliozzi GM, Goldsmith R, Ng SC, Banville SC, Zuckermann RN (1996) [25] synthesis of N-substituted glycine peptoid libraries. In: Methods in enzymology; combinatorial chemistry, vol 267. Academic Press, pp 437–447. https://doi.org/10.1016/S0076-6879(96)67027-X
54. Gibbons JA, Hancock AA, Vitt CR, Knepper S, Buckner SA, Brune ME, Milicic I, Kerwin JF, Richter LS, Taylor EW, Spear KL, Zuckermann RN, Spellmeyer DC, Braeckman RA, Moos WH (1996) Pharmacologic characterization of CHIR 2279, an N-substituted glycine peptoid with high-affinity binding for alpha 1-adrenoceptors. J Pharmacol Exp Ther 277(2):885–899
55. Kodadek T (2010) No splicing, no dicing: non-proteolytic roles of the ubiquitin-proteasome system in transcription*. J Biol Chem 285(4):2221–2226. https://doi.org/10.1074/jbc.R109.077883
56. Townsend J, Do A, Lehman A, Dixon S, Sanii B, Lam KS (2010) 3-Nitro-tyrosine as an internal quencher of autofluorescence enhances the compatibility of fluorescence based screening of OBOC combinatorial libraries. Comb Chem High Throughput Screen 13(5):422–429. https://doi.org/10.2174/138620710791292994
57. Olivos HJ, Baccawat-Sikder K, Kodadek T (2003) Quantum dots as a visual aid for screening bead-bound combinatorial libraries. Chembiochem 4:1242–1245
58. Lian W, Upadhyaya P, Rhodes CA, Liu Y, Pei D (2013) Screening bicyclic peptide libraries for protein–protein interaction inhibitors: discovery of a tumor necrosis factor-α antagonist. J Am Chem Soc 135(32):11990–11995. https://doi.org/10.1021/ja405106u
59. McEnaney P, Balzarini M, Park H, Kodadek T (2020) Structural characterization of a peptoid-inspired conformationally constrained oligomer (PICCO) bound to streptavidin. Chem Commun 56(72):10560–10563. https://doi.org/10.1039/D0CC02588G

60. Benhamou RI, Suresh BM, Tong Y, Cochrane WG, Cavett V, Vezina-Dawod S, Abegg D, Childs-Disney JL, Adibekian A, Paegel BM, Disney MD (2022) DNA-encoded library versus RNA-encoded library selection enables design of an oncogenic noncoding RNA inhibitor. Proc Natl Acad Sci U S A 119(6). https://doi.org/10.1073/pnas.2114971119
61. MacConnell AB, McEnaney PJ, Cavett VJ, Paegel BM (2015) DNA-encoded solid-phase synthesis: encoding language design and complex oligomer library synthesis. ACS Comb Sci 17:518–534. https://doi.org/10.1021/acscombsci.5b00106
62. Cochrane W, Malone ML, Dang VQ, Cavett VJ, Satz AL, Paegel BM (2019) Activity-based DNA-encoded library screening. ACS Comb Sci 21:425–435. https://doi.org/10.1021/acscombsci.9b00037
63. Doran TM, Gao Y, Mendes K, Dean S, Simanski S, Kodadek T (2014) The utility of redundant combinatorial libraries in distinguishing high and low quality screening hits. ACS Comb Sci 16:259–270. https://doi.org/10.1021/co50030f
64. Mikawa M, Wang H, Guo L, Liu R, Marik J, Takada Y, Lam K, Lau D (2004) Novel peptide ligands for integrin alpha 4 Beta 1 overexpressed in cancer cells. Mol Cancer Ther 3(10): 1329–1334
65. Peng L, Liu R, Marik J, Wang X, Takada Y, Lam KS (2006) Combinatorial chemistry identifies high-affinity peptidomimetics against Alpha4beta1 integrin for in vivo tumor imaging. Nat Chem Biol 2(7):381–389
66. Udugamasooriya DG, Dineen SP, Brekken RA, Kodadek T (2008) A peptoid "Antibody Surrogate" that antagonizes VEGF receptor 2 activity. J Am Chem Soc 130(17):5744–5752. https://doi.org/10.1021/ja711193x
67. Udugamasooriya DG, Dunham G, Ritchie C, Brekken RA, Kodadek T (2008) The pharmacophore of a peptoid VEGF receptor 2 antagonist includes both side chain and main chain residues. Bioorg Med Chem Lett 18(22):5892–5894
68. Udugamasooriya DG, Ritchie C, Brekken RA, Kodadek T (2008) A Peptoid antagonist of VEGF receptor 2 recognizes a "hotspot" in the extracellular domain distinct from the hormone-binding site. Bioorg Med Chem 16(12):6338–6343
69. Matharage JM, Minna JD, Brekken RA, Udugamasooriya DG (2015) Unbiased selection of peptide-peptoid hybrids specific for lung cancer compared to normal lung epithelial cells. ACS Chem Biol 10(12):2891–2899. https://doi.org/10.1021/acschembio.5b00592
70. Raymond AC, Gao B, Girard L, Minna JD, Gomika Udugamasooriya D (2019) Unbiased peptoid combinatorial cell screen identifies plectin protein as a potential biomarker for lung cancer stem cells. Sci Rep 9(1):14954. https://doi.org/10.1038/s41598-019-51004-3
71. Desai TJ, Toombs JE, Minna JD, Brekken RA, Udugamasooriya DG (2016) Identification of lipid-phosphatidylserine (PS) as the target of unbiasedly selected cancer specific peptide-peptoid hybrid PPS1. Oncotarget 7(21):30678–30690. https://doi.org/10.18632/oncotarget.8929
72. Gorske BC, Bastian BL, Geske GD, Blackwell HE (2007) Local and tunable N-->pi* interactions regulate amide isomerism in the peptoid backbone. J Am Chem Soc 129(29): 8928–8929. https://doi.org/10.1021/ja071310l
73. Gorske BC, Blackwell HE (2006) Tuning peptoid secondary structure with pentafluoroaromatic functionality: a new Design paradigm for the construction of discretely folded peptoid structures. J Am Chem Soc 128(44):14378–14387. https://doi.org/10.1021/ja065248o
74. Kirshenbaum K, Barron AE, Goldsmith RA, Armand P, Bradley EK, Truong KT, Dill KA, Cohen FE, Zuckermann RN (1998) Sequence-specific polypeptoids: a diverse family of heteropolymers with stable secondary structure. Proc Natl Acad Sci U S A 95:4303–4308
75. Armand P, Kirshenbaum K, Goldsmith RA, Farr-Jones S, Barron AE, Truong KT, Dill KA, Mierke DF, Cohen FE, Zuckermann RN, Bradley EK (1998) NMR determination of the major solution conformation of a peptoid pentamer with chiral side chains. Proc Natl Acad Sci U S A 95:4309–4314

76. Wu CW, Kirshenbaum K, Sanborn TJ, Patch JA, Huang K, Dill KA, Zuckermann RN, Barron AE (2003) Structural and spectroscopic studies of peptoid oligomers with alpha-chiral aliphatic side chains. J Am Chem Soc 125(44):13525–13530
77. Gellman SH (1998) Foldamers: a manifesto. Acc Chem Res 31(4):173–180
78. Shin SB, Yoo B, Todaro LJ, Kirshenbaum K (2007) Cyclic peptoids. J Am Chem Soc 129(11): 3218–3225
79. Gao Y, Kodadek T (2015) Direct comparison of linear and macrocyclic compound libraries as a source of protein ligands. ACS Comb Sci 17(3):190–195. https://doi.org/10.1021/co500161c
80. Morimoto J, Kodadek T (2015) Synthesis of a large library of macrocyclic peptides containing multiple and diverse N-alkylated residues. Mol Biosyst 11:2770–2779. https://doi.org/10.1039/c5mb00308c
81. Boehm M, Beaumont K, Jones R, Kalgutkar AS, Zhang L, Atkinson K, Bai G, Brown JA, Eng H, Goetz GH, Holder BR, Khunte B, Lazzaro S, Limberakis C, Ryu S, Shapiro MJ, Tylaska L, Yan J, Turner R, Leung SSF, Ramaseshan M, Price DA, Liras S, Jacobson MP, Earp DJ, Lokey RS, Mathiowetz AM, Menhaji-Klotz E (2017) Discovery of potent and orally bioavailable macrocyclic peptide-peptoid hybrid CXCR7 modulators. J Med Chem 60(23): 9653–9663. https://doi.org/10.1021/acs.jmedchem.7b01028
82. Kodadek T, McEnaney PJ (2016) Towards vast libraries of scaffold-diverse, conformationally constrained oligomers. Chem Commun (Camb) 52(36):6038–6059. https://doi.org/10.1039/c6cc00617e
83. Sarkar M, Liu Y, Morimoto J, Peng H, Aquino C, Rader C, Chiorazzi N, Kodadek T (2014) Recognition of antigen-specific B cell receptors chronic lymphocytic Leukemia patients by synthetic "Antigen Surrogates". Chem Biol 111:1670–1679
84. Sarkar M, Pascal BD, Steckler C, Micalizio GC, Kodadek T, Chalmers MJ (2013) Decoding split and pool combinatorial libraries with electron transfer dissociation tandem mass spectrometry. J Am Soc Mass Spec 24:1026–1036
85. Roy A, Koesema E, Kodadek T (2021) High-throughput quality control assay for the solid-phase synthesis of DNA-encoded libraries of macrocycles*. Angew Chem Int Ed Engl 60(21): 11983–11990. https://doi.org/10.1002/anie.202100230
86. Cochrane WG, Fitzgerald PR, Paegel BM (2021) Antibacterial discovery via phenotypic DNA-encoded library screening. ACS Chem Biol. https://doi.org/10.1021/acschembio.1c00714
87. Heinis C, Rutherford T, Freund S, Winter G (2009) Phage-encoded combinatorial chemical libraries based on bicyclic peptides. Nat Chem Biol 5(7):502–507. nchembio.184 [pii]. https://doi.org/10.1038/nchembio.184
88. Heinis C, Winter G (2015) Encoded libraries of chemically modified peptides. Curr Opin Chem Biol 26:89–98. https://doi.org/10.1016/j.cbpa.2015.02.008
89. Baeriswyl V, Rapley H, Pollaro L, Stace C, Teufel D, Walker E, Chen S, Winter G, Tite J, Heinis C (2012) Bicyclic peptides with optimized ring size inhibit human plasma Kallikrein and its orthologues while sparing paralogous proteases. ChemMedChem 7(7):1173–1176. https://doi.org/10.1002/cmdc.201200071
90. Carle V, Wu Y, Mukherjee R, Kong XD, Rogg C, Laurent Q, Cecere E, Villequey C, Konakalla MS, Maric T, Lamers C, Diaz-Perlas C, Butler K, Goto J, Stegmayr B, Heinis C (2021) Development of selective FXIa inhibitors based on cyclic peptides and their application for safe anticoagulation. J Med Chem 64(10):6802–6813. https://doi.org/10.1021/acs.jmedchem.1c00056
91. Maola K, Wilbs J, Touati J, Sabisz M, Kong XD, Baumann A, Deyle K, Heinis C (2019) Engineered peptide macrocycles can inhibit matrix metalloproteinases with high selectivity. Angew Chem Int Ed Engl 58(34):11801–11805. https://doi.org/10.1002/anie.201906791
92. Urech-Varenne C, Radtke F, Heinis C (2015) Phage selection of bicyclic peptide ligands of the Notch1 receptor. ChemMedChem 10(10):1754–1761. https://doi.org/10.1002/cmdc.201500261

93. Middendorp SJ, Wilbs J, Quarroz C, Calzavarini S, Angelillo-Scherrer A, Heinis C (2017) Peptide macrocycle inhibitor of coagulation factor XII with Subnanomolar affinity and high target selectivity. J Med Chem 60(3):1151–1158. https://doi.org/10.1021/acs.jmedchem.6b01548
94. Mudd GE, Brown A, Chen L, van Rietschoten K, Watcham S, Teufel DP, Pavan S, Lani R, Huxley P, Bennett GS (2020) Identification and optimization of EphA2-selective bicycles for the delivery of cytotoxic payloads. J Med Chem 63(8):4107–4116. https://doi.org/10.1021/acs.jmedchem.9b02129
95. Chen S, Morales-Sanfrutos J, Angelini A, Cutting B, Heinis C (2012) Structurally diverse cyclisation linkers impose different backbone conformations in bicyclic peptides. Chembiochem 13(7):1032–1038. https://doi.org/10.1002/cbic.201200049
96. Roberts RW, Szostak JW (1997) RNA-peptide fusions for the in vitro selection of peptides and proteins. Proc Natl Acad Sci U S A 94:12297–12302
97. Josephson K, Ricardo A, Szostak JW (2014) MRNA display: from basic principles to macrocycle drug discovery. Drug Discov Today 19(4):388–399. https://doi.org/10.1016/j.drudis.2013.10.011
98. Josephson K, Hartman MC, Szostak JW (2005) Ribosomal synthesis of unnatural peptides. J Am Chem Soc 127(33):11727–11735. https://doi.org/10.1021/ja0515809
99. Ohta A, Murakami H, Suga H (2008) Polymerization of alpha-hydroxy acids by ribosomes. Chembiochem 9(17):2773–2778. https://doi.org/10.1002/cbic.200800439
100. Ohuchi M, Murakami H, Suga H (2007) The flexizyme system: a highly flexible TRNA aminoacylation tool for the translation apparatus. Curr Opin Chem Biol 11(5):537–542. S1367-5931(07)00109-3 [pii]. https://doi.org/10.1016/j.cbpa.2007.08.011
101. Goto Y, Suga H (2021) The RaPID platform for the discovery of pseudo-natural macrocyclic peptides. Acc Chem Res 54(18):3604–3617. https://doi.org/10.1021/acs.accounts.1c00391
102. Kawakami T, Ohta A, Ohuchi M, Ashigai H, Murakami H, Suga H (2009) Diverse backbone-cyclized peptides via codon reprogramming. Nat Chem Biol 5(12):888–890. nchembio.259 [pii]
103. Kawakami T, Murakami H, Suga H (2008) Messenger RNA-programmed incorporation of multiple N-methyl-amino acids into linear and cyclic peptides. Chem Biol 15(1):32–42. S1074-5521(07)00447-4 [pii]. https://doi.org/10.1016/j.chembiol.2007.12.008
104. Nagano M, Huang Y, Obexer R, Suga H (2021) One-pot in vitro ribosomal synthesis of macrocyclic depsipeptides. J Am Chem Soc 143(12):4741–4750. https://doi.org/10.1021/jacs.1c00466
105. Goto Y, Ohta A, Sako Y, Yamagishi Y, Murakami H, Suga H (2008) Reprogramming the translation initiation for the synthesis of physiologically stable cyclic peptides. ACS Chem Biol 3(2):120–129. https://doi.org/10.1021/cb700233t
106. Reid PC, Goto Y, Katoh T, Suga H (2012) Charging of TRNAs using ribozymes and selection of cyclic peptides containing thioethers. Methods Mol Biol 805:335–348. https://doi.org/10.1007/978-1-61779-379-0_19
107. Fleming SR, Himes PM, Ghodge SV, Goto Y, Suga H, Bowers AA (2020) Exploring the post-translational enzymology of PaaA by MRNA display. J Am Chem Soc 142(11):5024–5028. https://doi.org/10.1021/jacs.0c01576
108. Peacock H, Suga H (2021) Discovery of de novo macrocyclic peptides by messenger RNA display. Trends Pharmacol Sci 42(5):385–397. https://doi.org/10.1016/j.tips.2021.02.004
109. Ottl J, Leder L, Schaefer JV, Dumelin CE (2019) Encoded library technologies as integrated lead finding platforms for drug discovery. Molecules 24(8):1629. https://doi.org/10.3390/molecules24081629
110. Buyanova M, Pei D (2022) Targeting intracellular protein-protein interactions with macrocyclic peptides. Trends Pharmacol Sci 43(3):234–248. https://doi.org/10.1016/j.tips.2021.11.008

111. Buller F, Mannocci L, Zhang Y, Dumelin CE, Scheuermann J, Neri D (2008) Design and synthesis of a novel DNA-encoded chemical library using Diels-Alder cycloadditions. Bioorg Med Chem Lett 18(22):5926–5931. https://doi.org/10.1016/j.bmcl.2008.07.038
112. Mannocci L, Zhang Y, Scheuermann J, Leimbacher M, De Bellis G, Rizzi E, Dumelin C, Melkko S, Neri D (2008) High-throughput sequencing allows the identification of binding molecules isolated from DNA-encoded chemical libraries. Proc Natl Acad Sci U S A 105(46): 17670–17675. https://doi.org/10.1073/pnas.0805130105
113. Nielsen J, Brenner S, Janda KD (1993) Synthetic methods for the implementation of encoded combinatorial chemistry. J Am Chem Soc 115(21):9812–9813. https://doi.org/10.1021/ja00074a063
114. Halpin DR, Harbury PB (2004) DNA display I. sequence-encoded routing of DNA populations. PLoS Biol 2(7):e173. https://doi.org/10.1371/journal.pbio.0020173
115. Dumelin CE, Trüssel S, Buller F, Trachsel E, Bootz F, Zhang Y, Mannocci L, Beck SC, Drumea-Mirancea M, Seeliger MW, Baltes C, Müggler T, Kranz F, Rudin M, Melkko S, Scheuermann J, Neri D (2008) A portable albumin binder from a DNA-encoded chemical library. Angew Chem Int Ed Engl 47(17):3196–3201. https://doi.org/10.1002/anie.200704936
116. Melkko S, Scheuermann J, Dumelin CE, Neri D (2004) Encoded self-assembling chemical libraries. Nat Biotechnol 22(5):568–574
117. Kinoshita Y, Nishigaki K (1995) Enzymatic synthesis of code regions for encoded combinatorial chemistry (ECC). Nucleic Acids Symp Ser 34:201–202
118. Li X, Liu DR (2004) DNA-templated organic synthesis: nature's strategy for controlling chemical reactivity applied to synthetic molecules. Angew Chem Int Ed 43(37):4848–4870

Top Med Chem (2022) 40: 41–64
https://doi.org/10.1007/7355_2022_146

Published online: 11 October 2022

Barcoding Strategies for the Synthesis of Genetically Encoded Chemical Libraries

Alice Lessing, Elena Detta, Jörg Scheuermann, and Andreas Brunschweiger

Contents

Abstract In this chapter, different encoding strategies for DNA- or PNA-encoded chemical libraries are presented. In general, DNA encoding can be distinguished into a DNA-recorded regimen, which encodes the individual chemical reaction steps, and which is most used in practice today, and a DNA-templated regimen where DNA templates guide the chemical synthesis. Various barcoding strategies have been employed most often for the generation of split-and-pool- or DNA-templated single-pharmacophore DELs, but also for dual-pharmacophore DELs displaying fragments/compounds on both ends of a DNA heteroduplex, both in a static and a dynamic setup. In addition to solution-phase protocols, encoded one-bead-one-

A. Lessing and J. Scheuermann (✉)
Department of Chemistry and Applied Biosciences, ETH Zürich, Institute of Pharmaceutical Sciences, Zürich, Switzerland
e-mail: joerg.scheuermann@pharma.ethz.ch

E. Detta and A. Brunschweiger (✉)
Department of Chemistry and Chemical Biology, TU Dortmund University, Dortmund, Germany
e-mail: andreas.brunschweiger@tu-dortmund.de

compound libraries and solid phase-initiated DNA-encoded library synthesis strategies are reported.

Keywords DNA-encoded libraries, DNA-routing, DNA-templated synthesis (DTS), Double-stranded DEL, Dual-pharmacophore DEL, Dynamic combinatorial DELs, Encoded combinatorial chemistry, Encoded self-assembled chemical (ESAC) libraries, Encoded solid-phase chemistry, Genetic barcode, Genetic tag, One-bead-one-compound libraries, PNA-encoded libraries, Single-pharmacophore DEL, Single-stranded DEL, Splint ligation, T4 ligation, yoctoReactor

1 Introduction: Encoding Molecules with Genetic Tags

Encoded chemical library technologies are based on the covalent linkage of chemically synthesized compounds to synthetic oligonucleotides that serve as barcode identifiers for the compound structure (Fig. 1a). This approach to screening library handling is commonly called phenotype–genotype coupling. It has its biological precedent in the phage-display, ribosome-display, or mRNA-display libraries which are widely used for screening for bioactive peptides and proteins. Like their biological relatives, genetically tagged chemically synthesized libraries can be screened for bioactive compounds on targets as complex mixtures, without the need for the large infrastructure needed to screen discrete libraries of compounds (high-throughput screening/HTS).

A seminal publication by Richard A. Lerner and Sydney Brenner in 1992 described for the first time the concept to synthesize molecular chimeras consisting of a genetic tag that was covalently connected to a chemically synthesized molecule [1]. In those early times of encoded chemistry, libraries were synthesized on beads that contained two linkers, one allowed for chemical coupling of amino acid building blocks by peptide coupling, the other for chemical coupling of DNA nucleotides by phosphoramidite nucleic acid chemistry [2, 3]. Coupling of amino acid building blocks and of DNA nucleotide building blocks was performed iteratively, and in combinatorial manner, so that each cycle of synthesis and encoding led to an exponential growth of the library, yet the different chemistries need for peptide synthesis and for phosphoramidite synthesis posed serious constraints to the initial setup. Since the infant days of encoded combinatorial chemistry, the field of encoded library technologies has branched out, and today several barcoding strategies are available for the synthesis of encoded libraries, which will be discussed in this chapter. They can broadly be divided into DNA-recorded techniques where the genetic tag serves to record the synthetic steps and techniques that use the DNA barcode for templating chemical reactions [4]. The class of DNA-recorded libraries comprises split-and-pool encoded solution-phase combinatorial synthesis [5–7], solid phase-initiated DNA-encoded combinatorial chemistry, and DNA-encoded solid-phase chemistry [8], as well as hybrid DNA/PNA-encoded chemistry

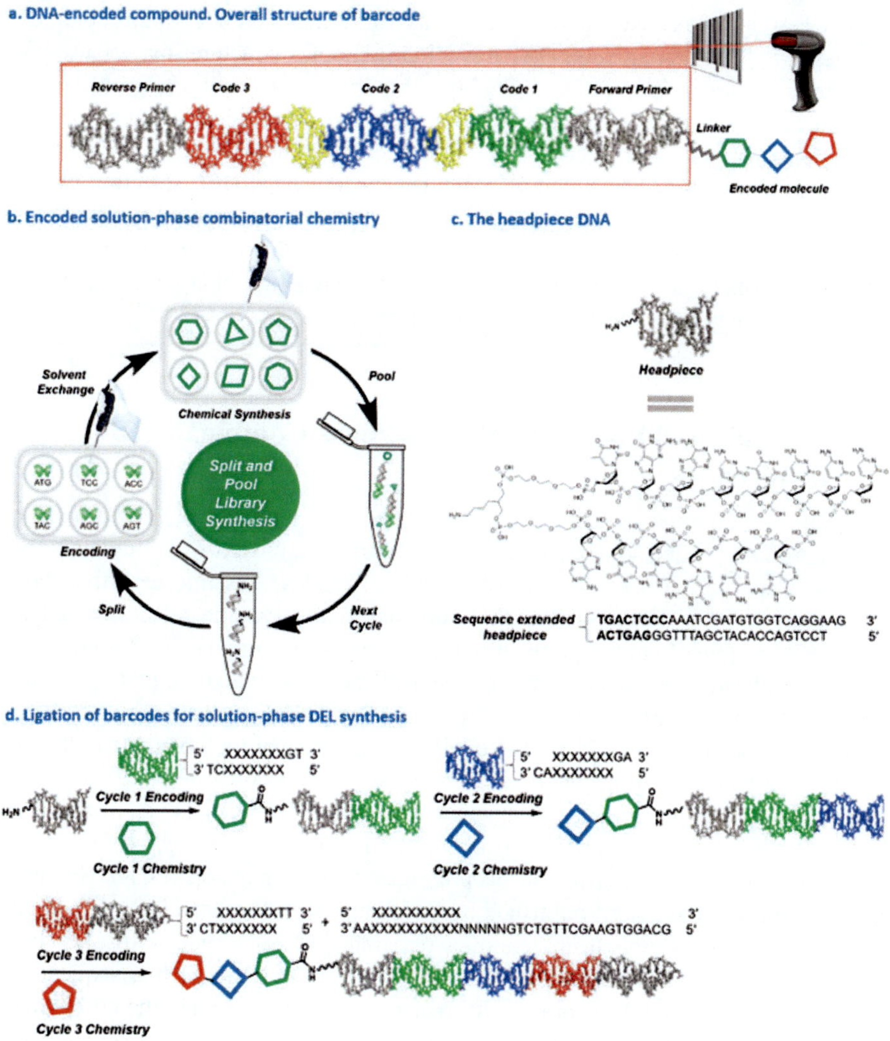

Fig. 1 Solution-phase synthesis of DNA-encoded libraries (DEL) by iterative cycles of chemical synthesis and barcode ligation. (**a**) Schematic presentation of a DNA-encoded molecule. (**b**) Split-and-pool encoded combinatorial chemistry workflow. (**c**) Structure and sequence of the "headpiece" DNA oligonucleotide. (**d**) Generic scheme of the synthesis of a DNA-encoded library initiated with the short hairpin duplex "headpiece" DNA

[9]. Further techniques that are based on recording reactions with DNA are dual-pharmacophore technologies and DNA-encoded dynamic combinatorial chemistry which enable both fragment screening and screening of complex molecules [10]. DNA-templated approaches comprise DNA-templated DEL synthesis which were mainly used for the construction of macrocyclic structures, and the yoctoReactor (TM) design, both using the DNA templates to direct the

oligonucleotide-linked reagents into proximity to the reaction site within complex mixtures [11, 12]. Alternatively, DNA/PNA hybrid duplex structures may be used in templated reaction strategies [9]. Notably, also the DNA-routing approach uses immobilized DNA strands to direct DNA templates to individual reaction vessels [13]. The use of single-stranded designs further enables the pairing with additional modular functionalities which can be exploited for advanced DEL selection strategies [1], such as photocrosslinking [2], affinity maturation of existing ligands [3], or the use of covalent or reversible-covalent warheads [4]. Such strategies may facilitate the identification of valuable compounds that bind to a biological target and serve as a starting point for drug development programs.

2 Split-and-Pool Encoded Solution-Phase Combinatorial Chemistry

Split-and-pool encoded solution-phase combinatorial chemistry, which today is the most used approach to construct DNA-encoded libraries, was first described in a publication in 1995 that described the combinatorial enzymatic ligation of DNA codes (Fig. 1) [14]. Seminal publications from the Neri group (ETH Zurich), and from a team headed by Barry Morgan, then at Praecis, described the first split-and-pool encoded library approaches [15, 16]. Morgan et al. designed a double-stranded DNA to encode compounds [15]. The overall architecture of their barcode is shown in Fig. 1a. It consisted of the encoded compound that was connected via a polyethylene glycol chain to both terminal nucleotides of the duplex DNA, forming a hairpin structure. Terminal primer regions of the DNA construct allowed for amplification of the genetic information by polymerase chain reaction (PCR). Suitable primer sequences have been published by different research groups [15, 16]. Alternatively, validated primers from genomics research, SELEX, or phage-display may be used, if researchers entering the DEL field wish to use other primer sequences. The number of internal degenerate 8–10 mer regions that contained the compound barcode mirrors the number of chemical reaction steps, and the number of barcodes per reaction step mirrors the number of chemical building blocks. Barcode sequences may be designed using freely available statistics software such as R, in order to avoid long homopurine or homopyrimidine sequences that may lead to sequencing artifacts and to define the Hamilton distance of barcodes. The Hamilton distance indicates the minimal number of different nucleosides between any two barcodes of a coding region. It is advisable to set the Hamilton distance to at least two for unambiguous compound identification by sequencing. The design of smaller libraries and/or longer coding regions would allow for a higher Hamilton distance. In the coding scheme published by Morgan et al. [16] any barcode sequences were connected by two-base regions that are needed for enzymatic barcode ligation with T4 ligase.

The workflow of encoded library synthesis was initiated with a short hairpin duplex DNA, called the "headpiece" that contained a polyethylene linker connecting the two arms of the hairpin (Fig. 1c). This linker carries a functional group for attaching the encoded molecule. The hairpin consisted of a duplex DNA sequence of six nucleotides and a two-base overhang that allowed for ligating a set of duplex (ds)DNA oligonucleotides with T4 ligase. This set of dsDNA oligonucleotides contained forward primer sequence, barcodes for the first set of chemical building blocks, and a two-base overhang for the second sticky-end ligation. Following barcode ligation, chemical building blocks were coupled to the linker part of the DNA constructs. As the linker moiety most often contained a primary amino group as a functional handle, any building blocks need to be substituted with a carboxylic acid, and at least one further functional group for subsequent encoded compound synthesis steps. In order to diversify the reagent space in this first cycle of library synthesis, alternative alkyl linkers were developed that contained a leaving group for substitution by nucleophiles [17]. Furthermore, chloroacetic acid or allyl-protected glycine was coupled to the linker, which allowed for synthesis of secondary amines from abundantly available primary amines and aldehydes, respectively [18, 19]. This process of encoded compound synthesis is called a cycle of encoded chemistry and concluded with a purification and pooling step (Fig. 1b). The last step is usually carried out by precipitation of the DNA conjugate from ethanolic solution. The cycles of synthesis and encoding are iterated as defined by library design. In the publication by Clark et al. [16], three and four cycles of encoded library synthesis have been described that yielded 7,000,000 and 800,000,000 encoded molecules, respectively (Fig. 1d). Libraries undergoing more combinatorial cycles have been synthesized to reach higher compound numbers. A library of encoded macrocycles was synthesized through six cycles of encoded peptide chemistry [20]. Today, libraries that are designed by three or even only two cycles [19] are increasingly being designed, likely, because higher numbers of synthesis steps tend to give rise to molecules with higher molecular weight and functional group compositions that may prove more challenging to develop toward drugs.

An alternative barcoding strategy developed by the Neri group used a long single-stranded (ss)DNA that contained a 5′-terminal linker for the encoded compound, the forward DNA primer, a compound barcode and region that was partially complementary to a set of ssDNA oligonucleotides containing the second barcode [16]. After coupling of compounds to the DNA oligonucleotides, all conjugates were pooled and split for the second cycle of encoded synthesis. Unlike the hairpin strategy that used T4 ligase and sticky-end ligation to concatenate barcode oligonucleotides prior to compound synthesis, the Neri group first performed compound synthesis, and then annealed a set of ssDNA barcode oligonucleotides to the first set of barcode ssDNA for primer elongation by Klenow fill-in (Fig. 2a-1). This strategy can be used to synthesize two-cycle libraries. A third cycle of library synthesis to reach a million-sized encoded library was demonstrated following digest of the construct with a restriction endonuclease and sticky-end ligation with T4 ligase of a third DNA barcode [21]. A more versatile encoding strategy for concatenating (theoretically) any number of ssDNA barcodes was developed by the Scheuermann

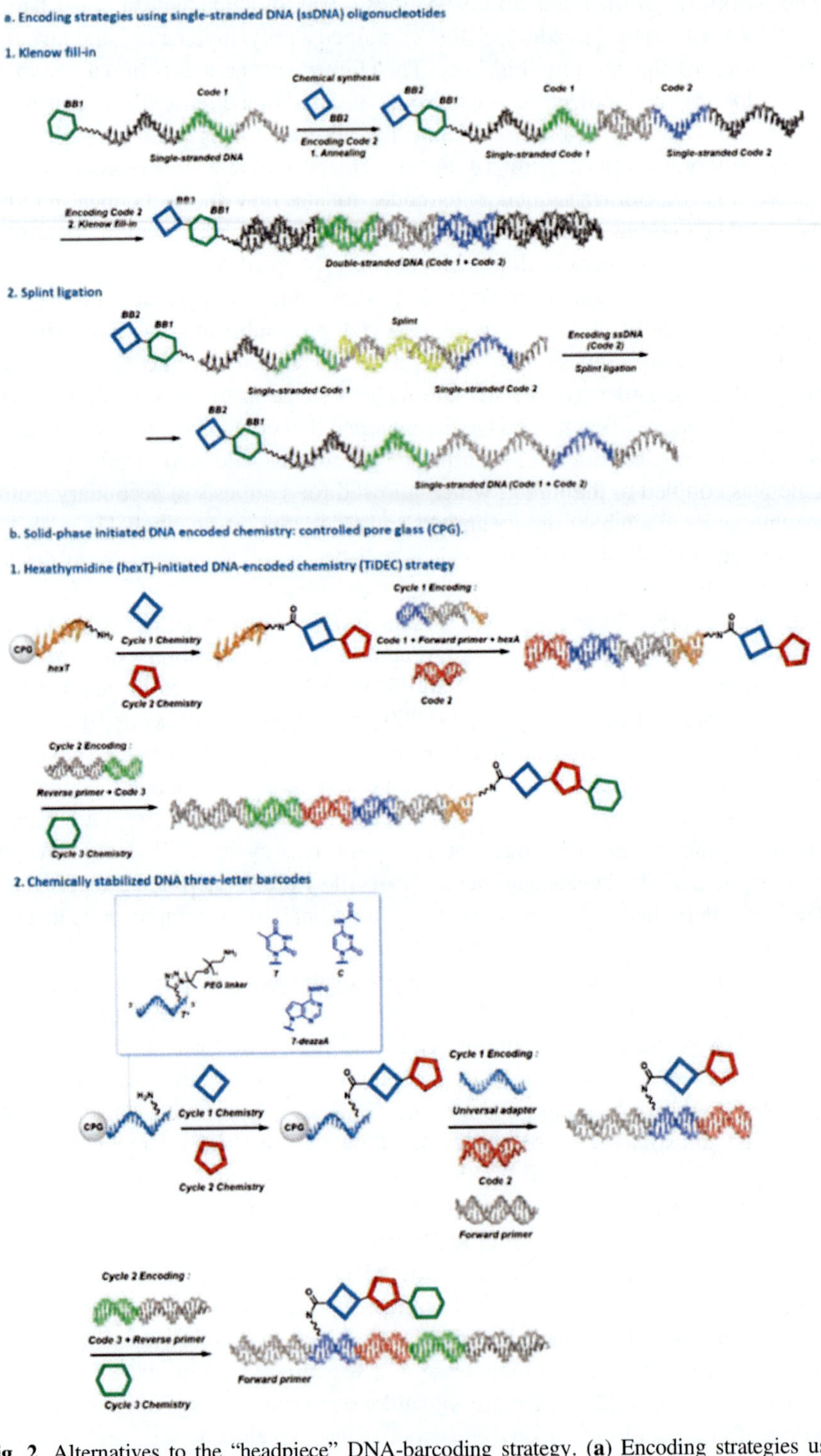

Fig. 2 Alternatives to the "headpiece" DNA-barcoding strategy. (**a**) Encoding strategies using Klenow fill-in or splint ligation for concatenation of single-stranded DNA barcodes. (**b**) Barcoding strategies initiated by DNA oligonucleotides coupled to controlled pore glass (CPG) solid phase

and Neri groups and used 20-mer splint oligonucleotides that were partially complementary to barcode oligonucleotides. Here, the internal compound barcodes were accordingly flanked by longer 10-mer sequences that were annealed to the splint ssDNA [22, 23] (Fig. 2a-2).

All solution-phase encoded library strategies require chemical building blocks and chemical reactions which are adapted to the process. The building blocks need to be functionalized to enable the encoded compound process, and there also needs to be a diverse set of them available to enable large split sizes, e.g. numbering hundreds of building blocks. The less cycles are used for DEL design, the more important is the ready availability of a class of building blocks. Library chemistry must be compatible with aqueous solvents and with the DNA chemical reactivity. Furthermore, methodology needs to be uniformly high-yielding and to perform robustly on a broad variety of substrates. The available reactions for library design and DNA damage reactions that need to be avoided are discussed in greater detail in chapter "Advancements in DEL-Compatible Chemical Reactions" of this volume.

In order to expand the scope of DEL design, the first cycles of library synthesis may optionally be carried out on DNA sequences covalently connected to a solid-phase matrix, for instance, on controlled pore glass (CPG) (Fig. 2b). This material is a standard solid phase for automated chemical oligonucleotide synthesis and allows for working under strictly dry conditions. Features of this approach are the option to perform reactions under strictly dry conditions, DNA nucleobase protection that prevents chemical deamination, and the possibility to remove reagent excess, and especially potentially harmful contaminants such as metal ions, under stringent washing conditions. Following on-DNA compound synthesis, the DNA oligonucleotides are ligated to further DNA barcodes. The DNA oligonucleotides may be ssDNAs that contain a partially primer sequence [24], primer and barcode [25, 26], short headpiece adapters such as the hexathymidine (*hexT*; Fig. 2b-1) headpiece [27, 28], or barcodes that are optionally chemically stabilized (*csDNA*; Fig. 2b-2) for greater reaction tolerance [29]. The latter two barcoding strategies used 4-mer overhangs to concatenate barcodes [30]. Both the hexT headpiece and chemically stabilized *csDNA* that is composed of pyrimidine nucleobases and 7-deazaadenine have been shown to tolerate reaction conditions that are hardly compatible with native DNA such as low pH and transition metal catalysis under forcing conditions. Thus, they expand the accessible chemical reaction space for solution-phase DEL design.

3 Split-and-Pool Encoded Solid-Phase Combinatorial Chemistry

The Kodadek and Paegel groups developed DNA-barcoding strategies for one-bead-one-compound (OBOC) synthesis (Fig. 3a) [31, 32]. Their DNA-encoded solid-phase synthesis (DESPS) approach employed a functionalized linker on Tentagel macrobeads (Fig. 3b). The bifunctional linker was functionalized with an azide to install a 6-mer dsDNA headpiece "HDNA" whose two strands were connected via an azide-substituted polyethylene glycol chain via copper-catalyzed azide alkyne cycloaddition (CuAAC). Like the headpiece solution-phase DEL strategy, the DESPS headpiece contained an overhang for ligation of the forward primer and first barcode (Fig. 3a). Following enzymatic ligation of the first dsDNA, the solvent could be exchanged by simple filtration, and a first encoded compound synthesis step was performed on the second arm of the bifunctional linker. This arm contained an amine as handle for encoded compound synthesis, and additionally a chromophore and an arginine residue to facilitate compound synthesis quality control by chromatographic and mass spectrometric analysis. Like the headpiece DEL strategy, encoded OBOC libraries were synthesized by iterative split-pool combinatorial cycles of compound synthesis and enzymatic ligation of DNA barcodes, here via three-base overhangs, resulting in oligomer compounds and DNA barcodes flanked by PCR primer sites. The reaction scope for this technology takes advantage of the option to exchange aqueous solvents by organic solvents and encompasses carbonyl chemistry, as well as cross-coupling reactions, aldol reactions, and a suite of aldehyde-diversifying reactions [8]. Encoded OBOC libraries may be screened by affinity-based selection screens [32]. In the future, they may offer the option to perform functional and even phenotypic screens, after cleavage of the encoded molecules from the DNA-encoded beads [33].

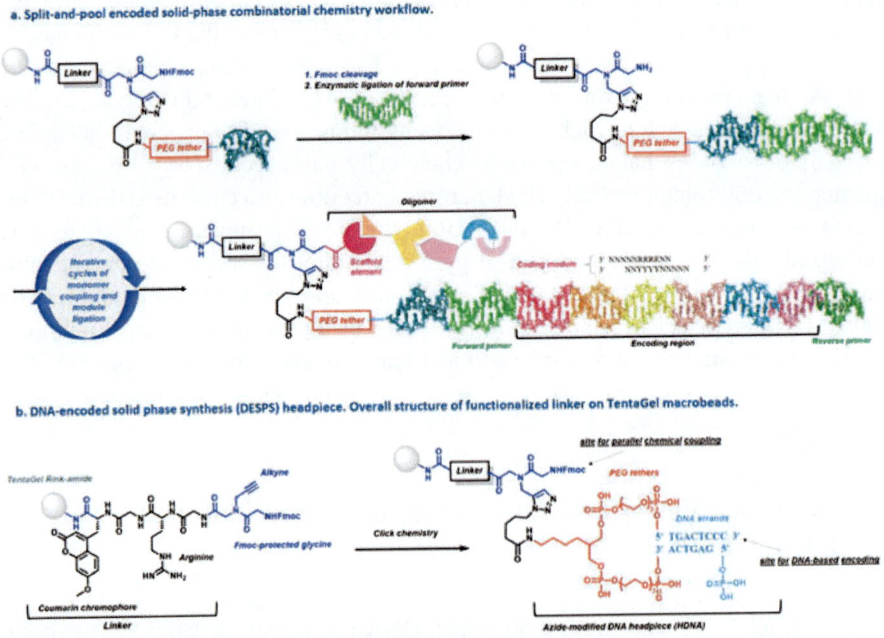

Fig. 3 DNA-encoding strategies for the synthesis of encoded one-bead-one compound (OBOC) libraries. (**a**) Generic workflow of the encoded OBOC library synthesis. (**b**) Structure of the solid phase for encoded OBOC libraries

4 PNA-Encoded Chemistry

Peptide-nucleic acid (PNA)-encoding strategies have been developed by the Winssinger group as an alternative to barcoding of compound synthesis with DNA [9]. Peptide-nucleic acids (PNAs) are artificially synthesized oligomers, having a backbone composed of repeating *N*-(2-aminoethyl)glycine units linked by peptide bonds, to which the various purine and pyrimidine bases are attached by a methylene bridge and a carbonyl group. PNA oligonucleotides specifically and strongly hybridize to complementary DNA or RNA sequences, forming duplexes that are resistant to degradation by either nucleases or proteases (Fig. 4a) [34].

Contrarily to unprotected DNA, PNA as coding oligomer does not easily undergo depurination, lacking the ribose sugar backbone and has a much higher thermal and pH-stability, enabling the use of a much broader spectrum of chemical methods for library preparation. Molecular diversity in PNA-encoded libraries can be achieved via a broad array of chemistries, including reactions catalyzed by transition metals, transformations yielding heterocyclic scaffolds, cyclization, and protic acid-promoted reactions [35, 36]. Currently, a considerable amount of peptide- and non-peptide-based PNA-encoded libraries have been reported, proving the broad applicability of the chemistry and the robustness of the encoding strategy [9].

However, the fundamental problem with PNA-encoding is that it does not benefit from the efficiency of enzymatic barcode ligation nor is the code amplifiable by PCR

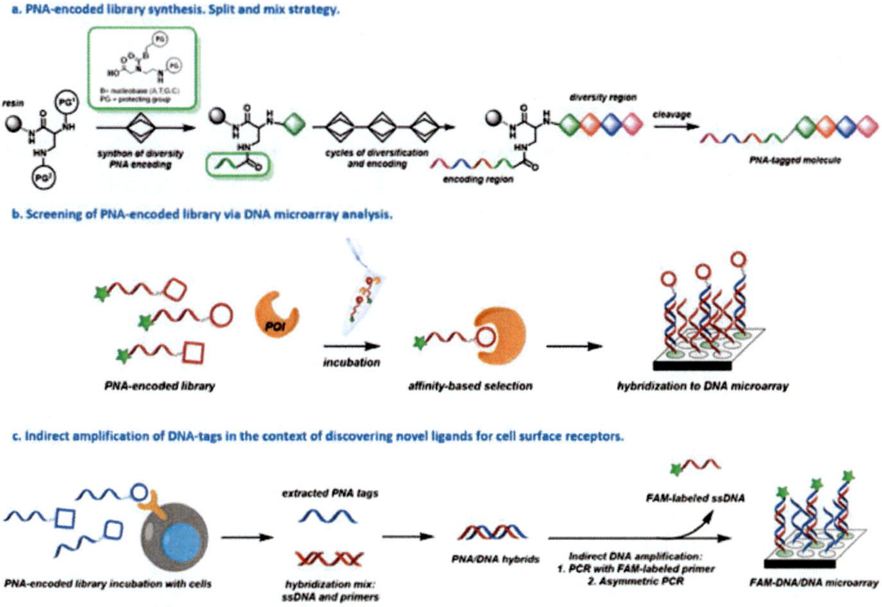

Fig. 4 Peptide-Nucleic Acid (PNA)-encoded chemistry. (**a**) Generic scheme of a PNA-encoded library synthesis. (**b**) Screening of PNA-encoded libraries on microarrays. (**c**) Indirect identification of PNA-encoded compounds by microarray analysis

or PCR-like techniques. The simplest format for the identification and quantification of selection outcomes from PNA-encoded libraries generally includes direct hybridization of the selected compounds, bearing a fluorophore on each PNA tag, to a DNA microarray (Fig. 4b) [37, 38]. Nevertheless, the success of this screening method strictly depends on the concentration of PNA tags, which should be sufficient for microarray-based detection [39].

To circumvent this limitation, novel screening strategies have been developed which offer the possibility to perform indirect DNA amplification of PNA tags prior to microarray analysis (Fig. 4c) [40, 41].

A merger of DNA- and PNA-encoding has been achieved with the development of a mating strategy that uses DNA template strands to direct the combinatorial ligation of PNA-encoded peptides. The resulting DNA/PNA-encoded peptides can be deconvoluted via amplification and sequencing of the DNA template strands [42].

5 Dual-Pharmacophore Encoding Strategies

In contrast to the DEL construction schemes described above featuring the display of just one chemical molecule on DNA (single-pharmacophore setup), dual-pharmacophore DELs feature the display of chemical entities (fragments and drug-like molecules) on the extremity of both the 5′ and 3′ end of DNA heteroduplexes. This approach had been pioneered by the Neri/Scheuermann group at ETH Zurich who reported the first such Encoded Self-Assembling Chemical (ESAC) library in 2004 [43]. Two sets of partially complementary sub-libraries, to which chemical compounds were conjugated at the 5′ and the 3′ terminus, respectively, were allowed to combinatorially assemble into stable DNA heteroduplexes (Fig. 5a). These fragment-like sub-libraries when assembled to a DNA heteroduplex would allow for a simultaneous screening of pairs of fragments for target binding, profiting from the chelate effect [44]. Initially, after PCR amplification ESAC libraries were read by hybridization to complementary oligonucleotides on microarrays, yet this read-out was performed individually for both sub-libraries, missing identification of pairs of synergistic binding moieties, initially limiting the potential of ESAC libraries to the lead expansion of known ligands [43]. The full potential of dual-display ESAC technology could only be uncovered using a novel setup that was compatible with the upcoming next-generation DNA sequencing. To this aim, a novel encoding methodology for enabling the transfer of the coding sequence from one strand to the other was proposed, based on the use of abasic DNA segments in one of the two sets of partially complementary oligonucleotides (Fig. 5a) [45]. The coding information of the 3′-strand is transcribed to the shorter 5′-strand by a DNA-polymerase assisted fill-in reaction (Fig. 5a). Together with the development of next-generation DNA sequencing, this method expanded the use of ESAC technology beyond affinity maturation and enabled the de novo discovery of pairs of synergistic binders [45]. A library consisting of 111′100 heteroduplexes obtained by the self-assembly of two mutually complementary sub-libraries of 550 × 202 fragments provided a

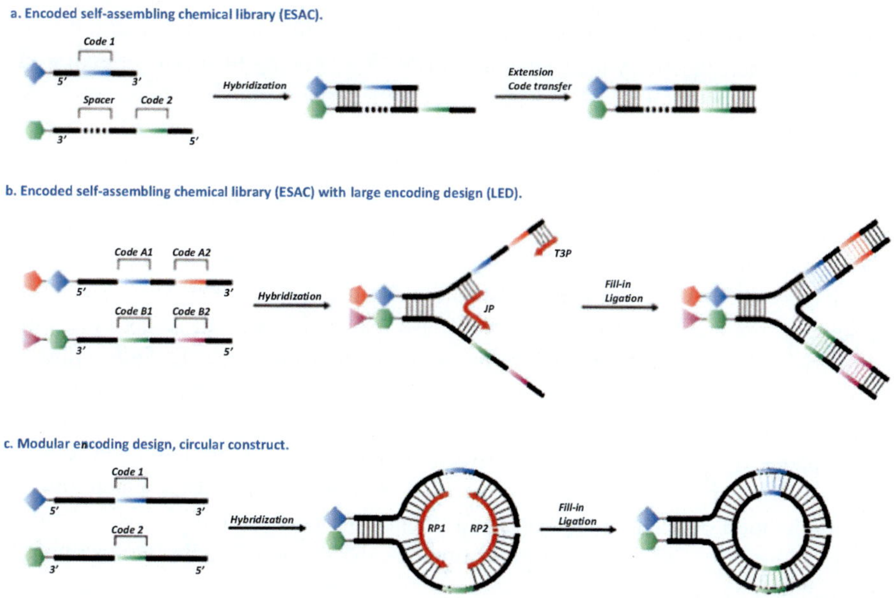

Fig. 5 Encoding strategies for dual-display DELs. (**a**) In encoded self-assembling chemical libraries (ESACs), two sub-libraries encoded with partially complementary DNA strands are constructed separately. Despite the variable sequences coding for the first fragment, hybridization of the two sub-libraries is possible owing to the introduction of an abasic spacer region. The extension of the shorter strand by a DNA-polymerase transfers the second code onto the first DNA strand. (**b**) A large-encoding design enables the construction of ESACs displaying more than three fragments. Two individually synthesized sub-libraries are stably annealed through a short hybridization region. After a fill-in reaction aided by annealing of a junction primer (JP) and terminal primer (T3P) and ligation, a single strand containing all coding sequences is generated. (**c**) A circular encoding construct can be generated through stable hybridization of two sub-libraries. Two relay primers assist the fill-in reaction followed by ligation, yielding a single strand containing all coding sequences

single pair of fragments which was strongly enriched against alpha-1-acid glycoprotein (AGP), while the individual fragments did not show significant enrichment in the selection and did not display a measurable Kd. Furthermore, in ESAC selections against carbonic anhydrase IX, a validated tumor-associated antigen, a highly enriched pair of acetazolamide (a known 20 nM CAIX ligand) and a phenolic compound could be identified with 40-fold improvement in affinity compared with acetazolamide alone, which is currently in clinical development. In order to further promote the applicability of dual-pharmacophore DEL technology, novel encoding methodologies have recently been postulated: "large-encoding design" (LED) (Fig. 5b) features the assembly and encoding of multi-building block sub-libraries, which allows for the construction and screening of DELs of unprecedented sizes and designs. This technology features sub-libraries with a stable hybridization domain and non-hybridizing encoding parts in a Y-shape manner. After combinatorial assembly and selection on a target of interest, the codes of the selected heteroduplexes can be PCR-amplified using a junction primer and a terminal primer and further ligated to a single PCR product containing all coding information and which can then be interrogated by next-generation DNA sequencing [46]. The selection of dual-pharmacophore DELs is generally performed on purified and immobilized proteins, but recent work on the screening of dual-pharmacophore DELs against carbonic anhydrase IX (CAIX) expressing tumor cells revealed the potential of on-cell selection protocols profiting from a substantial increase of ligand recovery and selectivity through the bivalent display of ligands [47].

A formidable challenge associated with the discovery of bidentate ligands from dual-display DELs relies on the identification of an optimal linker which connects the selected fragments/molecules. In most cases, several linkers differing in length, geometry, and rigidity may have to be tested, to identify the optimal solution for the scaffolding of synergistic pharmacophores. The choice of optimal linkers for the coupling of synergistic pharmacophores may be incorporated into future DNA-encoded library design. Recently, a strategy for "dimerizing" non-related single-stranded split-and-pool DELs into dual-pharmacophore DELs which allows for a modulation of fragment spacing and orientation has been devised as a circular construct (Fig. 5c), which may further broaden the scope of dual-pharmacophore DEL technology [10].

6 Conversion from Double-Stranded to Single-Stranded Library Format

A single-stranded library format is required for many of the recently reported advanced selection strategies. Apart from the deliberate single-stranded DEL synthesis two approaches have recently been proposed to convert double-stranded DELs into a single-stranded format. Selective digestion of one strand by exonuclease [48] or the use of a reversible-covalent DNA-headpiece designed to enable reversible interstrand photocrosslinking of a 3-cyanovinyl carbazole nucleoside on one strand with a thymine base on the opposite strand. Irradiation with either UV-A or UV-B light results in reversible conversion from the double- to the single-stranded DEL format [49]. Both methods hold the promise to enable transitioning between the robust double-stranded format and the versatile single stranded DEL format.

7 Dynamic Dual-Pharmacophore DEL Strategies

Dynamic combinatorial chemistry enables the identification of compounds binding to target structures with high affinity by connection of smaller fragments in the binding site [50]. In the context of DNA-encoded chemical libraries dynamic approaches rely on the incubation with the target to guide fragment assembly. In contrast to static dual-pharmacophore DNA-encoded libraries where a stable DNA heteroduplex is preformed, dynamic approaches feature a transiently stable DNA duplex upon incubation of the DNA-tagged fragment sub-libraries with the target. Tagging of fragments with short complementary DNA sequences allows for continuous shuffling of such tagged fragments, until pairs of fragments are binding to a target structure. After the equilibrium is reached, fragment pairs are locked to retain the information on binding pairs for decoding. A first pilot report was reported by Hamilton et al. in 2005, where a stable heteroduplex was heated above the melting temperature thus permitting the reshuffling of fragment pairs, which eventually lead to the enrichment of duplexes displaying the binding fragments [51]. About a decade later the dynamic concept in dual-pharmacophore DEL technology was elaborated independently by the groups of Yixin Zhang and Xiaoyu Li. A first dynamic proof-of-principle library was constructed and subjected to heating above the duplex melting temperature after the first round of selections. The resulting exchange of strands was found to lead to an enrichment of high-affinity duplexes [52]. A Y-shaped encoding setup with a very short hybridization region was proposed (Fig. 6a) where the locking of the equilibrium was accomplished by enzymatic ligation [53]. Further methods were established in order to lock the equilibrium state, e.g., by photocrosslinking of the two DNA strands [54], or by using short, complementary anchor DNA sequences [54] (Fig. 6b). More recently, chemical crosslinking by using *p*-stilbazole- or psoralen-modified DNA strands was proposed and tested [55, 56]. An interesting approach to facilitate ligand development after the identification of binding small molecule pairs, by linking of the small molecules rather than the two DNA strands was recently devised. Fragment conjugation was performed by imine formation of a single-stranded DEL containing a primary amine and a known ligand containing an aldehyde group, resulting in a dynamic imine library. After incubation with the target, the equilibrium could be locked by reduction of the imine [57]. Recently, an interesting PNA-based dynamic combinatorial

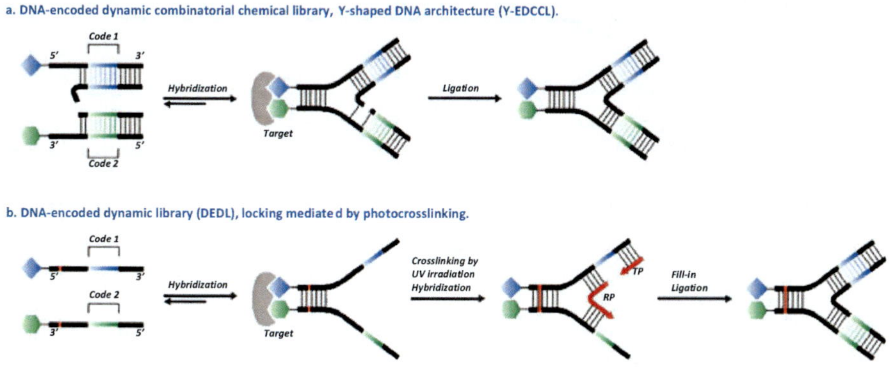

Fig. 6 In DNA-encoded dynamic combinatorial chemical libraries, the duplexes are unstable and duplexes binding with high affinity are stabilized by incubation with the target. (**a**) For Y-EDCCL, the equilibrium reached after incubation with the target was locked by ligation of both strands. (**b**) After incubation with the target, the equilibrium can be locked by crosslinking prompted by UV irradiation. A relay-primer-bypass strategy relying on a terminal primer and a relay primer and subsequent ligation affords a single DNA strand comprising all coding sequences

supramolecular approach has been reported which is based on the use of a constant short PNA tag to direct the combinatorial pairing of fragments [58]. While the concept of dynamic DELs is very promising, it must be mentioned that, until today, no real, large dynamic DEL, numbering hundreds of thousands, if not millions of compounds, has been reported in literature.

8 DNA-Templated Synthesis

An alternative strategy for the assembly of DNA-encoded chemical libraries features the use of DNA-templated synthesis (DTS) [24, 59, 60] (Fig. 7a). In this case, pre-defined DNA templates are employed both for library encoding and for directing the library construction. This setup was pioneered by David R. Liu et al. in 2001 when they demonstrated that chemical reactions are promoted by bringing DNA-linked reagents into close proximity through Watson–Crick base-pairing [24]. The effective molarity of such reactions may be very high, thus allowing to conduct reactions which are otherwise considered difficult or impossible to implement with conventional chemistry. The concept was used not only for reaction discovery [61] but also for the construction of DNA-encoded, macrocyclic libraries. Small organic compounds chemically linked to short single-stranded biotinylated oligonucleotides as donors are transferred upon hybridization to suitable

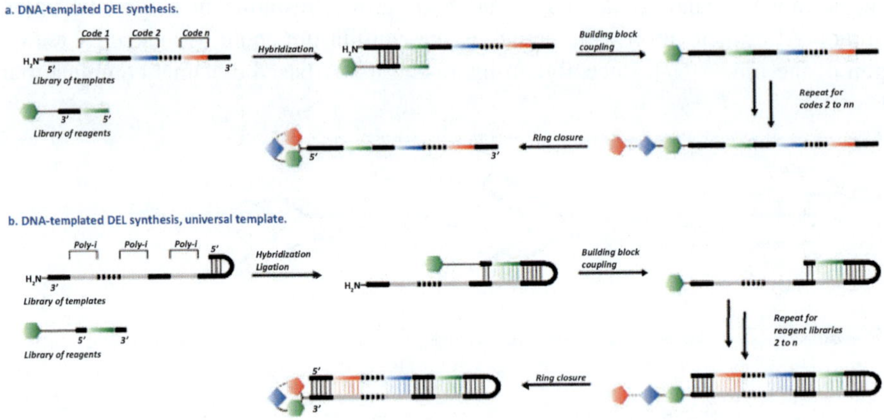

Fig. 7 DNA-templated synthesis uses pre-defined DNA template strands to direct library construction. (**a**) The hybridization of a library of DNA templates with a library of DNA-linked reagents brings reactants into proximity. Once the coupling of the first building block is complete, the hybridization and reaction steps are sequentially repeated for the remaining reagent libraries. (**b**) A universal template featuring tri-deoxyinosine anticodons enables hybridization with encoded reagents irrespective of their coding DNA sequence. After hybridization, the reagent DNA is ligated to the universal template and the chemical reaction takes place. This process is repeated for the remaining reagent libraries

complementary DNA sequences on the template, followed by cleavage and avidin-assisted "donor" strand removal [24]. In a stepwise fashion the template library is annealed with a first set of code-complementary oligonucleotides bearing each a different reactant, which is then chemically attached to the template (Fig. 7a). The linker between the short oligonucleotide and the reactant is subsequently cleaved and this process could be repeated two more times leading to a trimeric linear library. Finally, the library may be cyclized and used for selection experiments against target proteins. The first reported DTS-DEL comprised 65 library members made from three sets of four amino acids, followed by cyclization through Wittig olefination (see Fig. 4).

This proof-of-principle study was successful in selections against carbonic anhydrase. After technical optimization, a similar, yet larger macrocyclic library of 13,824 DNA-encoded compounds was constructed and reported in 2008 [24, 62], yielding hits for Src kinase and VEGFR2 [63], and six macrocycles against insulin degrading enzyme (IDE), which were also co-crystalized [64, 65].

In 2018, Usanov, Liu, and coworkers published a second-generation DNA-templated library of 256,000 macrocyclic members [66]. To do so, the DNA templates were computationally optimized regarding the orthogonality of each hybridization sequence and template assembly was improved through a polymerase-mediated strategy. In addition, macrocyclic Kihlberg rules [67] were applied to the design. With this DEL, selections were performed against IDE. One macrocyclic hit showed high potency with an IC_{50} of 40 nM [66]. The company Ensemble Therapeutics, co-founded by David R. Liu, used DTS for the construction of macrocyclic DELs in an industrial setting. In 2015, the company reported a DNA-templated macrocyclic DEL in collaboration with Bristol Meyer Squibb, consisting of five sets of building blocks which were eventually cyclized by a copper-catalyzed azide/alkyne cycloaddition (CuAAC) reaction [67]. Four libraries of 40,000 members were screened against XIAP BIR2 and BIR3 domains [68] and inhibitors with the ability to displace bound pro-apoptotic caspases were identified [69]. The positive aspects of DTS for DEL construction certainly encompass high chemical yields, access to unusual chemical reactions, one-pot reactions with reagent oligonucleotide conjugates, as well as specific purification methods [59]. On the other hand, template library generation, code-specificity for a larger set of building blocks, as well as the creation of larger sets of reagent oligonucleotide conjugates, to reach split sizes of, e.g., a few hundred building blocks can prove challenging. In 2013, Li and coworkers reported a smart solution for omitting the tedious template library generation by proposing a "universal template" consisting of a DNA hairpin containing tri-deoxyinosine anticodons which allow for the indiscriminate base-pairing and subsequent ligation with 3-base encoded reagent oligonucleotide conjugates (Fig. 7b). The feasibility of this DTS approach could be demonstrated with proof-of-principle DELs of 64*64*28 3-base codons [70].

9 DNA-Routing

The group of P. Harbury described "DNA-routing" [71] as a different DNA-template based strategy for library construction. This approach enables combinatorial chemistry by iterative sequence-specific immobilization-reaction and purification–elution step. A library of single-stranded DNA templates comprising the codes for the routing is subjected to resin columns with the immobilized complementary DNA code sequences and subsequently chemically modified (Fig. 8). After repooling, this split-and-pool step is then repeated for all diversity elements. In 2004, Halpin and Harbury described the construction of a combinatorial peptide library using a 340-mer oligonucleotide combinatorial template library consisting of alternating 20 base coding and 20 base constant regions obtained by PCR assembly of overlapping complementary 40-mer oligonucleotides [72]. To serve the purpose of routing on complementary DNA resin the double-stranded template library then was converted into single-stranded DNA by reverse-transcription and base-mediated hydrolysis of the RNA strand.

The routing anticodon resin is followed by a chemistry step performed on weak anion-exchange columns, varying high and low salt conditions [73]. The combined eluates after the reaction step can subsequently undergo further split-and-pool cycles. The routing can be performed sequence-specifically and efficiently with overall yield of 0.85 per hybridization round. Using this DNA-routing strategy an *N*-acylated pentapeptide library with a complexity of 10^6 was constructed [71], using 10 different amino acid building blocks for the first positions and nine carboxylic acids for the *N*-acylation step. The library included acylated leucine-enkephalin pentapeptides as a control and was subjected to affinity-based selections against

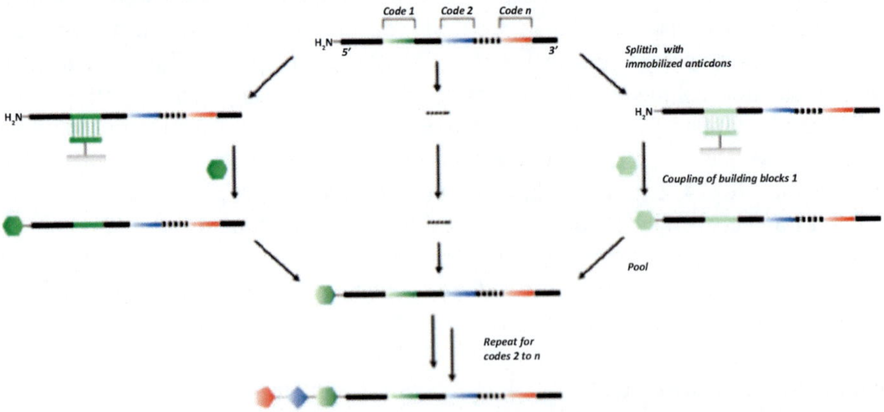

Fig. 8 A library of single-stranded DNA templates serves as the starting point for the construction of DELs by DNA-routing. DNA templates are separated by hybridization with anticodons immobilized on resin columns before being chemically modified. After pooling of all eluates, further rounds of sequence-specific separation and reaction can be carried out

the monoclonal antibody 3-E7 [74], a leucine-enkephalin binding antibody with 7.1 nM affinity.

The routing system was further optimized for peptide coupling yielding >90% conversion per reaction step and an 8-mer library of 100 Mio compounds was created and selected against the *N*-terminal SH3 domain of the proto-oncogene Crk [13]. Evolving over six generations, the hits converged to a small number of peptides with micromolar affinity. Further refinement allowed for a highly parallel performance in 384-well format [75] and the routing system, acting in analogy as an expanded genetic code, was tested for chemical evolution on protein kinase A (PKA) [76]. While the sole use of peptide bond chemistry and the sophisticated code design may pose restraints, the routing system allows for the generation of very large DNA-encoded libraries, numbering millions of compounds.

10 DNA-Directed Chemistry: The "yoctoReactor"

The final DNA-directed approach to be discussed in this chapter is the "yoctoReactor" or "yR" technology. The hallmark of the yR is a DNA sequence design that leads to self-assembly of two-four oligonucleotides which form a cavity with the name-giving volume of a yoctoliter, "yocto" being the lowest prefix in the SI system [12, 77]. Each of these DNA oligomers carries a functionalized chemical building block via a linker positioned within the sequence, and it is composed of a partial sequence for annealing to give the three- or four-way junction (the yR), and a partial sequence that encodes its building block (Fig. 9). The first set of encoded building blocks, i.e. the equivalent of the cycle 1 building blocks in encoded

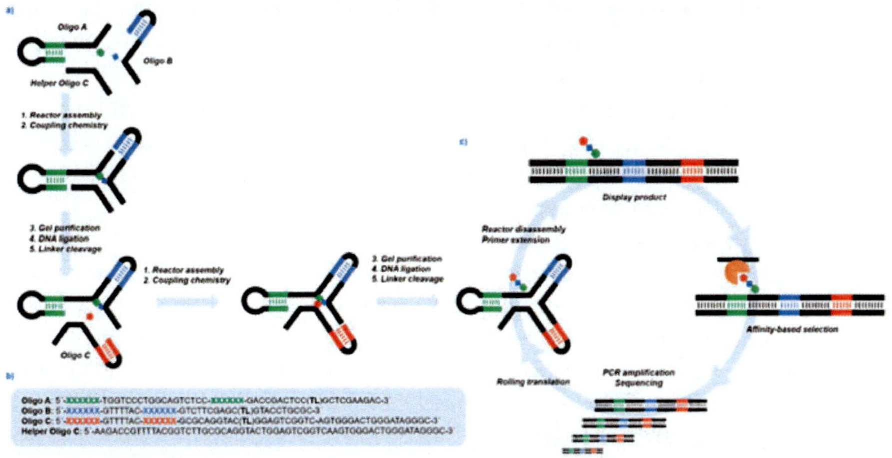

Fig. 9 Principles of the "yoctoReactor" (yR) technology. (**a**) Stepwise library assembly using the three-way form of the yR; (**b**) Sequences of the oligonucleotides; (**c**) Exemplary round of selection, amplification, and translation

combinatorial chemistry, is coupled to its DNA oligomers via a stable linker, while the cycle 2 and 3 building blocks are coupled to their barcodes via a scissile linker. Self-assembly of the DNA oligomers to the yR directs the chemical building blocks into proximity for a chemical reaction that couples them to the target molecule. Compound synthesis thus leads in effect to covalent coupling of the DNA barcodes. These constructs can be purified from any non-reacted starting materials by denaturing polyacrylamide gel electrophoresis (PAGE). Thus, the purification step of the anneal-react-purify sequence provides high fidelity in library synthesis, as only the successfully synthesized molecules will enter the encoded library. This fidelity constitutes a major advantage versus DNA-recorded chemistry. On the other hand, the rules for DNA-compatible chemistry apply to the yR, too, thus, carbonyl chemistry and nucleophile substation are prevalent in library design; the choice of building blocks for DEL design is limited to those that contain at least two functional groups, precluding tapping into the vast pool of monofunctionalized chemicals; and an additional chemical step is needed to couple building blocks to the barcodes. Following PAGE, the barcodes of the purified products are ligated to concatenate the genetic information of the compounds. Then, the scissile linker is cleaved to provide the final encoded molecules of a reaction step. As a last step, prior to screening, the complementary strand of the barcodes is formed by primer extension. As the yR can be constructed by annealing the DNA oligomers in combinatorial cycles, this technology furnished millions of encoded compounds, too [78].

11 Conclusion

Taken together the various encoding strategies for DELs have enabled the broad field of DEL technology of today. While DNA-templated synthesis and DNA-routing have led to various DELs, split-and-pool-based methods nowadays are mostly used in practice (Table 1).

While for DNA-recorded DELs most care is generally taken for the optimization on the "chemical" side, i.e., by employing high-yielding DNA-compatible reactions to achieve diversity, the quality of the encoding also critically matters for the performance of DELs in selections.

Encoding in single-stranded DNA format may serve important advantages for DEL selection protocols, since such DELs can be paired with either further single-stranded libraries to form dual-display DELs or by pairing them with known leads, covalent warheads, or with photoreactive moieties displayed on the complementary strand, therefore strategies have been devised for both creating single-stranded DELs by ligation and converting a double-stranded DEL into a single-stranded one {Zhao, 2022 #40; Huang, 2022 #41; Gui, 2022 #42} [1, 11, 12]. Sophisticated selection protocols will expand the scope of DEL technology for the identification of small bioactive compounds.

Table 1 Comparison of barcoding strategies

Encoding technique	Pros	Cons
Solution-phase combinatorial chemistry with duplex DNA (Fig. 1d)	Efficiency of library synthesis	Low(er) fidelity, DNA- and water/moisture-compatible chemistry
Solution-phase combinatorial chemistry with single-stranded DNA (Fig. 2a)	Efficiency of library synthesis, single-stranded DNA can be paired with modular functionalized second DNA strand	Low(er) fidelity, DNA- and water/moisture-compatible chemistry
"hexT"-barcoding strategy (Fig. 2b)	Fidelity, broad scope of chemical reactions	Low throughput, HPLC purification step
"csDNA"-barcoding strategy (Fig. 2b)	Fidelity, broad scope of chemical reactions	Low(er) throughput, HPLC purification step, upscaling to larger library sizes to be proven
Encoded solid-phase chemistry (Fig. 3)	Organic solvents usable, functional screening and off-DNA screening conceptually possible	Low(er) fidelity, technology not robustly proven
PNA-encoded libraries (Fig. 4)	Chemical stability of the barcode, broad compound reaction scope	Barcode not directly amplifiable, lower throughput, cost of the barcode
Dual-display libraries (Fig. 5)	Fidelity, enables fragment screening	Linkage of pharmacophores to be explored
Dynamic combinatorial encoded libraries (Fig. 6)	Fidelity, enables fragment screening	Functionalized building blocks required, limited chemistry scope
DNA-templated chemistry (Fig. 7)	Fidelity	Tedious synthesis of DNA-conjugates, limited to bifunctional building blocks, limited chemistry scope, code fidelity for larger libraries
DNA-routing (Fig. 8)	Fidelity, organic solvents	Limited chemistry scope
"yoctoreactor" (Fig. 9)	Fidelity	Tedious synthesis of DNA-conjugates, limited to bifunctional building blocks, limited chemistry scope, code fidelity for larger libraries

Acknowledgments The authors would like to thank Lars Kolster Petersen (Vipergen ApS) for helpful discussions.

Compliance with Ethical Standards Elena Detta declares no conflict of interest.

Andreas Brunschweiger is co-founder of the company Serengen GmbH that offers DNA-encoded library services.

This chapter is a review and no humans or animals were involved in the preparation of the chapter.

References

1. Brenner S, Lerner RA (1992) Encoded combinatorial chemistry. Proc Natl Acad Sci U S A 89: 5381–5383
2. Needels MC, Jones DG, Tate EH, Heinkel GL, Kochersperger LM, Dower WJ, Barrett RW, Gallop MA (1993) Generation and screening of an oligonucleotide-encoded synthetic peptide library. Proc Natl Acad Sci U S A 90:10700–10704
3. Nielsen J, Brenner S, Janda KD (1993) Synthetic methods for the implementation of encoded combinatorial chemistry. J Am Chem Soc 115:9812–9813
4. Shi B, Zhou Y, Huang Y, Zhang J, Li X (2017) Recent advances on the encoding and selection methods of DNA-encoded chemical library. Bioorg Med Chem Lett 27:361–369
5. Goodnow Jr RA, Dumelin CE, Keefe AD (2017) DNA-encoded chemistry: enabling the deeper sampling of chemical space. Nat Rev Drug Discov 16:131–147
6. Gironda-Martínez A, Donckele EJ, Florent Samain F, Neri D (2021) DNA-encoded chemical libraries: a comprehensive review with successful stories and future challenges. ACS Pharmacol Transl Sci 4:1265–1279
7. Satz AL, Brunschweiger A, Flanagan ME, Gloger A, Hansen NJV, Kuai L, Kunig VBK, Lu X, Madsen D, Marcaurelle LA, Mulrooney C, O'Donovan G, Sakata S, Scheuermann J (2022) DNA-encoded chemical libraries. Nat Rev Methods Primers 2:3
8. Fitzgerald PR, Paegel B (2021) DNA-encoded chemistry: drug discovery from a few good reactions. Chem Rev 121:7155–7177
9. Zambaldo C, Barluenga S, Winssinger N (2015) PNA-encoded chemical libraries. Curr Opin Chem Biol 26:8–15
10. Oehler S, Plais L, Bassi G, Neri D, Scheuermann J (2021) Modular assembly and encoding strategies for dual-display DNA-encoded chemical libraries. Chem Commun 57:12289–12292
11. Kleiner RA, Dumelin CE, Liu DR (2011) Small-molecule discovery from DNA-encoded chemical libraries. Chem Soc Rev 40:5707–5717
12. Blakskjaer P, Heitner T, Hansen NJV (2015) Fidelty by design: Yoctoreactor and binder trap enrichment for small-molecule DNA-encoded libraries and drug discovery. Curr Opin Chem Biol 26:62–71
13. Wrenn SJ, Weisinger RM, Halpin DR, Harbury PB (2007) Synthetic ligands discovered by in vitro selection. J Am Chem Soc 129:13137–13143
14. Kinoshita Y, Nishigaki K (1995) Enzymatic synthesis of code regions for encoded combinatorial chemistry (ECC). Nucleic Acids Symp Ser 34:201–202
15. Clark MA, Acharya RA, Arico-Muendel CC, Belyanskaya SL, Benjamin DR, Carlson NR, Centrella PA, Chiu CH, Creaser SP, Cuozzo JW, Davie CP, Ding Y, Franklin GJ, Franzen KD, Gefter ML, Hale SP, Hansen NJV, Israel DI, Jiang J, Kavarana MJ, Kelley MS, Kollmann CS, Li F, Lind K, Mataruse S, Medeiros PF, Messer JA, Myers P, O'Keefe H, Oliff MC, Rise CE, Satz AL, Skinner SR, Svendsen JL, Tang L, van Vloten K, Wagner RW, Yao G, Zhao B, Morgan BA (2009) Design, synthesis and selection of DNA encoded small-molecule libraries. Nat Chem Biol 5:647–654
16. Mannocci L, Zhang Y, Scheuermann J, Leimbacher M, De Bellis G, Rizzi E, Dumelin C, Melkko S, Neri D (2008) High-throughput sequencing allows the identification of binding molecules isolated from DNA-encoded chemical libraries. Proc Natl Acad Sci U S A 105: 17670–17675
17. Sun Z-M, Yang S-G, Xue L-J, Zhang J, Yang K, Hu Y-J (2022) N-alkyl linkers for DNA-encoded chemical libraries. Chem Asian J 17:e202200016
18. Franzini RM, Samain F, Abd Elrahman M, Mikutis G, Nauer A, Zimmermann M, Scheuermann J, Hall J, Neri D (2014) Systematic evaluation and optimization of modification reactions of oligonucleotides with amines and carboxylic acids for the synthesis of DNA-encoded chemical libraries. Bioconjug Chem 25:1453–1461
19. Rianjongdee F, Atkinson SJ, Chung CW, Grandi P, Gray JRJ, Kaushansky LJ, Medeiros P, Messenger C, Phillipou A, Preston A, Prinjha RK, Rioja I, Satz AL, Taylor S, Wall ID, Watson

RJ, Yao G, Demont EH (2021) Discovery of a highly selective BET BD2 inhibitor from a DNA-encoded library technology screening hit. J Med Chem 64:10806–10833

20. Zhu Z, Shaginian A, Grady LC, O'Keeffe T, Shi XE, Davie CP, Simpson GL, Messer JA, Evindar G, Bream RN, Thansandote PP, Prentice NR, Mason AM, Pal S (2018) Design and application of a DNA-encoded macrocyclic peptide library. ACS Chem Biol 13:53–59
21. Buller F, Steiner M, Frey K, Mircsof D, Scheuermann J, Kalisch M, Bühlmann P, Supuran CT, Neri D (2011) Selection of carbonic anhydrase IX inhibitors from one million DNA-encoded compounds. ACS Chem Biol 6:336–344
22. Favalli N, Bassi G, Pellegrino C, Millul J, De Luca R, Cazzamalli S, Yang S, Trenner A, Mozaffari NL, Myburgh R, Moroglu M, Conway SJ, Sartori AA, Manz MG, Lerner RA, Vogt PK, Scheuermann J, Neri D (2021) Stereo- and regiodefined DNA-encoded chemical libraries enable efficient tumour-targeting applications. Nat Chem 13:540–548
23. Bassi G, Favalli N, Vuk M, Catalano M, Martinelli A, Trenner A, Porro A, Yang S, Lean Tham C, Moroglu M, Yue WW, Conway SJ, Vogt PK, Sartori AA, Scheuermann J, Neri D (2021) A single-stranded DNA-encoded chemical library based on a stereoisomeric scaffold enables ligand discovery by modular assembly of building blocks. Adv Sci 7:2001970
24. Gartner ZJ, Tse BN, Grubina R, Doyon JB, Snyder TM, Liu DR (2004) DNA-templated organic synthesis and selection of a library of macrocycles. Science 305:1601–1605
25. Franzini RM, Ekblad T, Zhong N, Wichert M, Decurtins W, Nauer A, Zimmermann M, Samain F, Scheuermann J, Brown PJ, Hall J, Gräslund S, Schüler H, Neri D (2015) Identification of structure-activity relationships from screening a structurally compact DNA- encoded chemical library. Angew Chem Int Ed 54:3927–3931
26. Yuen LH, Dana S, Liu Y, Bloom SI, Thorsell AG, Neri D, Donato AJ, Kireev DB, Schuler H, Franzini RM (2019) A focused DNA-encoded chemical library for the discovery of inhibitors of NAD^+-dependent enzymes. J Am Chem Soc 141:5169–5181
27. Klika Škopić M, Salamon H, Bugain O, Jung K, Gohla A, Doetsch LJ, dos Santos D, Bhat A, Wagner B, Brunschweiger A (2017) Acid- and Au(I)-mediated synthesis of hexathymidine-DNA-heterocycle chimeras, an efficient entry to DNA-encoded libraries inspired by drug structures. Chem Sci 8:3356–3361
28. Kunig VBK, Potowski M, Akbarzadeh M, Klika Škopić M, dos Santos SD, Arendt L, Dormuth I, Adihou H, Blaž Andlovic B, Karatas H, Shaabani S, Zarganes-Tzitzikas T, Neochoritis CG, Ran Zhang R, Groves M, Guéret SM, Ottmann C, Rahnenführer J, Fried R, Dömling A, Brunschweiger A (2020) TEAD–YAP interaction inhibitors and MDM2 binders from DNA-encoded indole-focused Ugi peptidomimetics. Angew Chem Int Ed 59:20338–20342
29. Potowski M, Kunig VBK, Eberlein L, Vakalopoulos A, Kast SM, Brunschweiger A (2021) Chemically stabilized DNA barcodes for DNA-encoded chemistry. Angew Chem Int Ed 60: 9744–19749
30. Potapov V, Ong JL, Kucera RB, Langhorst BW, Bilotti K, Pryor JM, Cantor EJ, Canton B, Knight TF, Evans TC, Lohman GJS (2018) Comprehensive profiling of four base overhang ligation fidelity by T4 DNA ligase and application to DNA assembly. ACS Synth Biol 7:2665–2674
31. MacConnell AB, McEnaney PJ, Cavett VJ, Paegel BM (2015) DNA-encoded solid-phase synthesis: encoding language design and complex oligomer library synthesis. ACS Comb Sci 17:518–534
32. Mendes KR, Malone ML, Ndungu JM, Suponitsky-Kroyter I, Cavett VJ, McEnaney PJ, MacConnell AB, Doran TM, Ronacher K, Stanley K, Utset O, Walzl G, Paegel BM, Kodadek T (2016) High-throughput identification of DNA-encoded IgG ligands that distinguish active and latent mycobacterium tuberculosis infections. ACS Chem Biol 12:234–243
33. Hackler AL, Fitzgerald FG, Dang VQ, Satz AL, Paegel BM (2020) Off-DNA DNA-encoded library screening. ACS Combi Sci 22:25–34
34. Pothukanuri S, Pianowski Z, Winssinger N (2008) Expanding the scope and orthogonality of PNA synthesis. Eur J Org Chem 18:3141–3148

35. Chouikhi D, Ciobanu M, Zambaldo C, Duplan V, Barluenga S, Winssinger N (2012) Expanding the scope of PNA-encoded synthesis (PES): Mtt-protected PNA fully orthogonal to fmoc chemistry and a broad array of robust diversity-generating reactions. Chemistry 18: 12698–12704
36. Daguer JP, Ciobanu M, Alvarez S, Barluenga S, Winssinger N (2011) DNA-templated combinatorial assembly of small molecule fragments amenable to selection/amplification cycles. Chem Sci 2:625–632
37. Harris JL, Winssinger N (2005) PNA encoding (PNA=peptide nucleic acid): from solution-based libraries to organized microarrays. Chemistry 11:6792–6801
38. Debaene F, Da Silva JA, Pianowski Z, Duran FJ, Winssinger N (2007) Expanding the scope of PNA-encoded libraries: divergent synthesis of libraries targeting cysteine, serine and metalloproteases as well as tyrosine phosphatases. Tetrahedron 63:6577–6658
39. Galloway WR, Spring DR (2011) PNA to DNA to microarray decoding facilitates ligand discovery. Chem Biol 18:1209–1210
40. Svensen N, Díaz-Mochón JJ, Bradley M (2011) Encoded peptide libraries and the discovery of new cell binding ligands. Chem Commun 47:7638–7640
41. Svensen N, Díaz-Mochón JJ, Bradley M (2011) Decoding a PNA encoded peptide library by PCR: the discovery of new cell surface receptor ligands. Chem Biol 18:1284–1289
42. Vummidi BR, Farrera-Soler L, Daguer JP, Dockerill M, Barluenga S, Winssinger N (2022) A mating mechanism to generate diversity for the Darwinian selection of DNA-encoded synthetic molecules. Nat Chem 14:141–152
43. Melkko S, Scheuermann J, Dumelin CE, Neri D (2004) Encoded self-assembling chemical libraries. Nat Biotechnol 22:568–574
44. Melkko S, Dumelin CE, Scheuermann J, Neri D (2006) On the magnitude of the chelate effect for the recognition of proteins by pharmacophores scaffolded by self-assembling oligonucleotides. Chem Biol 13:225–231
45. Wichert M, Krall N, Decurtins W, Franzini RM, Pretto F, Schneider P, Neri D, Scheuermann J (2015) Dual-display of small molecules enables the discovery of ligand pairs and facilitates affinity maturation. Nat Chem 7:241–249
46. Plais L, Lessing A, Keller M, Martinelli A, Oehler S, Bassi G, Neri D, Scheuermann J (2022) Universal encoding of next generation DNA-encoded chemical libraries. Chem Sci 13:967–974
47. Oehler S, Catalano M, Scapozza I, Bigatti M, Bassi G, Favalli N, Mortensen MR, Samain F, Scheuermann J, Neri D (2021) Affinity selections of DNA-encoded chemical libraries on carbonic anhydrase IX-expressing tumor cells reveal a dependence on ligand valence. Chem A Eur J 27:8985–8993
48. Gui Y, Wong CS, Zhao G, Xie C, Hou R, Li Y, Li G, Li X (2022) Converting double-stranded DNA-encoded libraries (DELs) to single-stranded libraries for more versatile selections. ACS Omega 7:11491–11500
49. Zhao GX, Zhong ST, Zhang G, Li YF, Li YZ (2022) Reversible covalent headpiece enables interconversion between double- and single-stranded DNA-encoded chemical libraries. Angew Chem Int Ed 61:e202115157
50. Mondal M, Hirsch AK (2015) Dynamic combinatorial chemistry: a tool to facilitate the identification of inhibitors for protein targets. Chem Soc Rev 44:2455–2488
51. Sprinz KI, Tagore DM, Hamilton AD (2005) Self-assembly of bivalent protein-binding agents based on oligonucleotide-linked organic fragments. Bioorg Med Chem Lett 15:3908–3911
52. Reddavide FV, Lin W, Lehnert S, Zhang Y (2015) DNA-encoded dynamic combinatorial chemical libraries. Angew Chem Int Ed 54:7924–7928
53. Reddavide FV, Cui M, Lin W, Fu N, Heiden S, Andrade H, Thompson M, Zhang Y (2019) Second generation DNA-encoded dynamic combinatorial chemical libraries. Chem Commun 55:3753–3756
54. Li G, Zheng W, Chen Z, Zhou Y, Liu Y, Yang J, Huang Y, Li X (2015) Design, preparation, and selection of DNA-encoded dynamic libraries. Chem Sci 6:7097–7104

55. Zhou Y, Li C, Peng J, Xie L, Meng L, Li Q, Zhang J, Li XD, Li X, Huang X et al (2018) DNA-encoded dynamic chemical library and its applications in ligand discovery. J Am Chem Soc 140:15859–15867
56. Zhou Y, Peng J, Shen W, Li X (2020) Psoralen as an interstrand DNA crosslinker in the selection of DNA-encoded dynamic chemical library. Biochem Biophys Res Commun 533: 215–222
57. Deng YQ, Peng JZ, Xiong F, Song YA, Zhou Y, Zhang JF, Lam FS, Xie C, Shen WY, Huang YR et al (2020) Selection of DNA-encoded dynamic chemical libraries for direct inhibitor discovery. Angew Chem Int Ed 59:14965–14972
58. Farrera-Soler L, Jean-Pierre Daguer J-P, Patrick Raunft P, Sofia Barluenga S, Anne Imberty A, Nicolas Winssinger N (2020) PNA-based dynamic combinatorial libraries (PDCL) and screening of lectins. Bioorg Med Chem 28:115458
59. Gartner ZJ, Liu DR (2001) The generality of DNA-templated synthesis as a basis for evolving non-natural small molecules. J Am Chem Soc 123:6961–6963
60. Li X, Liu DR (2004) DNA-templated organic synthesis: Nature's strategy for controlling chemical reactivity applied to synthetic molecules. Angew Chem Int Ed 43:4848–4870
61. Kanan MW, Rozenman MM, Sakurai K, Snyder TM, Liu DR (2004) Reaction discovery enabled by DNA-templated synthesis and in vitro selection. Nature 431:545–549
62. Tse BN, Snyder TM, Shen Y, Liu DR (2008) Translation of DNA into a library of 13 000 synthetic small-molecule macrocycles suitable for in vitro selection. J Am Chem Soc 130: 15611–15626
63. Kleiner RE, Dumelin CE, Tiu GC, Sakurai K, Liu DR (2010) In vitro selection of a DNA-templated small-molecule library reveals a class of macrocyclic kinase inhibitors. J Am Chem Soc 132:11779–11791
64. Maianti JP, McFedries A, Foda ZH, Kleiner RE, Du XQ, Leissring MA, Tang WJ, Charron MJ, Seeliger MA, Saghatelian A et al (2014) Anti-diabetic activity of insulin-degrading enzyme inhibitors mediated by multiple hormones. Nature 511:94–98
65. Maianti JP, Tan GA, Vetere A, Welsh AJ, Wagner BK, Seeliger MA, Liu DR (2019) Substrate-selective inhibitors that reprogram the activity of insulin-degrading enzyme. Nat Chem Biol 15: 565–574
66. Usanov DL, Chan AI, Maianti JP, Liu DR (2018) Second-generation DNA-templated macrocycle libraries for the discovery of bioactive small molecules. Nat Chem 10:704–714
67. Doak BC, Zheng J, Dobritzsch D, Kihlberg J (2016) How beyond rule of 5 drugs and clinical candidates bind to their targets. J Med Chem 59:2312–2327
68. Seigal BA, Connors WH, Fraley A, Borzilleri RM, Carter PH, Emanuel SL, Fargnoli J, Kim K, Lei M, Naglich JG et al (2015) The discovery of macrocyclic XIAP antagonists from a DNA-programmed chemistry library, and their optimization to give lead compounds with in vivo antitumor activity. J Med Chem 58:2855–2861
69. Wilkinson JC, Cepero E, Boise LH, Duckett CS (2004) Upstream regulatory role for XIAP in receptor-mediated apoptosis. Mol Cell Biol 24:7003–7014
70. Li Y, Zhao P, Zhang M, Zhao X, Li X (2013) Multistep DNA-templated synthesis using a universal template. J Am Chem Soc 135:17727–17730
71. Halpin DR, Harbury PB (2004) DNA display I. Sequence-encoded routing of DNA populations. PLoS Biol 2(E173)
72. Halpin DR, Harbury PB (2004) DNA display II. Genetic manipulation of combinatorial chemistry libraries for small-molecule evolution. PLoS Biol 2:E174
73. Halpin DR, Lee JA, Wrenn SJ, Harbury PB (2004) DNA display III. Solid-phase organic synthesis on unprotected DNA. PLoS Biol 2:E175
74. Meo T, Gramsch C, Inan R, Hollt V, Weber E, Herz A, Riethmuller G (1983) Monoclonal antibody to the message sequence Tyr-Gly-Gly-Phe of opioid peptides exhibits the specificity requirements of mammalian opioid receptors. Proc Natl Acad Sci U S A 80(13):4084–4088

75. Weisinger RM, Wrenn SJ, Harbury PB (2012) Highly parallel translation of DNA sequences into small molecules. PLoS One 7(3):e28056
76. Krusemark CJ, Tilmans NP, Brown PO, Harbury PB (2016) Directed chemical evolution with an outsized genetic code. PLoS One 11(8):e0154765
77. Hansen MH, Blakskjaer P, Petersen LK, Hansen TH, Højfeldt JW, Gothelf KV, Hansen NJV (2009) A yoctoliter-scale DNA reactor for small-molecule evolution. J Am Chem Soc 131: 1322–1327
78. Petersen LK, Blakskjær P, Chaikuad A, Christensen AB, Dietvorst J, Holmkvist J, Knapp S, Kořínek M, Larsen LK, Pedersen AE, Röhm S, Sløk FA, Hansen NJV (2015) Novel p38α MAP kinase inhibitors identified from yoctoReactor DNA-encoded small molecule library. Med Chem Commun 7:1332–1339

Top Med Chem (2022) 40: 65–122
https://doi.org/10.1007/7355_2022_151

Published online: 30 September 2022

Advancements in DEL-Compatible Chemical Reactions

Matthew J. Anderson, Isaline F. S. F. Castan, Jessica S. Graham, Haitham Hassan, Jake Odger, Catherine L. A. Salvini, Cameron Taylor, and Michael J. Waring

Contents

M. J. Anderson, I. F. S. F. Castan, J. S. Graham, H. Hassan, J. Odger, C. L. A. Salvini, C. Taylor, and M. J. Waring (✉)
Chemistry, School of Natural and Environmental Sciences, Newcastle University, Newcastle upon Tyne, UK
e-mail: Mike.Waring@newcastle.ac.uk

Abstract DNA-Encoded Libraries (DELs) use chemical reactions to build organic moieties on the coding DNA strand. Accordingly, the generation of a DEL requires robust chemical transformations that are compatible with DNA and the aqueous conditions required for its solubilisation. Reactions that damage DNA cannot be employed, neither can reactions that are hindered by the functionality in DNA (Malone and Paegel, ACS Comb Sci 18(4):182–187, 2016). The nature of DEL synthesis imposes further restrictions: reactions must be compatible with split-and-pool, and reactions must proceed predominantly to the desired product without excessive side products. A broad substrate scope is required using accessible reagents.

As a consequence, a great deal of effort has been expended in developing synthetic methodologies that are DNA-compatible, in order to increase the chemical space DELs can cover. These are, most often, adaptations of off-DNA synthesis methods extended to a DEL setting. The range of reactions available to DEL chemists is ever-expanding, covering an extensive range of reactions including, but not limited to amide couplings, cycloadditions, heterocycle syntheses, nucleophilic additions, reductive aminations, S_NAr reactions, and a wide variety of metal-catalysed cross-couplings, such as Suzuki-Miyaura, Buchwald-Hartwig, and Sonogashira couplings (Kunig et al., Biol Chem 399 (7):691–710, 2018; Shi et al., RSC Adv 11(4):2359–2376, 2021; Castan et al., Bioorg Med Chem 43:116273, 2021; Fair et al., Bioorg Med Chem Lett 51:128339, 2021).

This synthetic toolkit is not all-encompassing; however, new methodologies towards DEL-compatible chemical reactions are constantly being elucidated, broadening the available tools towards DEL synthesis. Herein, recent advances towards DNA-compatible synthetic methods are outlined, comprising DNA damage assessment procedures, technologies to prevent this damage, and applications of these techniques to DEL generation.

Keywords DNA compatible reactions, on-DNA chemistry, DNA-encoded library synthesis, DNA damage

Abbreviations

°C	Degree Celsius
3CR	3-Component reaction
4CR	4-Component reaction
A	Adenine
APTAC	Amphiphilic polymer-facilitated transformations under anhydrous conditions
Boc	*Tert*-butoxycarbonyl group
bp	Base pairs
bpy	2,2′-Bipyridine
C	Cytosine
CMC	Critical micelle concentration
COD	Cyclooctadiene
CPG	Controlled pore glass
Da	Dalton
DEAE	Diethylaminoethyl
DEL	DNA-encoded library
dF(CF3)ppy	2-(2,4-Difluorophenyl)-5-(trifluoromethyl)pyridine
DHP	Dihydropyridines
DMA	Dimethylacetamide
DMF	Dimethylformamide
DMSO	Dimethyl sulfoxide
DNA	Deoxyribonucleic acid
dtbbpy	4,4′-Di-tert-butyl-2,2′-dipyridyl
eq	Equivalent
G	Guanine
GOase	Galactose oxidase
h	Hours
HexT	Hexathymidine
HFIP	1,1,1,3,3,3-Hexafluoro-2-propanol
HPLC	High-performance liquid chromatography
ISC	Intersystem crossing
LC-MS	Liquid chromatography–mass spectrometry
LED	Light-emitting diode
L-TA	L-threonine aldolase
MALDI MS	Matrix-assisted laser desorption/ionisation mass spectrometry
MeCN	Acetonitrile
min	Minutes
mM	Millimolar

MOPS	4-Morpholinepropanesulfonic acid
NGS	Next generation sequencing
nm	Nanometer
nmol	Nanomole
NMP	*N*-Methyl-2-pyrrolidone
OPA	*o*-Phthaldialdehyde
PBS	Phosphate buffered saline
PCR	Polymerase chain reaction
p-cym	p-Cymene
PEG	Poly(ethylene glycol)
PES	PNA-encoded synthesis
pH	Potential of hydrogen
Phe	Phenyl
pKa	Acid strength
PLP	Pyridoxal phosphate
PNA	Peptide nucleic acid
ppy	2-Phenylpyridine
qPCR	Quantitative polymerase chain reaction
RASS	Reversible adsorption to solid support
ROS	Reactive oxygen species
rt	Room temperature
SET	Single-electron transfer
SnAP	Sn amino protocol
T	Thymine
TEM	Transition electron microscopy
TEMPO	(2,2,6,6-Tetramethylpiperidin-1-yl)oxyl
TFA	Trifluoroacetic acid
THF	Tetrahydrofuran
TMEDA	Tetramethylethylenediamine
TMHD	2,2,6,6-Tetramethyl-3,5-heptanedione
TPGS-750-M	D,L-α-tocopherol methoxypolyethylene glycol succinate
uL	Microliters
v/v	Volume per volume
λ	Wavelength

1 Measuring DNA Damage

1.1 Introduction

Successful in DEL screening is critically dependent on the use of chemistry in library construction that proceeds with efficient conversion across a wide range of substrates, without appreciable degradation of the coding DNA tags, and the use of synthesis schemes that result in lead-like library members [1–5]. Critical to DELS screening is the amplification and sequencing of the DNA barcodes of selected hits. Sequencing evaluates the number of times that each DNA barcode is detected [1, 6–8]. The relationship between the counts of a particular library member and its equilibrium dissociation constant with the target of interest has been explicitly described by Satz (when using low concentrations of protein and high fidelity DELs) [9]. Essential to the reliability of this analysis is avoiding significant DNA damage during library synthesis so that the amplification and sequencing is accurate.

1.2 Overview

Preserving the information stored within the DNA barcodes during DEL construction is an important requirement of DNA-compatible chemistry. The amount of

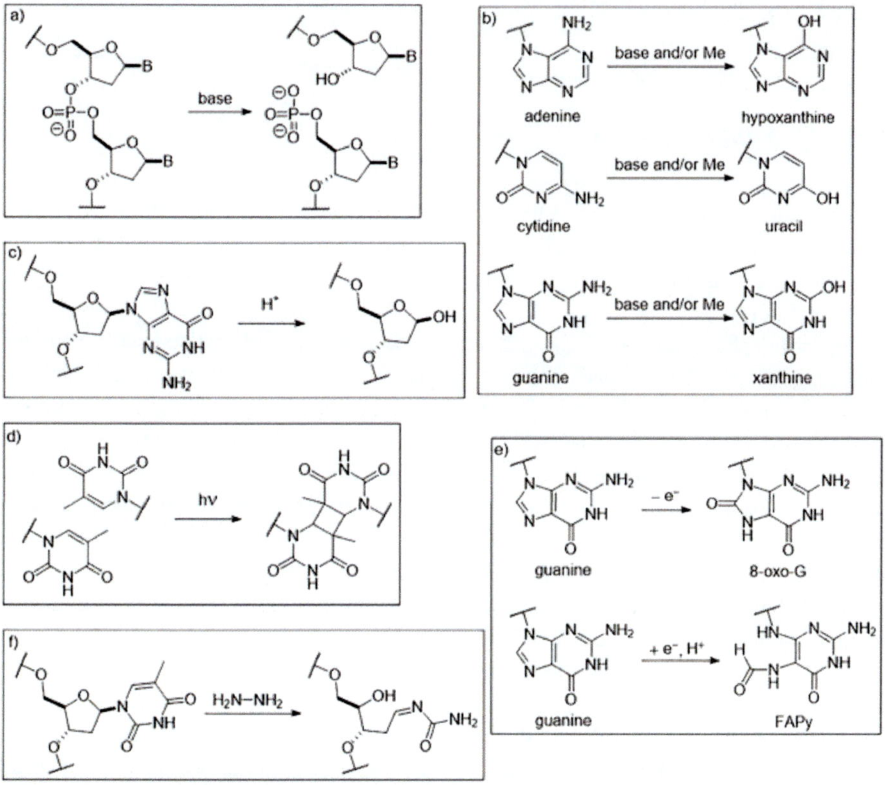

Fig. 1 Selected modes of DNA damage that can potentially be caused by DEL reaction conditions: (**a**) phosphate backbone hydrolysis (**b**) hydrolytic deamination (**c**) depurination (**d**) thymine dimerisation (**e**) oxidative radical damage (**f**) nucleophilic attack. The DNA nucleobases are coloured: red for adenine, green for cytosine, yellow for guanine, and blue for thymine. Reprinted with permission from Tetrahedron Letters, Volume 61, Issue 22, 28 May 2020, 151889. Copyright 2020 ScienceDirect

damage that can be tolerated has not been defined but, ideally, the majority of DNA within a library should remain amplifiable.

Several types of DNA damage are well known (Fig. 1). Depurination is relatively common, occurring predominantly at low pH [10]. Conversely, fragmentation may occur at extremely high pH. Metal ions are known to catalyse deamination, most often with cytosine but purines are also vulnerable [11, 12]. Purines are also susceptible to oxopurine formation in the presence of oxidants and radical species [13, 14]. Finally strong nucleophiles have been observed to add to the 6-position of pyrimidines resulting in nucleobase loss.

Several methods have been developed to measure some part of said DNA damage caused by chemical reactions. When using low molecular weight DNA tags, DNA damage can be monitored by mass spectrometry, whilst for higher molecular weight DNA tags, Sanger sequencing can evaluate DNA [15–17]. While Sanger sequencing can report on significant point mutations, it cannot quantify less frequent point mutations [18].

This section will describe the advancements in measuring DNA damage, namely qPCR and NGS.

1.3 Quantitative PCR

In the context of DNA-encoded libraries, qPCR reports on changes in DNA amplification caused by damaged primer regions or by polymerase stalling at sites of damage. qPCR as a means of measuring DNA damage has been used extensively in DEL research [1, 15, 19, 20].

For solid-phase DELs, qPCR uses magnetic beads functionalised with a model DNA tag [1, 20]. These beads are then subjected to the reaction of interest, and the amount of amplifiable DNA remaining on these beads, normalised to the unexposed sensor bead stock, is then assessed. This is done via a TaqMan polymerase qPCR assay. As well as measuring damage for new reaction development, this method has also been used to assess encoding tag integrity of previously published "DNA-compatible" reactions, which showed that $<0.2\%$ of the DNA remained after being subjected to reaction conditions in some cases [21].

qPCR has also been used for analysing solution-phase DEL post synthesis, again utilising a TaqMan polymerase qPCR assay. Through the use of the qPCR assay, Abel et al. discovered that the addition of dimethyl sulfoxide, a known radical scavenger, into the aqueous mixture dramatically suppressed DNA damage during the Cu(I)-catalysed azide-alkyne cycloaddition [19]. qPCR using SYBR Green has also been applied in the solution phase [22, 23]. The TaqMan assay is highly specific and detects specified amplification products, whereas SYBR Green is less specific and binds with any double-stranded DNA.

1.4 Characterisation of Chemical Reaction Compatibility

Because relatively long DNA sequences (i.e. high molecular weight) are required for qPCR, analysis of chemical conversions by mass spectrometry, and DNA fidelity by qPCR, is difficult. Through the incorporation of a ClaI restriction endonuclease sequence, Paegel et al. have developed a solution-phase platform that allows simultaneous mass fragment chemical analysis, whilst also allowing qPCR [24]. The authors hope that this all-in-one method will allow quantitative comparisons between different reactions.

In addition to this, ligation efficiency was measured to assess DNA damage, directly reporting issues for encoding. The ligation efficiency is a vital consideration when short overhangs are used. They are particularly vulnerable to significant damage which can impair the encoding [24].

1.5 NGS to Measure DNA Damage

The methods described above to quantify DNA damage are limited to write efficiency, where the ability to ligate DNA onto a working encoded library strand is measured, and qPCR to measure the amplifiability of the DNA. These measures read signal quantity and overall integrity, but do not report on specific changes to the encoded information. Some reagents, such as hydroxylamine, induce mutations that would damage information stored in DNA, yet minimally impact amplifiability in PCR and thus escape notice. To circumvent this problem, Sauter et al. have used NGS to measure the quality of the read signal in order to quantify the accuracy of the retrieved information [25].

Through data analysis, the authors identified reactions that caused specific transversions. In these cases, the authors suggest that the codons exposed to the damaging conditions should be designed to avoid damage. For example, having

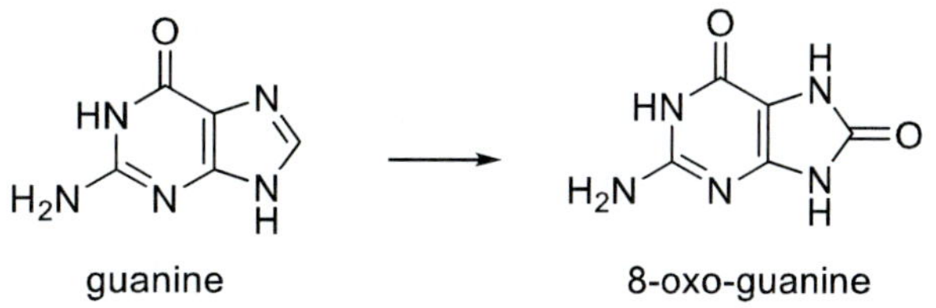

Fig. 2 Oxidation of guanine can result in subsequent G → T transversion

reduced numbers of G/C pairs, due to the risk of oxidation and the subsequent G → T transversion (Fig. 2). The authors go on to suggest that the DNA compatibility of a new reaction should be established not only by qPCR, but also by sequencing.

Despite the widespread adoption of various methods to measure DNA damage by the DEL community, there is no consensus as to what an acceptable level of damage for a single reaction is.

2 Modified DNA to Improve DNA Compatibility

2.1 Introduction

Many traditional organic chemistry reaction conditions can be detrimental when transferred to on-DNA syntheses. For example, metal-catalysed and Brønsted-acid mediated reaction conditions are commonly used in traditional library synthesis. When translating these conditions to on-DNA chemical transformations, reagents such as strong acids and metal catalysts can often interact, or even react, with DNA.

Common methods of DNA degradation include deamination, either spontaneously or via metal ion promotion; oxidative DNA damage, which can occur through various different mechanisms; alkylation of DNA, through the reaction of electrophiles with nucleophilic sites of the DNA bases; and metal- or acid-mediated depurination [11–15]. Purine bases are particularly susceptible to cleavage under organic reaction conditions.

2.2 Overview

Given the difficulties of performing on-DNA reactions due to depurination and degradation of DNA, modifications to the DNA barcodes to improve their stability and compatibility with various organic reagents are highly desirable to increase the range of applicable chemistry. Given that purine bases are particularly susceptible to side reactions leading to degradation, the removal or alteration of these purine nucleosides within the DNA sequence may lead to improved chemical stability, allowing for harsher conditions to be carried out on the DNA.

Thus far, there have been relatively few reports of modifications to improve DNA compatibility for DEL synthesis. Recent work by Brunschweiger et al. described a hexathymidine (HexT) DNA oligonucleotide [26–29]. This is proposed to be a more stable DNA sequence due to the absence of purine bases. Hence, this DNA tag can withstand harsher reaction conditions. It has been demonstrated that HexT can be recognised by T4 ligase and has been successfully ligated to coding DNA sequences. Moreover, HexT can be analysed by MALDI MS and analytical HPLC and can be purified using preparative HPLC.

More recently, the same group has investigated the replacement of HexT with a chemically stabilised DNA barcode [30]. This was hypothesised to withstand similar reaction conditions on encoded mixtures of starting materials, rather than individual HexT-tagged conjugates. The chemical modifications explored involved the

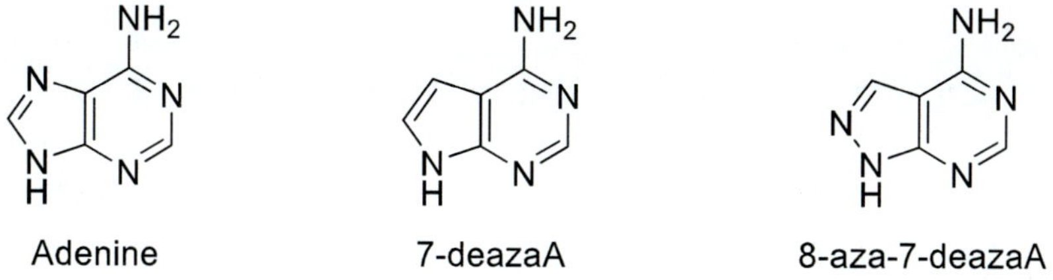

Fig. 3 Structures of nucleobase adenine, A (left), and its chemically modified analogues 7-deazaA (centre) and 8-aza-7-deazaA (right)

substitution of the vulnerable purine bases with 7-deazaA and 8-aza-7-deazaA (Fig. 3), in combination with the CPG solid support approach. It was experimentally shown that the 7-deazaA tolerates harsher reaction conditions, especially acid treatment [31].

The use of peptide nucleic acids (PNAs) is also a potential method of improving the chemical compatibility of DNA barcodes. PNAs were first reported in an encoded library in 2004 by Wissinger et al. [32] PNAs were initially investigated as encoding oligonucleotides due to several favourable properties, including their chemical robustness. Oligomerisation of PNAs for library synthesis would involve a mild amide coupling between peptides. Furthermore, a range of protecting groups on the N-terminus of PNA can be employed to accommodate diverse functional group tolerances. PNA-encoded synthesis (PES) can be performed in traditional organic solvents, with more forceful conditions compared to those utilised in DEL synthesis. However, PNA-encoded libraries are not compatible with PCR amplification. Therefore, hybridisation to complementary DNA fragments is required. Nevertheless, the lack of negative charges on PNA increases its affinity for DNA and reduces the influence of salt concentration on hybridisation.

2.3 HexT

Brunschweiger et al. have published several papers investigating the use of HexT for various chemical reactions that require harsh conditions. Encoding DNA barcodes can then be ligated after harsh reaction. These reactions include: a Brønsted acid-catalysed Pictet-Spengler reaction, for the synthesis of β-carbolines [26]; Au-mediated reactions, synthesising pyrazol(in)e and spiroheterocycle conjugates [27]; Yb-mediated Castangoli-Cushman reactions and Ag-mediated 1-3-cycloadditions, generating tetrahydroisoquinoline conjugates and a pyrrolidine core scaffold, respectively [28]; and a range of Ugi-type reactions, forming various heterocyclic core scaffolds [29]. These reactions would cause challenges for traditional DNA sequences due to the reagents present. The strong Brønsted- and Lewis-acids, and

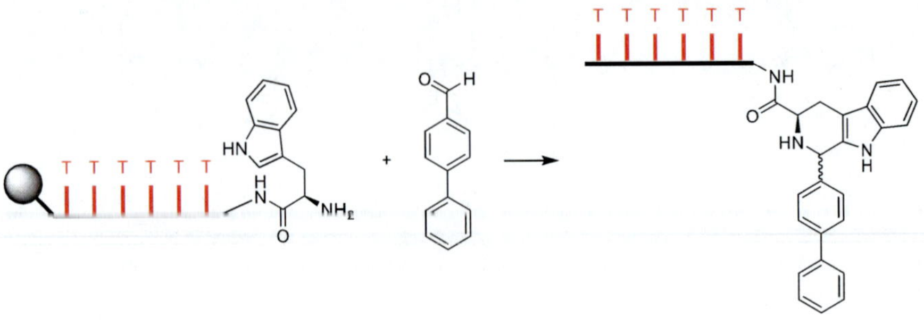

Fig. 4 Pictet-Spengler reaction facilitated using the modified DNA, HexT, protecting the DNA tag from the Brønsted acid, TFA

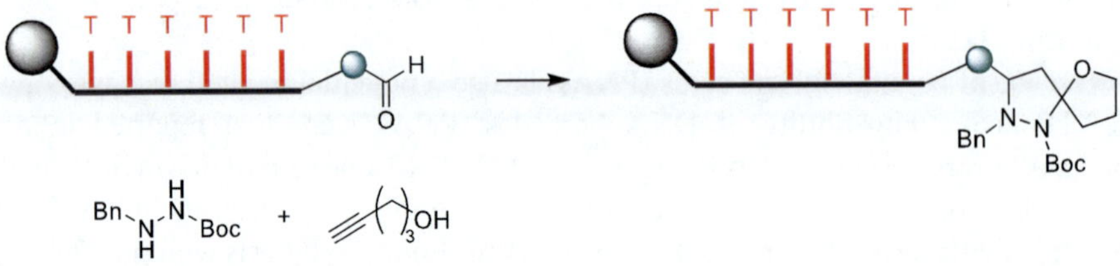

Fig. 5 Utilisation of modified DNA, HexT, to protect DNA tag from Au-based Lewis acid employed in the displayed spiroheterocycle formation 3CR, using [tris(2,4-di-tert-butylphenyl) phosphite]gold chloride and $AgSbF_6$

metal catalysts, used in these reactions are known to cause DNA degradation via depurination and deamination.

Skopic et al. found that the Pictet-Spengler reaction (Fig. 4), which used 1% TFA, gave a conversion of around 70% desired product with <5% DNA damage [26]. In fact, the use of 10% TFA also gave rise to <5% DNA degradation; however, this offered no improvement on the conversion to desired product. This was later revisited and trialled with ATC- and ATGC-containing 10mer sequences. Under the same reaction conditions, both the ATC- and ATGC-containing DNA sequences were degraded by >90% whilst no detectable product was observed.

The same group have also demonstrated that the two Au-catalysed reactions can be done successfully using CPG-bound, HexT-tagged substrates [27]. With respect to the gold-catalysed three-component reaction (3CR) to form spiroheterocycles (Fig. 5), it was found that HexT can tolerate the metal-mediated conditions with no reported DNA damage. However, this reaction was also tolerated by pyrimidine- and, surprisingly, adenine-containing DNA barcodes. Conversely, guanine-containing sequences were degraded under the gold conditions, attributed by the authors to the lower redox potential and greater susceptibility to electrophilic attack of guanine, compared to adenine. The reaction was also trialled using DNA-tagged hydrazides and alkynols; however, the use of the DNA-tagged aldehyde conjugate was the only viable option for library synthesis, with limited side product formation.

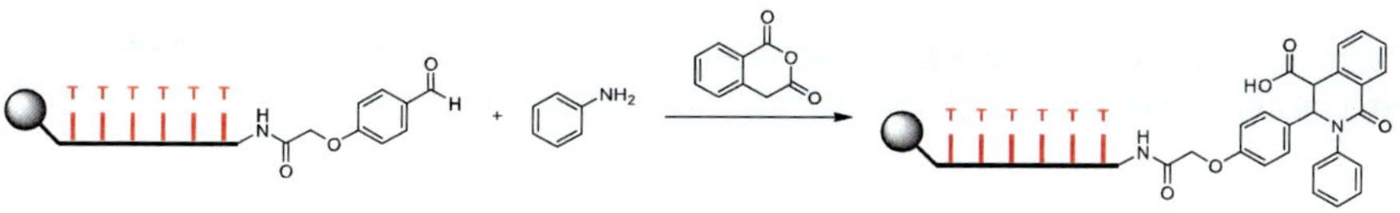

Fig. 6 $Yb(OTf)_3$-mediated Castagnoli-Cushman reaction facilitated by HexT, resulting in no DNA damage

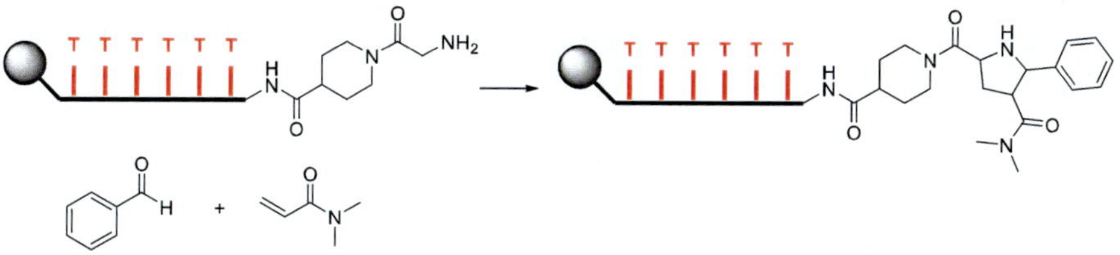

Fig. 7 AgOAc-mediated 1,3-cycloaddition, with no DNA damage, facilitated by HexT

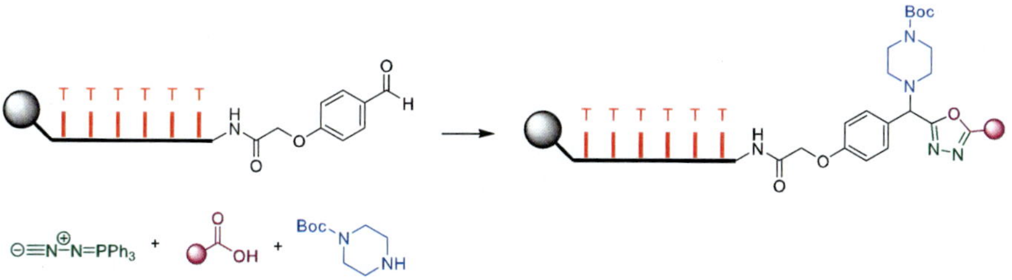

Fig. 8 Synthesis of 1,3,4-oxadiazoles via one-pot Ugi 4CR/aza-Wittig reaction, requiring HexT to prevent DNA damage

Work by Potowski et al. showed the tolerance of CPG-bound HexT conjugates to several Lewis acid reagents [28]. The Yb-mediated Castagnoli-Cushman reaction (Fig. 6) led to a 70% conversion to desired product under optimised conditions (using 50 equivalents of $Yb(OTf)_3$) with no reported DNA damage. Similarly, the Ag-mediated 1,3-cycloaddition reaction (Fig. 7) was also tolerated with similar conversions and no DNA degradation. However, both the aforementioned conditions were also tolerated by a CPG-bound ATGC 10mer, with conversions ranging from 46–72% for the Castagnoli-Cushman reaction and 10–70% for the cycloaddition reaction. There were no reports of DNA damage; therefore, the use of HexT was not essential for withstanding these conditions.

Kunig et al. demonstrated that several Ugi-type reactions can be performed on solid-supported coding DNA, including the Ugi and Ugi-azide 4CRs, and the Groebke-Blackburn-Bienayme 3CR [29]. However, the Ugi 4CR/aza-Wittig reaction to synthesise 1,3,4-oxadiazoles (Fig. 8) was found to be detrimental to the solid-supported DNA. The DNA damage caused was attributed, by the authors, to concomitant cleavage of purine bases. However, when this reaction was carried out on HexT-tagged substrates, the desired product was formed with no reports of DNA damage.

2.4 Chemical Modification

Potowski et al. investigated the chemical compatibility of modified DNA using 7-deazaA and 8-aza-7-deazaA bases in place of the vulnerable purine bases, compared to native and pyrimidine-only containing DNA sequences [30]. A stability screen against a wide range of metal salts, organic reagents, and protic acids was conducted. The pyrimidine-DNA tolerated most reagents with <20% DNA degradation observed. The native DNA sequence, however, was severely degraded (>61%) by strong acids, oxidants, and certain metal ions, including Pd, Ce, and Ru (41–60% degradation). Both chemically modified DNA sequences showed similarities to the stability profile of the pyrimidine-DNA; almost all reagents were tolerated with <20% DNA damage. Furthermore, there was little difference in stability profile between the two different modifications, 7-deazaA and 8-aza-7-deazaA.

The more readily available 7-deazaA was then subjected to a series of on-DNA chemical reactions. The Ugi reactions that were successful on native DNA were revisited, and all reactions gave almost quantitative conversions with <5% DNA damage. The Ugi 4CR/aza-Wittig reaction, which had previously required HexT to avoid DNA damage, was also explored. The reaction on CPG-bound 16mer 7-deazaATC DNA-encoded substrates led to desired product, with conversions ranging from 32–49%. However, some DNA damage was observed. The modified DNA resulted in 24% damage observed, which was a slight improvement on native DNA, which saw 32% degradation.

Brønsted acid-mediated Biginelli, Povarov, and Pictet-Spengler reactions were also investigated. The Biginelli and Povarov reactions gave >85% conversion to desired product, with <5% DNA damage observed. For the Pictet-Spengler reaction, which was proven to cause significant damage to native DNA, conversions to desired product for the modified DNA were >95% for most substrates, whilst DNA damage was limited to <17%.

Various metal-mediated reactions were studied, including aza-Diels-Alder, Petasis, Castangoli-Cushman, 1,3-cycloadditions, and Au-mediated pyrazol(in)e syntheses. Most of these reactions saw <5% DNA degradation, whilst giving conversions to desired product of 49–90%.

Finally, a Yb-promoted pyrazole synthesis utilising aryl hydrazines was investigated. Using a native 10mer sequence, this resulted in 69% DNA damage. When transferred to the modified DNA sequence, the degradation was reduced to 33%.

3 Non-covalent Reversible Solid Supports in DEL Synthesis

3.1 Introduction

While the suite of available chemistry for the synthesis of DELs has greatly expanded in the last 10 years, currently established transformations have mostly been limited to water-compatible reactions [3]. The primary approach to the development of new transformations for DELs has been the adaptation of standard organic protocols, replacing typical organic solvents in single-substrate transformations for aqueous media, applicable to sets of hundreds of compounds. While this tactic has proved successful in yielding new synthetic protocols, DEL synthetic methods lag behind those of traditional synthetic medicinal chemistry. The demand to increase the range of available chemistry for the assembly of DELs, giving access to a wider chemical space, provides impetus to increase the speed of discovery of DEL-compatible conditions away from previously well-worn approaches.

Protocols for on-DNA transformations, in which the DNA is covalently conjugated on to a solid support, have been used to facilitate a broad range of chemical transformations [33, 34]. However, it is known that biomacromolecules can be reversibly adsorbed and subsequently eluted from solid supports, without the covalent linkage. This phenomenon has been exploited by a number of groups in the modification of peptides and DNA-peptide conjugates [35–37].

The application of this approach to DEL synthesis proves an attractive route to enable the use of solvents and conditions traditionally unamenable to DNA. Herein, the recent development of reversible solid supports for DELs synthesis is outlined.

3.2 Overview

The earliest examples of adsorbing DNA onto solid support for DEL synthesis employ the carbohydrate anion-exchange resin DEAE-Sepharose. Work by Halpin et al. used a 20 bp modified DNA headpiece. The supported DNA species were used to produce libraries of peptides via sequential acylations using established chemistry (Fig. 9) [37].

More recently, efforts by separate groups simultaneously developed further solid support materials, allowing the expansion of applicable chemistry. Flood et al. employed a similar strategy, coining the method reversible adsorption to solid support (RASS) (Fig. 10). They identified a polystyrene-based strong-anion exchange resin, Strata-XA/XAL, as a suitable support [38].

The support was selected for its favourable properties with regard to avoiding reactive supports and premature elution, highlighting that it allows for the inclusion of organic solvents. The ensuing solid-phase DNA was used to present first-in-class chemistry techniques: electrochemical amination and decarboxylative

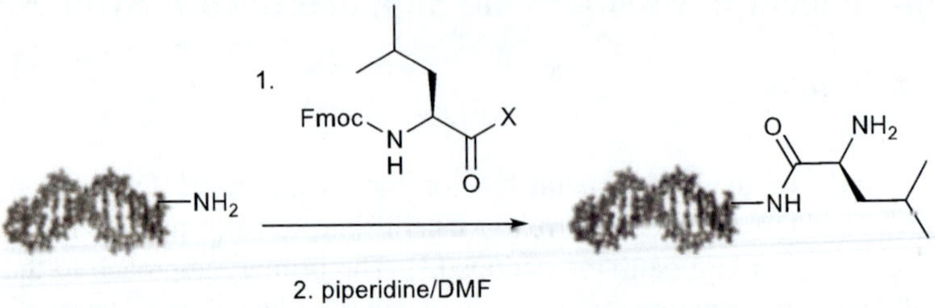

Fig. 9 Amino acid (leucine) coupling to amine-functionalised DNA headpiece via an acylation reaction, facilitated by DEAE-Sepharose. X represents a succinimidyl or EDC/HOAt-activated ester

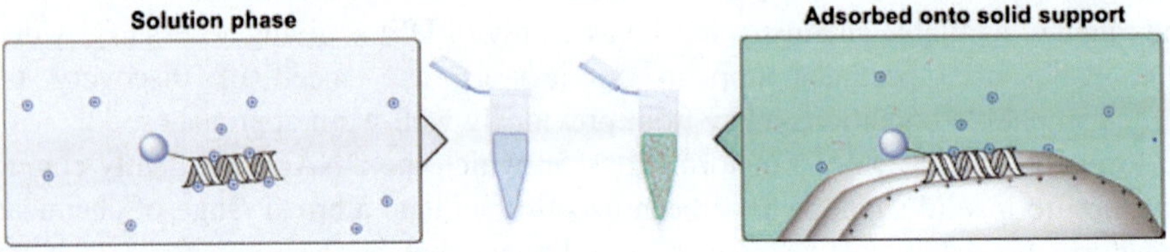

Fig. 10 Schematic illustrating RASS method of adsorbing DNA onto solid support (right) and comparison with "normal" solution-phase DEL chemistry

cross-coupling, as well as reductive amination, sulfonamide, and sulphide formation, which are already known in the literature [39].

Similar work by Ruff et al. from Novartis employed an amphiphilic cationic polymer support to specifically access sensitive anhydrous conditions, terming them amphiphilic polymer-facilitated transformations under anhydrous conditions (APTAC). In contrast to the previous, this work used PEG-based resins [40].

With the APTAC methodology, three proof-of-concept transformations were described: umpolung chemistry; water-intolerant tin-catalysed SnAP chemistry, affording sp^3 rich spirocyclic scaffolds; and photoredox chemistry, namely a radical-mediated cross-coupling (Fig. 11).

3.3 Support Material

During their assessment of suitable resins, Halpin et al. excluded for unsuitable bind and release properties; DEAE-silica, macro-prep ceramic hydroxyapatite, and quaternary amine anion-exchange resins were hence eliminated. Other resins that were excluded were macro-porous methacrylate, for poor reswelling behaviours during organic to aqueous phase transition, and reverse phase resins, as they were expected to give poor retention in organic media. DEAE-Sepharose is composed of crosslinked 6% agarose beads with diethylaminoethyl (DEAE) weak anion-exchange groups (Fig. 12) [37]. Whist it has been successfully used in the literature for both click and acylation chemistry, other groups have highlighted its deficiencies

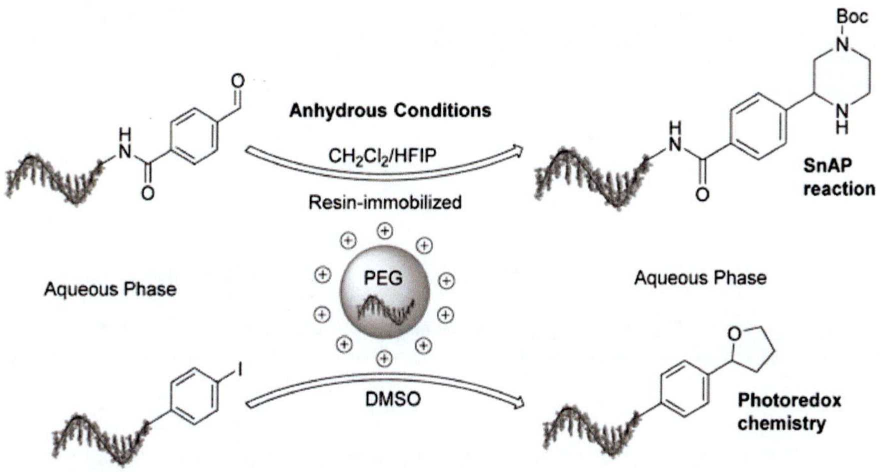

Fig. 11 Application of APTAC methodology, utilising a PEG-based resin, in SnAP chemistry (top) and photoredox cross-coupling (bottom), yielding bicyclic compounds

Fig. 12 Structure of DEAE-Sepharose, comprised of sugar moieties and protonated tertiary amine weak anion-exchange groups

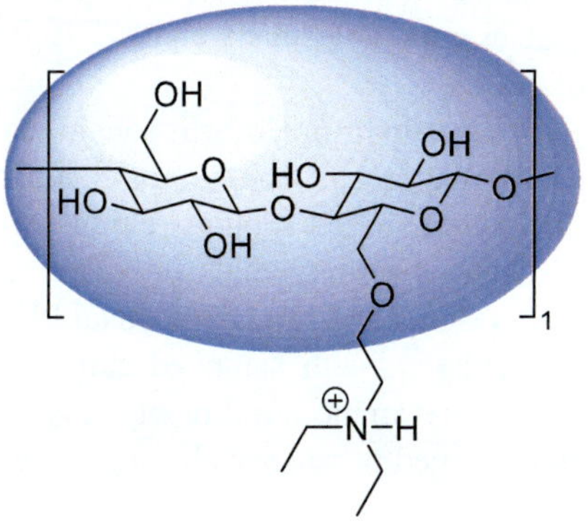

related to the tertiary amines present, which could interfere in desirable transformations such as photoredox chemistry and metal-catalysed cross-couplings.

Flood et al. came to similar conclusions in their investigations into suitable solid-phase media; however, weak anion-exchange resins were ruled out due to bearing nucleophilic, or otherwise reactive, functionalities [38]. Whilst this may be compatible with the acylation transformations carried out in the synthesis of peptidic species, it was speculated that the presence of abundant hydroxylated moieties and tertiary amines has curtailed the scope of their use.

The use of strong-anion exchange resins overcomes the issues presented by the presence of basic and nucleophilic moieties in weak exchange resins. Initially, the polystyrene-based Phenomenex Strata-XA strong-anion exchange resin, containing quaternary ammonium species, eluted with high concentration salt buffer, was chosen (Fig. 13). In later work the similar Phenomenex Strata XAL resin was preferred, due to superior DNA retention and capacity properties (vide infra).

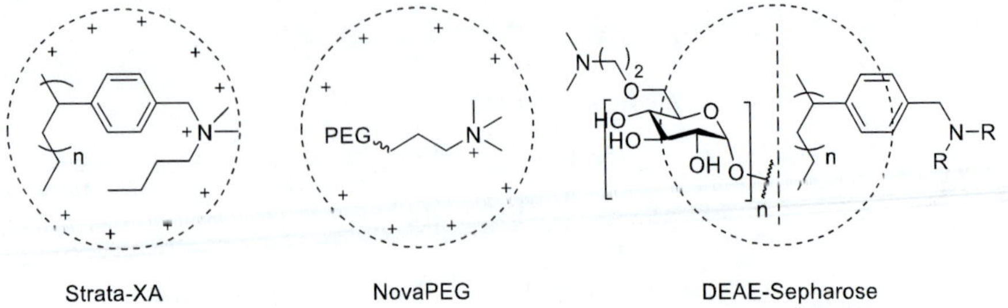

Fig. 13 Structures of three commonly used resins: Strata-XA (strong-anion exchange resin), NovaPEG (PEG-based strong-anion exchange resin, and DEAE-Sepharose (weak anion-exchange resin)

THF, DMF, DMSO, and MeCN were all found to be compatible with this approach, with yields comparable to those observed off-DNA with a similar dependence on solvent. These strong-anion exchange resins are designed to absorb and release macromolecules carrying negative charges. Strata XA possesses a particle size of 30 μm containing 90 Å pores, whereas the Strata XAL was designed to adsorb species with molecular weights >10,000, comprising particles of 100 μm with pores of 300 Å.

The group from Novartis sought to move away from carbohydrate-based resins to access capricious anhydrous chemistry. Differing from Flood, they discounted polystyrene-based supports citing solvent swelling concerns; they instead selected PEG-based resins, which provide excellent swelling in both organic and aqueous media (Fig. 13) [40]. Commercial PEG resin H-rink amide NovaPEG was modified by acylation with saturated carbon-chain carboxylic acid possessing a terminal quaternary ammonium moiety (Fig. 14). Critically, these resins could be successfully purged of water with successive solvent washing.

3.4 DNA Support Retention and Capacity

Flood et al. monitored DNA absorption onto the resin by loss of absorbance at 260 nm of the supernatant; DNA headpieces of differing structures were also assessed [39]. DNA sequences of 6–57 bp were evaluated via incubation in PBS. It was found that complete absorbance had occurred after 30 min at all sequence lengths (Fig. 15a). The alternative resin, Strata XA, performed poorly in comparison, as did the adoption in deionised water. From these results, it was suggested that the pH and ionic strength of the adsorbing solution are key, as well as the pore size of the resin.

Furthermore, the capacity of the resins for DNA loading was investigated. During the synthesis of DELs it is typical to work on scales of 10–50 nmols. It was shown that Strata XAL could be loaded with 50 nmols of 40mer double-stranded DNA in 30 min. However, when the resin was incubated with 100 nmol of DNA, it was

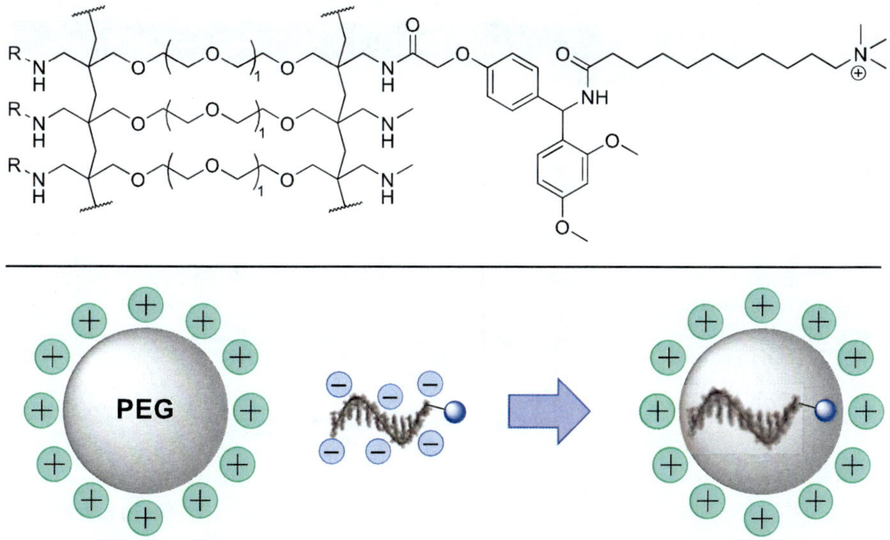

Fig. 14 Full structure of NovaPEG resin (top) and schematic representation displaying capture of DNA oligonucleotide (bottom)

reported that 25% of the DNA remained in solution. Authors described, therefore, that 50 nmol of DNA can be safely adsorbed to 100 μL of resin with recovery after elution of between 77 and 95% (Fig. 15b).

Similarly, the loading capacity of the so-called PEG+ supports implemented by the Novartis group estimated loading capacity by incubation of 5 mg of resin with increasing amounts of DNA, while measuring absorbance of supernatant. The APTAC resin loading capacity was estimated to be 11.5 nmol/mg for short (20 bp) single-stranded DNA sequences and 0.6 nmol/mg of longer (85 bp) and double-stranded DNA constructs. Elution affected with high salt concentration buffer yielded reported recoveries between 59 and 71% across all sequence lengths (Fig. 15c). Alternative elution buffers were screened at 40 bps with 1 M NH_4PF_6, 20% MeOH, and 40 mM Tris at pH 8.5 being identified as the optimal conditions.

3.5 DNA Protection

The DNA-protective potential of the RASS approach was investigated in a qualitative capacity using dsDNA intercalator SYBR [38]. It was observed that the fluorescence was maintained while bound; it was therefore proposed that the DNA duplex remained intact whilst adsorbed onto the solid phase, and in the presence of organic solvent (Fig. 16). The impact of adsorption on ligation efficiency is also of major importance for the use of the resins in DEL synthesis. It was reported that, for the RASS system, ligation efficiency between DNA exposed to RASS cycle and DNA not exposed was "identical".

Fig. 15 DNA binding and elution: (**a**) DNA adsorption kinetics to Strata XAL with various DNA constructs (**b**) Loading capacity of DNA onto Strata XAL using 40mer dsDNA C) DNA elution from Strata XAL using various DNA constructs. Reprinted with permission from Angew. Chem. Int. Ed. 2020, 59, 7377. Copyright 2020 John Wiley & Sons, Inc.

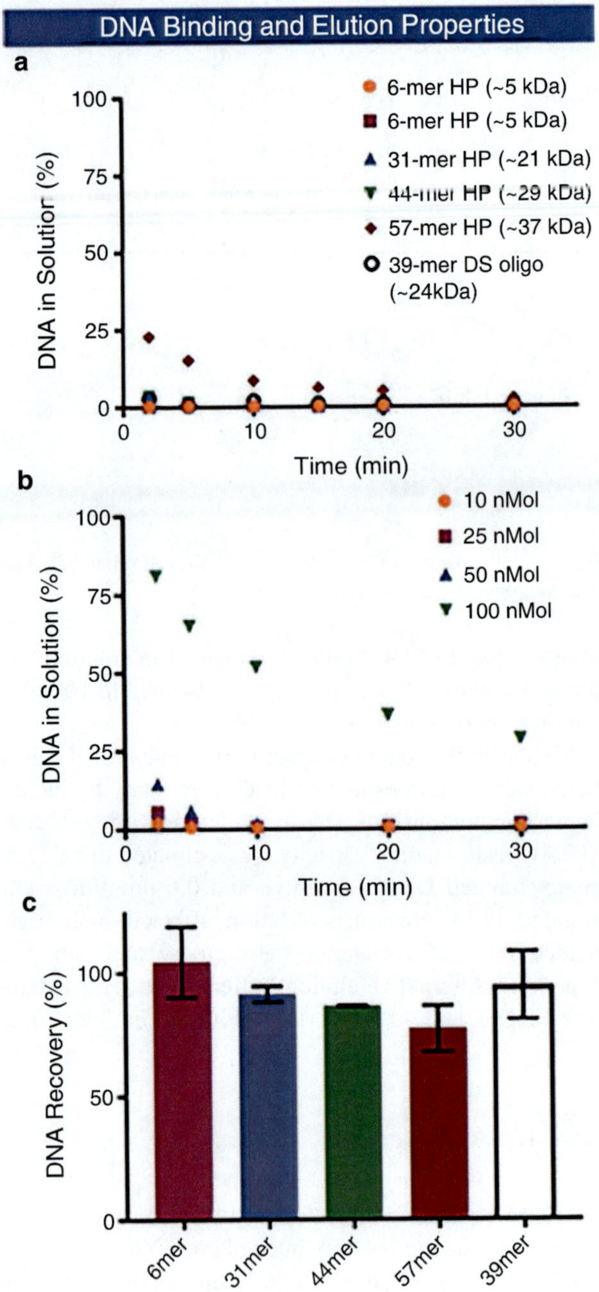

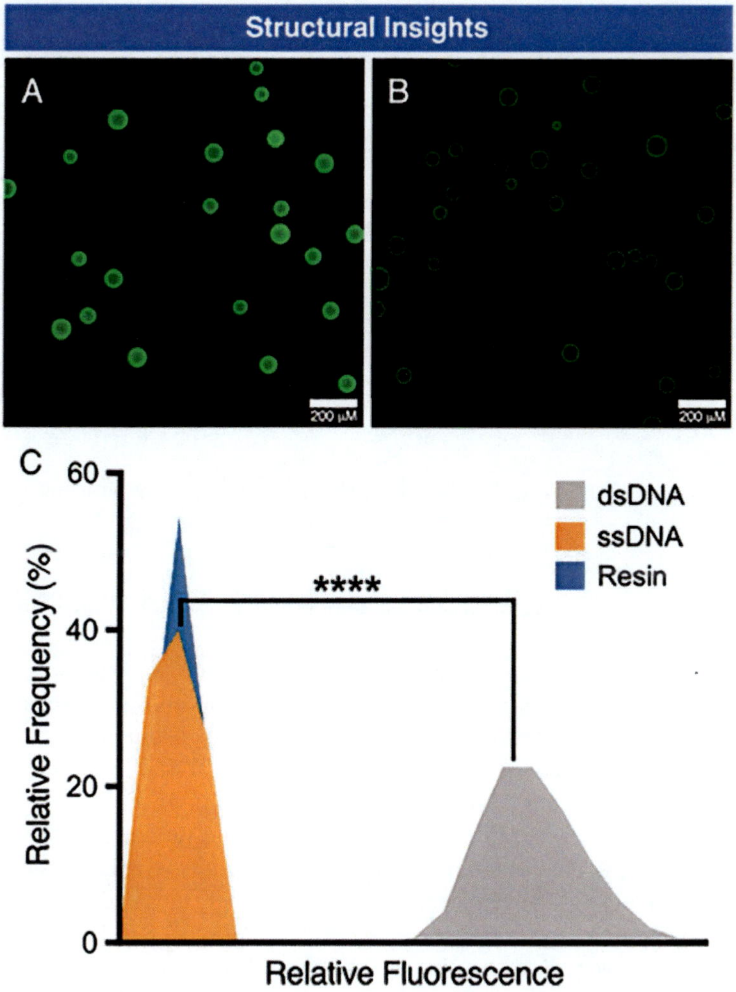

Fig. 16 Confocal microscopy images of resin with: (**a**) double-stranded DNA (**b**) Single-stranded DNA adsorbed and stained with SYBR green. (**c**) Quantification of resin fluorescence, comparing dsDNA, ssDNA, and resin. Reprinted with permission from J. Am. Chem. Soc. 2019, 141, 25, 9998–10006. Copyright 2019 American Chemical Society

3.6 New Chemistry Frontiers

While the range of transformations available for the construction of DELs has been greatly expanding in recent years [5], these transformations have predominantly been limited to aqueous solvent preparations, precluding DEL construction from accessing the anhydrous chemistry typical of traditional organic synthesis. To this end, Ruff et al. used their ABTAC methodology in 3 proof-of-concept transformations: umpolung chemistry; water-intolerant tin-catalysed SnAP chemistry,

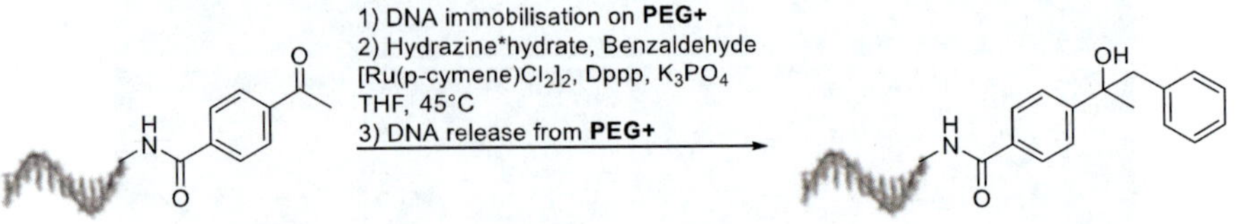

Fig. 17 Umpolung reaction using PEG+ to attain sufficiently anhydrous conditions allowing for hydrazone addition to carbonyls

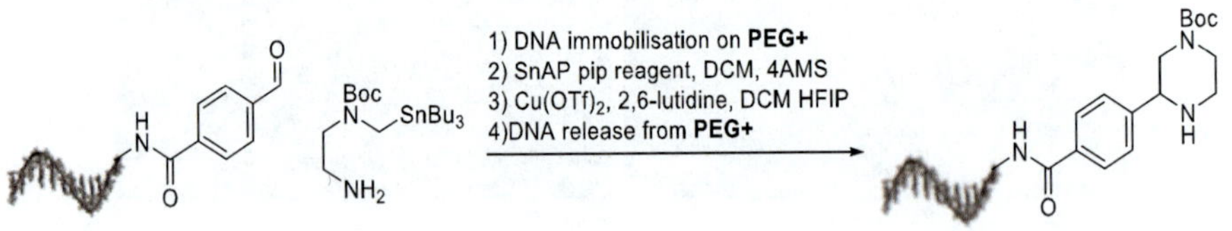

Fig. 18 Utilisation of PEG+ to facilitate tin-catalysed SnAP chemistry, resulting in bicyclic compounds

affording sp^3-rich spirocyclic scaffolds; and photoredox radical-mediated cross-coupling [40].

Once adsorbed onto the solid phase, sufficiently anhydrous conditions could be attained by sequential rinsing of the resin with the desired anhydrous solvent. Tertiary alcohols were accessed through hydrazone additions to carbonyls (Fig. 17). The work demonstrated that the ruthenium-catalysed Umpolung reaction preferentially takes place in THF under anhydrous conditions. Control experiments in aqueous solvent showed no conversion to the desired product. Their optimised conditions gave 75% conversion using PEG+. These results were contrasted against constructs supported on DEAE-Sepharose, delivering only the condensation product hydrazone derivative.

Furthermore, water-intolerant tin-catalysed SnAP chemistry could be achieved, giving up to 87% conversion with optimised conditions (Fig. 18). Additionally, the use of photoredox in a decarboxylative cross-coupling was demonstrated via this methodology, achieving 17–92% conversions. The authors commented that the use of successive 15-min reaction cycles improved conversions, without the need for work-up in between.

Historically, electrochemistry has not been available to unsupported DNA, due to the charged nature of the macromolecule being incompatible with the electrodes, with contact realising irreversible absorption onto the cathode surface. Flood et al. demonstrated, using the RASS methodology, electrochemical aminations of iodo-functionalised DNA constructs were achievable in 26–74% yield, generating the aniline derivatives (Fig. 19) [38]. Other transformations established using this paradigm included nickel-catalysed decarboxylative cross-couplings (49–97% conversion) and reductive aminations (16–82% conversion). However, the substrate scope for the latter was limited, for the most part, to ketones as the few aldehydes shown proceeded with more modest conversions (23–40%).

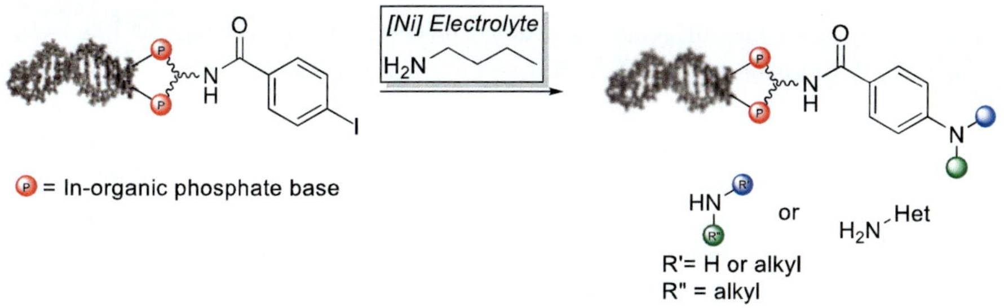

Fig. 19 Electrochemical aminations of DNA-conjugated aryl iodide, using alkyl (1° or 2°) or heterocyclic (1°) amines

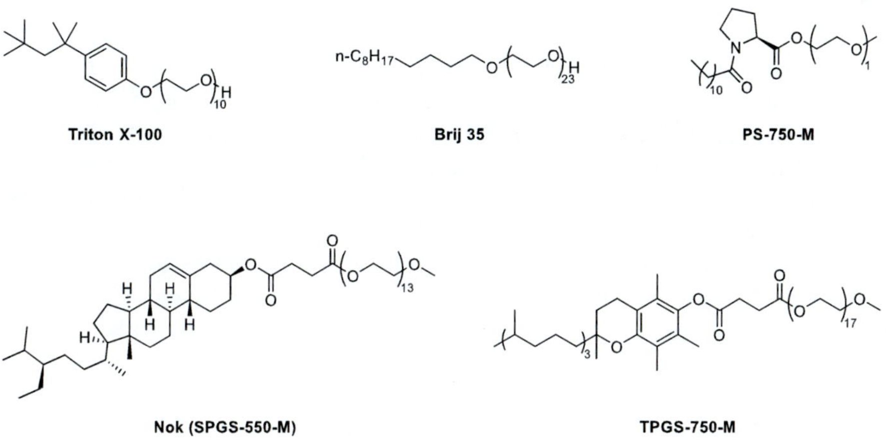

Fig. 20 Representative micelle-forming surfactants utilised in aqueous reactions

Recently, Flood et al. expanded the application of RASS to the generation of sulfonamides – a moiety not widely reported in DELs synthesis due to the instability of the substrates [39]. DNA-conjugated sulfonyl chlorides are not stable to library building transformations including ethanol precipitations and enzymatic ligations.

4 Micellar Catalysis for DEL Synthesis

4.1 Introduction

One possible solution to overcome the issues associated with carrying out reactions in water is to use micelle-forming surfactants (Fig. 20) [41, 42]. The surfactants used are amphiphilic compounds consisting of a hydrophilic head and hydrophobic tail. Micelles are formed at their critical micelle concentration (CMC), which is the lowest concentration at which spherical micelles form. Micellar catalysis takes advantage of the hydrophobic effect that exists within the lipophilic core of micelles upon dissolution of the surfactant in water. The use of micellar catalysis has been extensively demonstrated off-DNA in transition metal-catalysed reactions, such as olefin metathesis [43], Heck reactions [44], and Suzuki-Miyaura couplings [45].

Micellar catalysis is associated with high efficiency and mild reaction temperatures in an aqueous reaction environment. Consequently, this technology is an extremely attractive area to apply to DEL synthesis. The use of surfactants to promote on-DNA reactions emerged in 2019, with Brunschweiger and Waring being pioneers in utilising this approach to DEL synthesis [46, 47].

4.2 Overview

Micellar catalysis is one advancing technology employed within DELs to promote on-DNA reactions that use conditions normally incompatible with DNA. Most reagents and catalysts used in on-DNA transformations are typically insoluble in water. Hence, the addition of the surfactant helps to solubilise the reagents by localising them within the lipophilic core, protecting the DNA from damage by such reagents.

Strongly acidic conditions are typically forbidden for on-DNA transformations because they lead to protonation of the purine bases that may lead to depurination [41]. As a solution to this issue, amphiphilic block copolymers, covalently modified with sulfonic acid residues, have been synthesised [46]. The copolymers assemble in water and locate the Brønsted acid catalyst within the cores of the micelles. These acid nanoreactors have been utilised to promote the conversion of DNA-conjugated aldehydes to diversely substituted tetrahydroquinolines and aminoimidazopyridines by Povarov and Groebke–Blackburn–Bienaymé reactions, respectively. Furthermore, the cleavage of Boc protecting groups from amines, as well as the oxidation of DNA-conjugated alcohols to the corresponding aldehydes, have also been reported (Fig. 21). Prior to this work, these classes of reactions were very challenging to carry out under normal aqueous conditions, for the aforementioned reasons. The work described presents an alternative advancement to carry out organic transformations on DNA conjugates that may otherwise be classed as forbidden.

The block copolymers reported require bespoke synthesis. However, commercially available micelle-forming surfactant, TPGS-750-M, has been shown to promote a variety of on-DNA metal-catalysed reactions: Suzuki-Miyaura reactions [47], hydrogenations [23], and Buchwald-Hartwig couplings of (hetero)arylamines [48], as well as forward and reverse amide couplings (Fig. 22) [49]. The micellar-catalysed reactions reported, in all cases, are highly efficient with most substrates proceeding with >90% conversion to the desired products for all four reactions developed. The reaction set-up is no different to other on-DNA reactions; the commercially available surfactant used is simply in place of a regular solvent. The work described provides an alternative to the use of non-aqueous solvents, that has the potential to be applied to various synthetic approaches and expand the available chemistries that can be carried out on DNA.

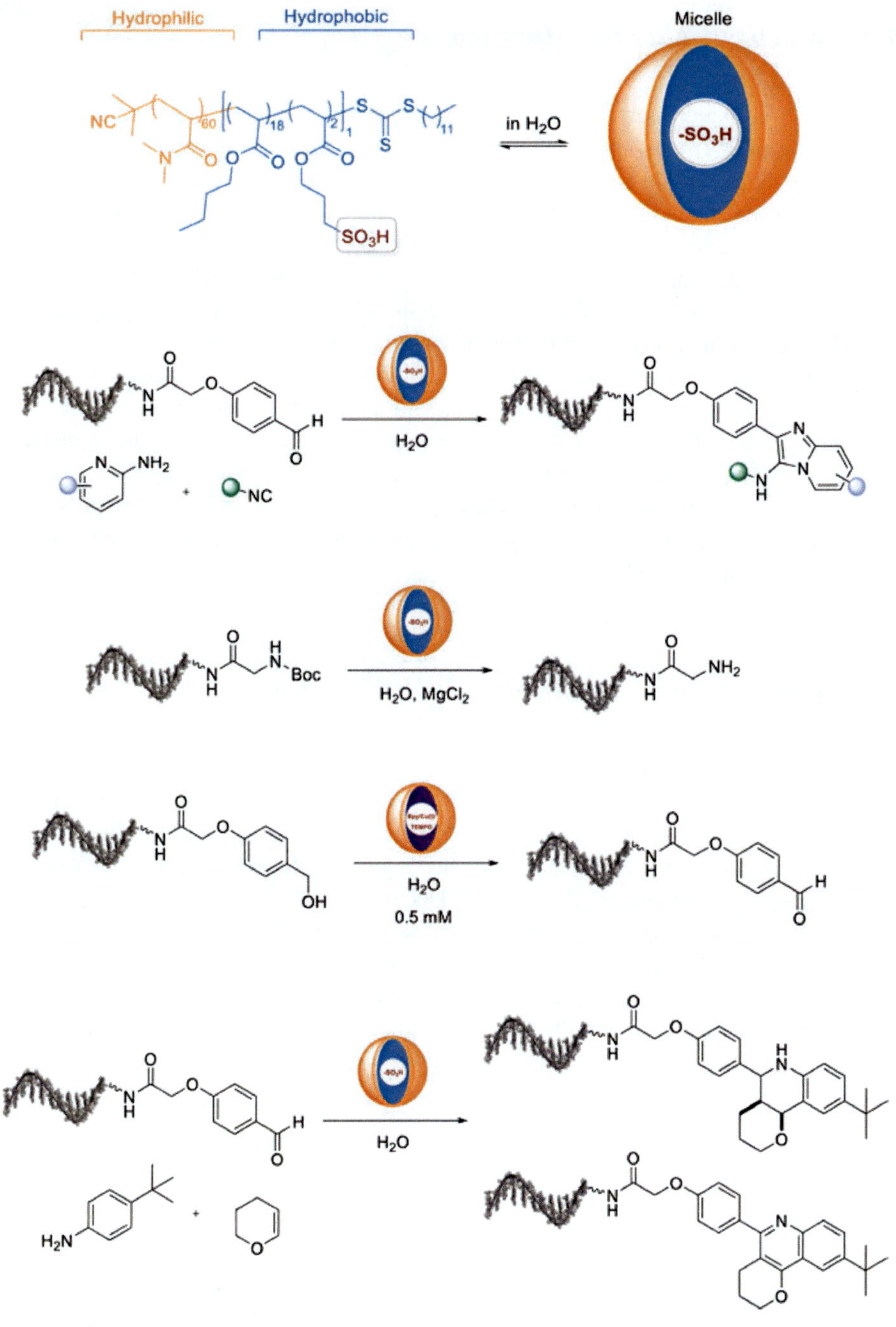

Fig. 21 Structure of the amphiphilic block copolymer and reactions facilitated by the amphiphilic block copolymeric micelles: aminoimidazopyridine formation via Groebke–Blackburn–Bienaymé reaction (top scheme), Boc deprotection of amines (second), primary alcohol oxidation to the corresponding aldehyde (third), and tetrahydroquinoline formation via Povarov reaction (bottom). Conditions: DNA-aldehyde conjugate, 2-amino pyridine, isocyanide, block copolymer, H_2O, EtOH, 40°C, 54 h (top scheme), DNA-*t*Boc protected conjugate, $MgCl_2$, copolymer, H_2O, 50°C, 4 h (second); DNA-alcohol conjugate, copolymer, CuBr, N-methylimidazole, TEMPO, H_2O,

4.3 Micelles Under the Microscope

To provide an insight into on-DNA micelle-promoted reactions, transition electron microscopy (TEM) was used. When the copolymer surfactants were used, it was shown that the structural composition of the micelles was not altered by the addition of DNA, or the alkene utilised in the reaction, and the same spherical shape was observed (Fig. 23a–c). However, the addition of the DNA, alkene, and amine showed the aggregation of micelles into large conglomerates, where the micelles appear rod-like compared to the spherical shape observed on their own (Fig. 23d). The block copolymer on its own is 20 nm swelling to 100 nm in the reaction mixture, suggesting organic substrates are being encompassed within the micelle [46].

When the commercially available surfactant TPGS-750-M was used, TEM was used to demonstrate the effect that different surfactant concentrations have on the micelle composition (Fig. 24). Micelles were shown to form at 2.5% TPGS-750-M with a diameter of 50 nm, swelling to 200 nm at 3.5% TPGS-750-M in the presence of DNA. This suggests that the DNA associates with the micelles and has an effect on their shape and size [49].

4.4 Linker Choice for Micellar DEL Chemistry

For the development of the micelle-promoted reverse amide coupling reaction, initial investigations were carried out using a PEG linker that is typically used in DEL synthesis; however, poor conversion was observed using this linker [49]. It was hypothesised that the use of a hydrophobic linker would promote the reaction by encouraging the organic portion of the DNA-conjugated headpiece to associate within the micelle core. This theory was confirmed when a more hydrophobic hexadecanoic acid linked headpiece (Fig. 25) was used and coupling was observed for a subset of amines. Similarly, the work with the block copolymer surfactant was carried out using an all-carbon hydrophobic linker. The authors observed no difference in conversions when the coupling was performed with a more hydrophilic PEG linker [49].

Fig. 21 (continued) rt. 24 h (third); DNA-aldehyde conjugate, aniline, olefin, block copolymer, H_2O, EtOAc, 25°C, 18 h (bottom)

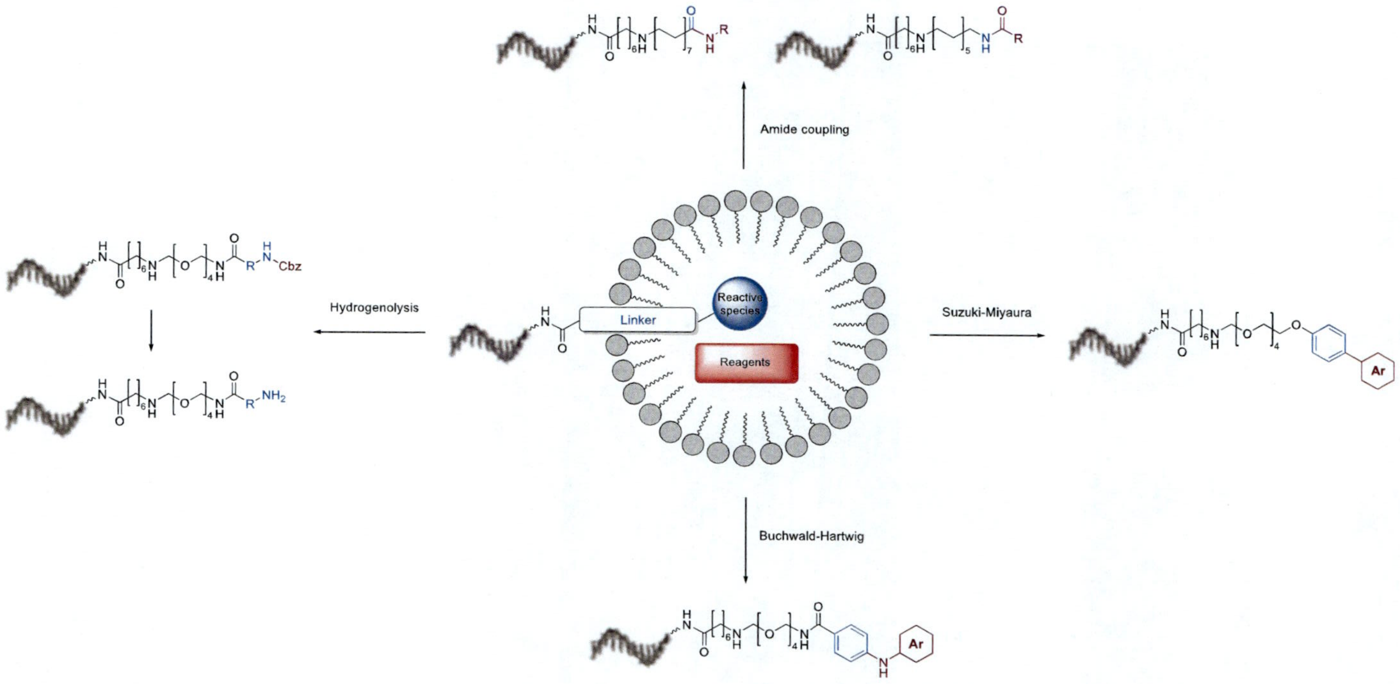

Fig. 22 Reactions promoted by micelle-forming surfactant TPGS-750-M: forward and reverse amide couplings (top), Suzuki-Miyaura cross-couplings (right), Buchwald Hartwig couplings of (hetero)arylamines (bottom), and hydrogenations (left)

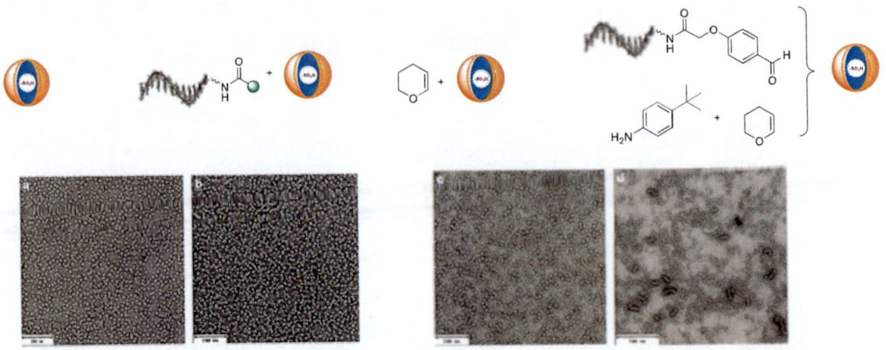

Fig. 23 TEM images depicting (**a**) micelles (**b**) micelles + DNA (**c**) micelles + olefin (**d**) micelles, DNA, olefin, and amine, showing the effects of reagents on the micellar size and aggregation. Reprinted with permission from J. Am. Chem. Soc. 2019, 141, 26, 10546–10555. Copyright 2019 American Chemical Society

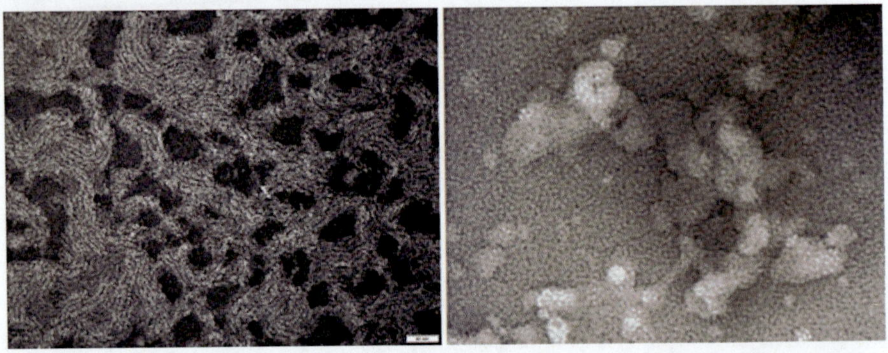

Fig. 24 TEM images displaying: 2.5% aqueous TPGS-750-M (left) and 0.1 mM amino-C_{11}-hexylamidoDNA in 3.5% aqueous TPGS-750-M (right), with the swelling of the micelles apparent. Reprinted with permission from Chem. Sci., 2021, 12, 9475. Copyright 2021 Royal Society of Chemistry

5 Enzymatic Transformations for DEL Synthesis

5.1 *Introduction*

Enzyme-mediated transformations have long been utilised in organic synthesis. Revolutionary technologies in genome sequencing, bioinformatics, and gene synthesis in the past 20 years have allowed limitations in stability, activity, or selectivity of enzymes from Nature to be efficiently addressed in laboratories, via enzyme engineering strategies [50]. Common transformations used in organic synthesis employ enzymes including oxidoreductases, transferases, hydrolases, lyases, and isomerases [51].

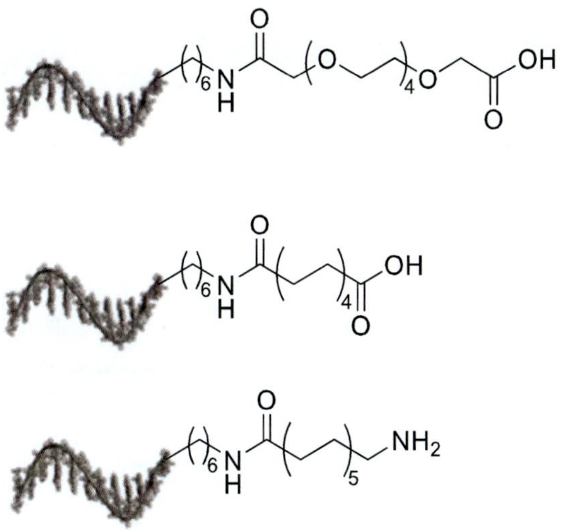

Fig. 25 Structures of DNA-linker conjugates utilised in micellar DNA-based reactions: unsuccessful PEG linker (top), acid-bearing hydrophobic linker (middle), and amine-bearing hydrophobic linker (bottom)

Enzyme-catalysed transformations are associated with very high selectivity, and such transformations are quite orthogonal to traditional chemical modifications because of the mild aqueous reaction conditions used. The combination of these two factors makes enzyme-mediated reactions an extremely attractive area to apply to DEL synthesis. However, the application of enzymes for DNA-encoded libraries has only recently begun to be explored.

Because of the demand to incorporate a diverse range of building blocks in DELs [52], substrate specificity remains one of the large barriers in applying enzymatic reactions to DNA-encoded library technology. This section outlines the advancements made in overcoming this problem.

5.2 *Overview*

To date there have been only two examples of biocatalysis applied to DEL synthesis, incorporating either carbohydrates or ß-hydroxy-α-amino acids (both of which have been largely absent from previous DELs).

Work by Thomas et al. created carbohydrate-based libraries via both enzymatic glycosylation and oxidation [53]. Established chemical synthesis was used to append synthetic glycans to a DNA headpiece via a urea linkage. The on-DNA glycans were then used as substrates for glycosylation enzymes, followed by site-specific oxidation of the primary sugar hydroxyl group. The ensuing aldehyde was then capped by reductive amination or hydrazone ligation (Fig. 26).

Chai et al. have also employed a similar strategy using L-threonine aldolase to create a β-hydroxy-α-amino acid-containing DEL [54]. A range of carboxylic acid-containing aldehydes were chemically attached to a DNA headpiece via an amide

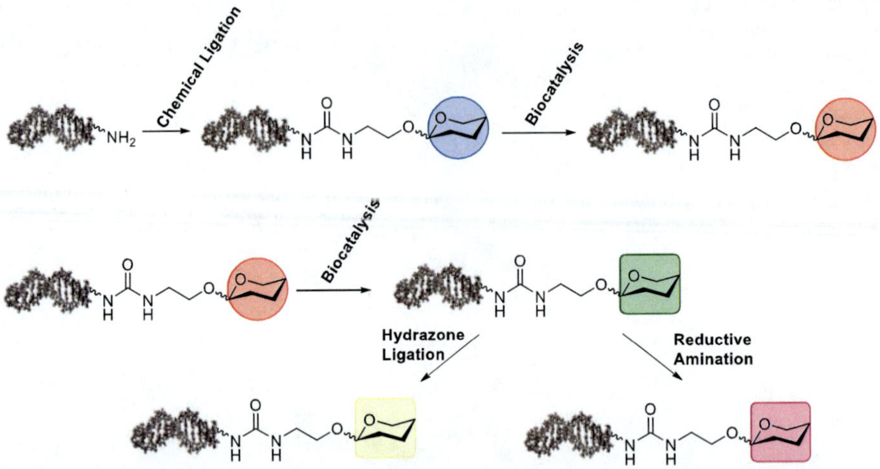

Fig. 26 Carbohydrate-based DEL synthesis by Thomas et al., incorporating sugar rings via chemical ligation and the subsequent enzyme-mediated transformations and the ultimate capping

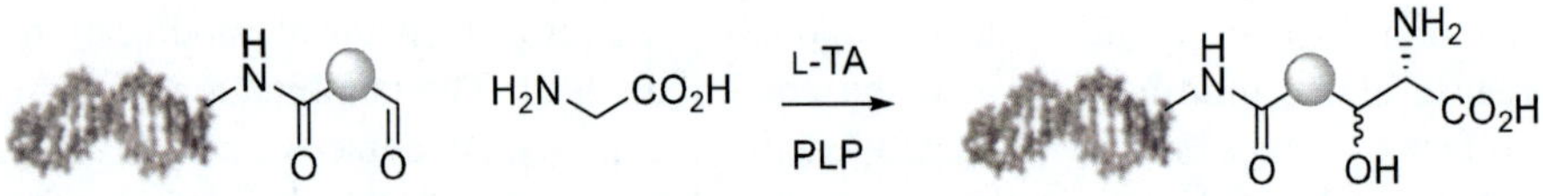

Fig. 27 Threonine aldolase-mediated β-hydroxy-α-amino acid DEL synthesis by Chai et al., involving donor amino acid (glycine) and PLP

coupling. The resultant on-DNA aldehydes were then used as substrates for the L-threonine aldolases, with glycine as the donor amino acid (Fig. 27). A range of chemical methods were then trialled for the subsequent diversification of the β-hydroxy-α-amino acid scaffold.

In both instances, neither reactivity nor isolation of the product was reported as an issue. The main problem encountered was enzyme specificity. The authors also had some difficulty with product analysis due to the novel nature of the products formed in these reactions.

5.3 *Enzyme Specificity*

In the case of the glycosylations, different enzymes, each with unique conditions, were required to transfer different glycans. Bovine β1,4-galactosyltransferase was used in the transfer of Gal-β1,4 and two different enzymes were used to produce α2,6- and α2,3-linked sialosides (Fig. 28).

In contrast, galactose oxidase (GOase) accepted a reasonably wide substrate scope, allowing the oxidation of multiple glycoconjugates. The specific activity of

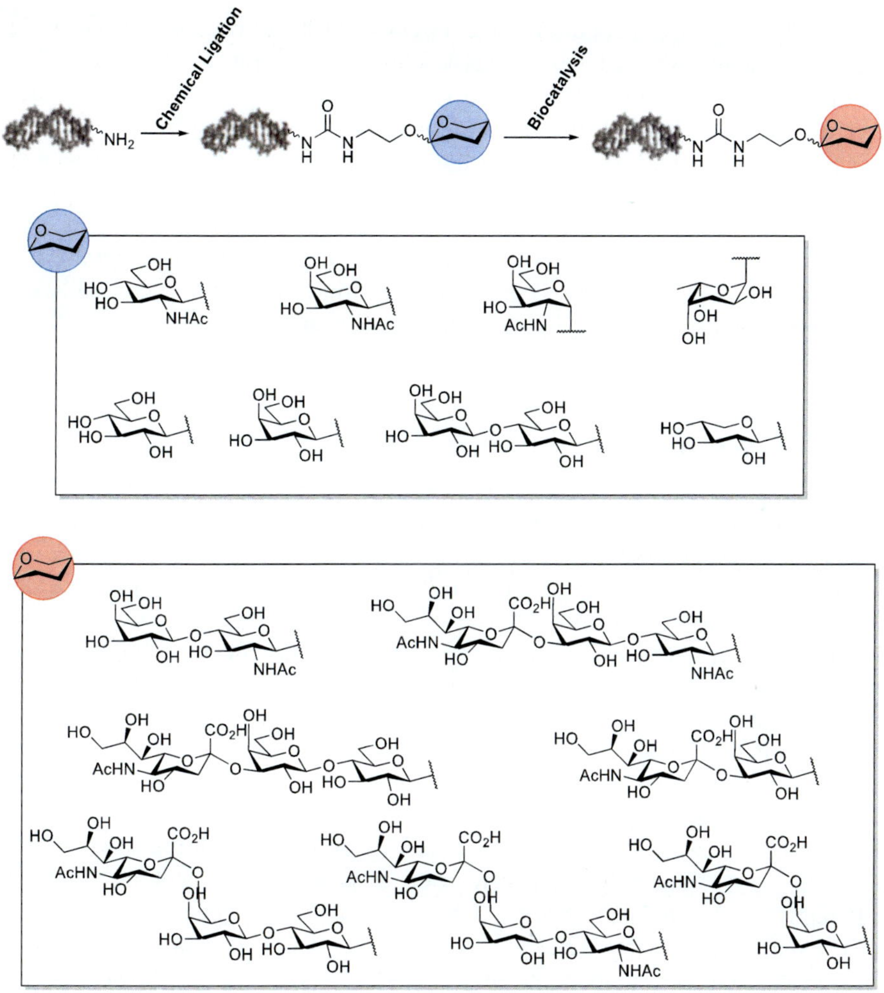

Fig. 28 Structures of glycoconjugates generated from the chemical coupling of the displayed DNA headpiece with a range of 2-aminoethyl glycosides (top box) and subsequent enzymatic glycosylation yielding the shown sialosides (bottom box). Bovine β1,4-galactosyltransferase was used in the transfer of Gal-β1,4 (top box, top left to bottom box, top left) and two different enzymes were used to produce α2,6- (bottom box, top right and middle row) and α2,3-linked (bottom box, bottom row) sialosides from the corresponding precursors

two variants of GOase (M_1 and F_2) towards a range of aldehydes was quite different with no ascertainable pattern. The authors did note that neither variant accepted αGalNAc or αMan as a substrate, indicating that the β configuration of the DNA glycoconjugates was an essential requirement.

Chai et al. found optimised conditions for the conversion of the 4-formylbenzoic acid derived substrate (Fig. 27, left), finding three L-threonine aldolases that showed robust conversion under various conditions, with reasonable enzyme loading

[54]. With the optimised conditions, the on-DNA aldehyde scope was investigated using all three threonine aldolases. Heterocycles gave moderate to high conversion, but bicyclic substrates resulted in no to low conversion. Two aliphatic aldehydes were tested; the cyclohexyl substrate resulting in high conversion, while the linear substrate did not react. It is perhaps such discrepancies between chemically similar structures that will remain an issue for enzymatic DEL synthesis. Further to this, the three enzyme varieties seemed to differ in performance, comparatively, for one substrate relative to another, leaving no obvious choice of optimal enzyme.

In addition to exploring the aldehyde scope, serine and alanine were trialled as amino acid donors, with both showing very low conversion.

5.4 *Product Analysis*

Whilst conducting the research, both laboratories were forced to create new methods to analyse on-DNA products.

Traditionally, on-DNA reactions have been monitored by MALDI or LC-MS. However, Thomas et al. monitored reaction progress by reversed-phase HPLC, as direct analysis of sialosides by mass spectrometry was difficult due to fragmentation of the product [53]. The MS indicated a much lower conversion than was observed by HPLC trace analysis.

In a separate problem, owing to the high molecular weight of the glycoconjugates (ca. 5,000 Da) and the relatively low mass change upon oxidation (mass shift of −2), determining the activity of the enzyme by LC-MS proved to be challenging. To circumvent this issue, a subsequent hydrazone ligation was performed, which was shown to be a fast, highly efficient reaction. This allowed the authors to readily differentiate between the aldehyde and the non-oxidised compounds.

Threonine aldolase introduces two stereogenic centres and Chai et al. wanted to measure the stereoselectivity of the enzyme [54]. All aldol products showed a single peak on achiral LC-MS columns; however, using *o*-phthaldialdehyde (OPA) and *N*-Boc L-cystine as amino acid derivatisation reagents the diastereomers could be eluted as four distinct peaks. Consequently, it was then determined that all three Threonine aldolases give excellent α-carbon stereoselectivity and moderate to good β-carbon selectivity, consistent with literature examples of the enzyme [55]. There is no current method to establish the absolute configuration of the samples to date.

5.5 *Linker Importance on Selectivity*

Chai et al. noted that all aldehyde substrates showed β-carbon selectivity (>83.0% de) except in the below substrates (Fig. 29) [54]. It was postulated that extending the linker between the DNA and aldehyde building block reduced reaction selectivity dramatically.

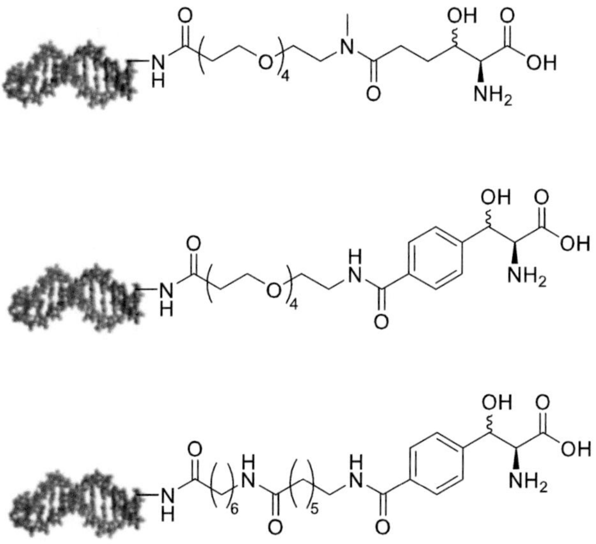

Fig. 29 Consequent products of substrates which exhibited poor diastereoselectivity in the L-threonine aldolase-catalysed aldol reaction

6 Electrochemistry and Radicals in a DEL Context

6.1 Introduction

C(sp^3)-C(sp^3) bond formation is one of the reactions absent from the chemical toolbox for DEL syntheses. These types of reactions primarily proceed through radical chemistry, formerly avoided in DEL-based synthesis. Generating radicals in a DEL context was hypothesised, for a long time, as an unapproachable area. The obligatory aqueous medium to perform DEL-based syntheses, alongside the very low concentrations, represents a major hurdle for radical reactions. Generated radicals could also damage the functional group rich DNA-based barcode.

The importance of C(sp^3)-C(sp^3) bond formation as a chemical step in DEL synthesis is attributed to its potential to create unprecedented diversity from simple building blocks. To figure out how to bring C(sp^3)-C(sp^3) bond formations through radical processes into a DEL context, a general approach should be developed to overcome the hurdles mentioned above. In this section, how to bring organic reactions, in general, to DEL settings and how radical chemistry could be compatible with DNA and its idiosyncratic requirements, is discussed.

6.2 Overview

Introducing C(sp^3)-C(sp^3) and C(sp^2)-C(sp^3) bond formation within a DEL synthesis enhances the 3D character of the library and hence enables deeper exploration of chemical space. Although C(sp^3)-C(sp^3) bond formation is well explored in

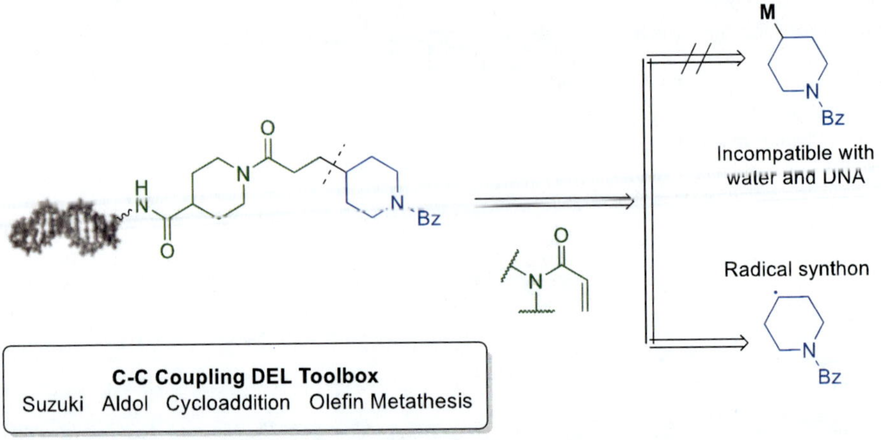

Fig. 30 Scheme displaying C-C Coupling methods, with the application of radical chemistry to C(sp^3)-C(sp^3) couplings due to metal-based couplings of the same kind being incompatible with DEL-like aqueous conditions

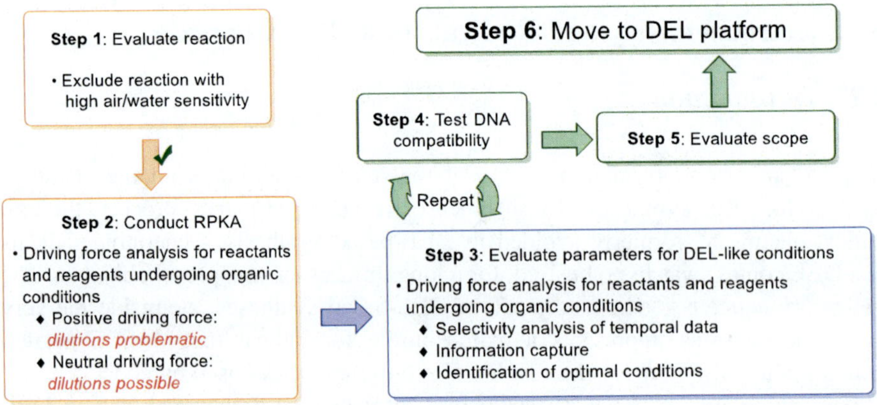

Fig. 31 Flow chart exemplifying the method of bringing organic reactions to a DEL setting, centred on reaction kinetics. *RPKA* Reaction Progress Kinetic Analysis

conventional organic synthesis, applying the required radical chemical conditions in DEL synthesis settings is less straightforward (Fig. 30).

To bring organic reactions into aqueous DEL-like conditions, Wang et al. developed a mechanism-guided protocol dependent on analysis of reaction kinetics, with a proposed six-step flowchart (Fig. 31) [56]. Initially, reactions that are very sensitive to air and/or moisture are avoided; this is followed by selection of reactions that should proceed at high dilution. To provide insight into this information, Reaction Progress Kinetic Analysis protocols (RPKA) were applied. As a rule, reactions which show zero- or low positive-order kinetics in the reactants have the potential to be transferred to dilute concentrations, as required for DELs, without affecting the reaction rate. Guided by the obtained information, an initial set of DNA-compatible reaction conditions are identified. In step 3, empirical evaluation of all reaction parameters over time (temperature, solvent, additives, concentrations) gives myriad valuable data. Such data, when graphically presented, reveals the different patterns of the reaction progress, subsequently leading to better understanding of reaction mechanisms and improved product conversions. It is important, in this step, to assess the reaction deriving forces by considering absolute concentrations, rather than numbers of equivalents. After analysing these data, initial conditions compatible with DEL synthesis are selected. In the final two steps, the success of the selected conditions leads to scope evaluation, which in turn will give access to DEL platform synthesis.

6.3 Giese-Type Reaction

As an application of this protocol, Wang et al. applied it to a Giese-type reaction (Fig. 32) [56]. After applying RPKA in organic conditions, the reaction rate is independent of the concentration of the reactants (redox-active ester and acrylate). This Giese reaction, hence, has zero-order kinetics in relation to reactant concentrations. Consequently, it could be potentially translated to a DEL-like environment. In addition, concentration of Ni had no significant impact on the reaction rate; the only two important parameters were the concentration of LiCl and Zn powder. These findings suggested that the reaction is dependent on the available surface area of Zn, as well as the concentration of LiCl, which served as an electrolyte to facilitate the electron transfer.

The authors then used high LiCl concentrations and large surface areas of Zn (either through higher amount of Zn or higher surface area nanoparticles) in DEL-based synthesis. Various conditions and variables were screened: reactant and catalyst concentrations, different metal powders, buffers, surfactants, and solvent ratios. The incorporation of the wide-reaching data into a graphical form led to elucidation of an ideal set of conditions for DEL synthesis. One of the important findings from these optimisations was that the kinetic patterns are transferrable from the parent organic conditions to DEL conditions, providing evidence that the reaction has the same mechanism in organic and DEL-like conditions.

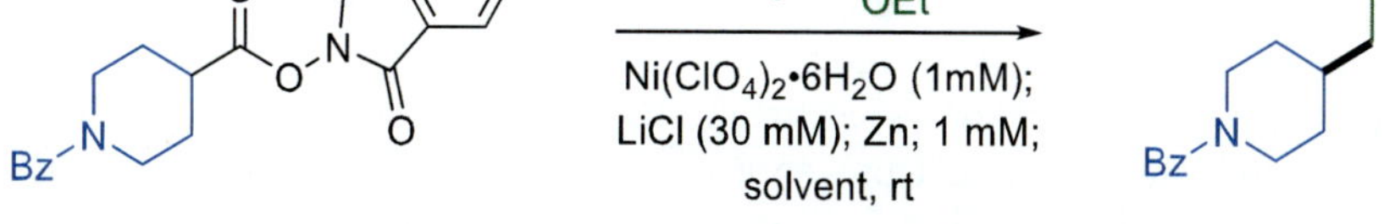

Fig. 32 Off-DNA Giese-type reaction to investigate reaction kinetics before transferring to an on-DNA application

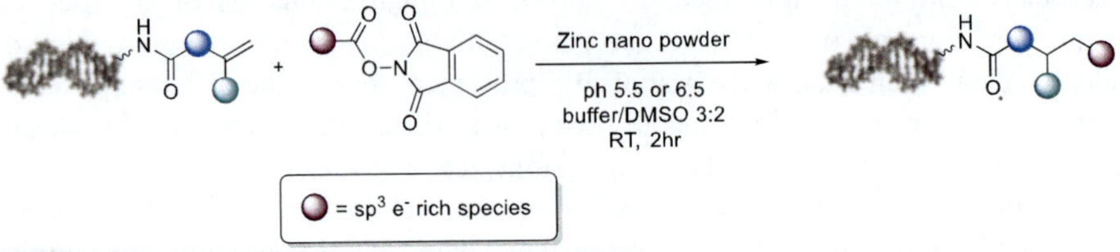

Fig. 33 Optimised conditions for on-DNA Giese-type reaction yielding a diverse range of C(sp^3)-rich structures

With the optimised DEL-like conditions in hand, Wang et al. explored the scope of the Giese reaction in DEL settings with medicinally relevant building blocks rich with sp^3 carbons, including primary, secondary, and tertiary carboxylic acids (to form the redox-active ester) [56]. The Michael acceptor scope included acrylates, acrylamides, vinyl ketones, acrylonitriles, and vinyl sulfones. The reaction provided the desired products in 60–90% yields, from α- and β-substituted Michael acceptors. The reaction tolerated several functional groups including aldehydes, epoxides, and acids, which enable further library synthesis cycles.

Finally, the authors applied the optimal conditions in a DEL synthesis. Whilst it was viable to attach either the acrylate or the redox-active ester to DNA, to avoid side products, an amine bound to 14 bp DNA headpiece was used and coupled to the corresponding acid-bearing acrylate motif. After fine-tuning of solvents and buffer, the conditions were translated successfully to DEL-based settings to give wide, diverse, C(sp^3)-rich structures (Fig. 33).

6.4 *Electrochemistry in an Aqueous Environment*

Organic electrochemistry gives high and precise selectivity through controlling redox potentials [57]. Although this technology is still struggling to find its way into the organic chemist toolbox, Flood et al. applied it to DELs [38]. At first, it was assumed that the aqueous reaction conditions used in DEL synthesis would not be compatible with electric current applied through the aqueous medium. The electricity-water compatibility was not the only hurdle; the charged nature of DNA, which will be absorbed irreversibly to the electrode surface in the reaction cell, was also problematic.

RASS technology is a possible solution to these hurdles (see section "Utilisation of Non-Covalent Reversible Solid Supports DEL Synthesis"). RASS technology was postulated to provide site protection to the DNA during electrochemical redox transformations. It also provides practically anhydrous conditions. To test this concept, the authors chose Ullmann-Buchwald-Hartwig type amination reactions to perform reaction optimisations and investigate the compatibility with electrochemistry (Fig. 34). The first attempt in aqueous conditions resulted in no recoverable DNA. However, once RASS technology was applied using the standard

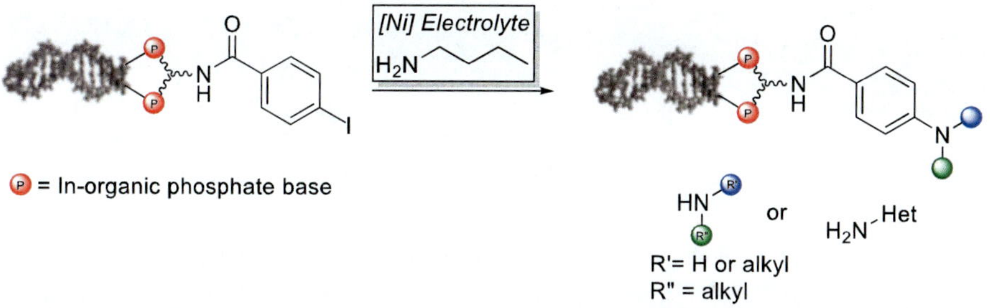

Fig. 34 RASS-enabled on-DNA electrochemical Ullmann-Buchwald-Hartwig type amination with alkyl (1° and 2°) and heterocyclic (1°) amines. Standard conditions: $Ni(bpy)_3Br_2$ precatalyst (50 mM), DBU (300 mM), and 4 mA of current for 3 h

conditions, the DNA conjugate could be isolated in 37% yield, along with the corresponding phenol as the major side product. Optimising the conditions to avoid this side product led to desired product in 74% yield. The scope of the electrochemical reaction covered several alkyl and heteroaryl amines, along with one example for an amide. Anilines and piperazines were not compatible with this reaction. The yields are comparable to those observed with off-DNA substrates (20–70%).

7 C-H Activation and DELs

7.1 Introduction

Selectively functionalising C-H bonds in drug-like molecules could achieve high molecular diversity. Several advancements have been achieved with respect to the activation of C-H bonds in organic media [58], but they are proving difficult to apply to DEL chemistry. The activation of C-H bonds could address the structural gaps in the presently explored chemical space. It could also allow the incorporation of C (sp^3)-rich structures to designed libraries. The power of C-H activation is the ability to convert almost any nonfunctional species to a bifunctional one. For example, a simple, aromatic carboxylic acid can be used as a monofunctional building block to undergo amide formation reactions, then using a C-H activation strategy this could be converted into a bifunctional one, allowing for further transformations (Fig. 35).

Despite the potential for CH-activation in DEL synthesis, there are major challenges, such as aqueous reaction medium compatibility, dilute concentrations, and the presence of DNA tags, which all stand as a barrier to selectively activate C-H bonds. In addition, in this specific case, it is essential to have predictable regiochemical outcomes since in DEL synthesis, there are multiple C-H bonds that could be activated (or multiple couplings could occur on a substrate) and there are no analytical methods that would allow determination of the product identities. In this

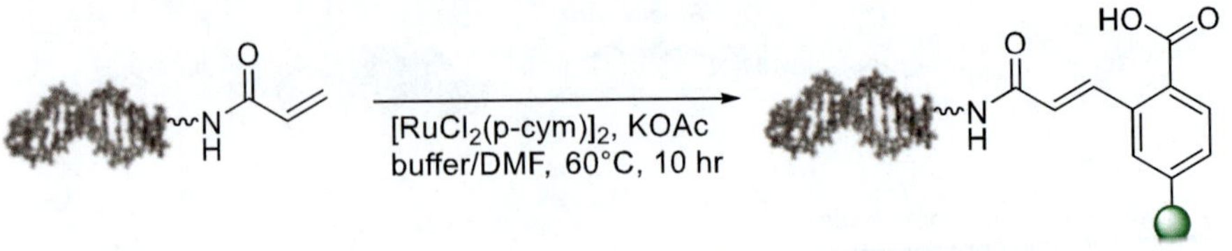

Fig. 35 C-H activation coupling functionalised benzoic acids to a DNA-conjugated acrylamide to yield bifunctional building blocks

section, circumvention of these challenges and the revelation of unexplored chemical space by C(sp$^{3/2}$)-H will be outlined.

7.2 Overview

Previously, it was not possible to perform C-H activation reactions in DEL aqueous media due to the challenges mentioned above. To circumvent this, Gerry et al. designed a library of ca. 100,000 DNA-barcoded saturated *N*-heterocycles, featuring under-represented properties like defined stereocentres and topographic complexity, through the Pd-mediated C-H arylation of azetidines and pyrrolidines [59]. This C-H arylation reaction, however, could not be performed in the presence of a DNA conjugate.

Each of D-azetidine-2-carboxylic acid and D-proline was converted to the corresponding aminoquinolines – the precursors for the C-H arylation reaction with 1,4-diiodobenzene (Fig. 36) [60]. All four stereoisomers of both the azetidine and pyrrolidine scaffolds were obtained after amine protection and removal of the aminoquinoline directing group. DNA was then attached through reacting the amine-functionalised DNA headpiece with the activated NHS esters to give the DNA conjugates ready for library synthesis. The library synthesis was continued after removal of the Boc protecting groups and functionalising the secondary amine with either aldehydes (through reductive amination) or with sulfonyl chlorides. In addition, the iodo-aryl group could be exploited for further functionalisation through Suzuki coupling.

7.3 C(sp^2)-H Activation

Despite the previous approach benefiting from traditional off-DNA chemistry and DNA-encoding technology, it considerably loses the power of encoding every single reagent in every synthetic step during the split-pool cycles, leading to limited variety of building blocks and, consequently, limited library size. In contrast, using the C-H activation as a synthetic step in the synthesis of a DEL will give access to diverse chemical space with larger library sizes. The first reported C(sp^2)-H activation

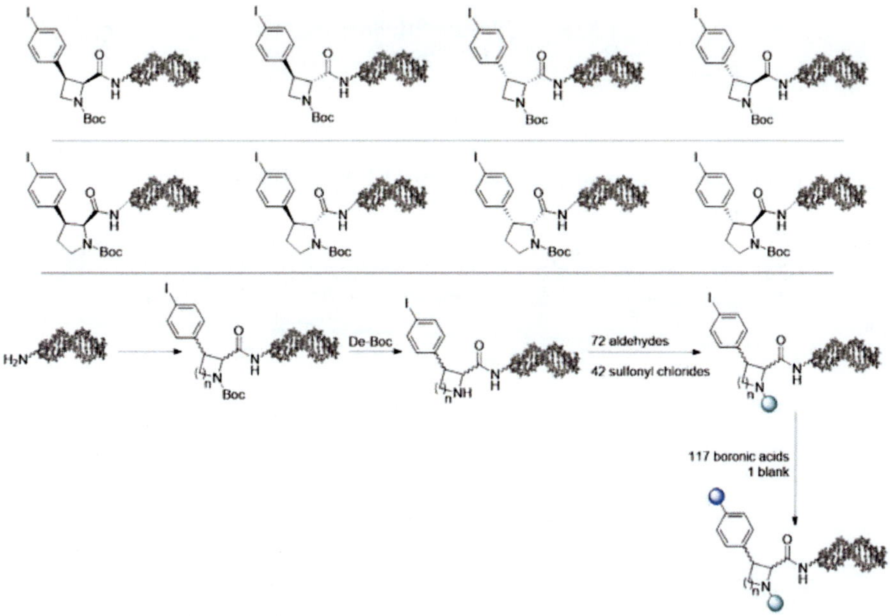

Fig. 36 Four stereoisomers of azetidine (top) and pyrrolidine (middle) aminoquinoline scaffolds. Formation of DNA conjugates of functionalised saturated azacycles, deprotections, and subsequent library synthesis

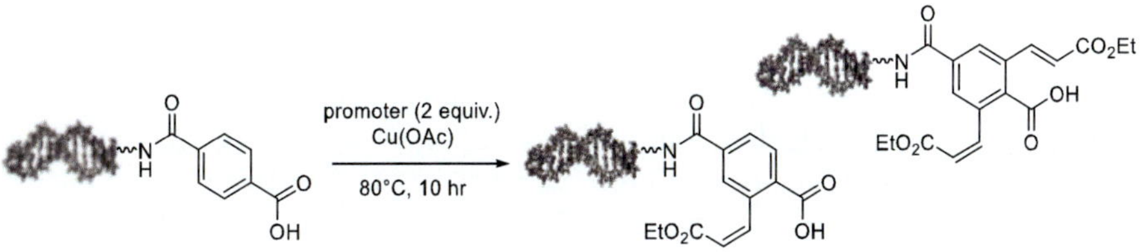

Fig. 37 C-H activation of DNA-conjugated benzoic acids at the *ortho* position, trapped with acrylic esters to yield both mono- and di-alkylated products

compatible with DEL conditions was by Wang et al. [15] The authors developed ruthenium-promoted C(sp^2)-H bond activation in the *ortho* position of aromatic acids, trapped by acrylic esters (Fig. 37).

To begin with, the reaction between the simple DNA-conjugated aromatic acid and acryl esters was investigated. After a series of optimisations related to catalyst loading, $Cu(OAc)_2$ equivalents, and solvent ratios, the desired mono-addition product was obtained (72%). As a control experiment, the DNA-conjugated arene, lacking carboxylic acid, was allowed to react with the acrylate ester to ensure that the C-H activation was directed by the carboxylic acid (Fig. 38). Fortunately, the authors found that there was no product formed (Table 1). In parallel, the reaction of

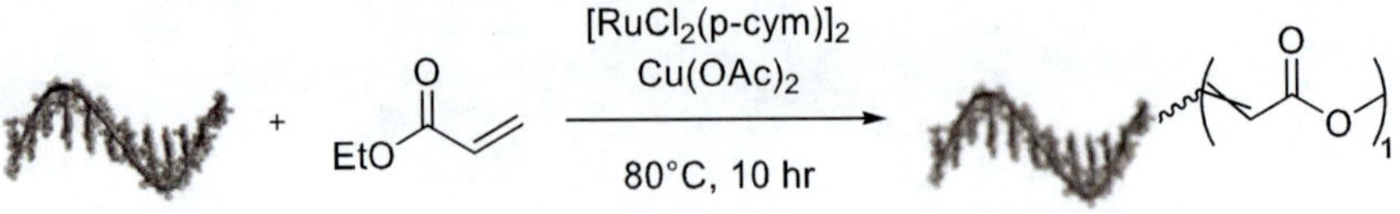

Fig. 38 Control experiment exposing unfunctionalised DNA to C-H activation conditions

Table 1 DNA constructs exposed to C-H activation conditions, all displaying no DNA adduct

Entry	DNA	DNA adduct[a]
1	HeadPiece—NH 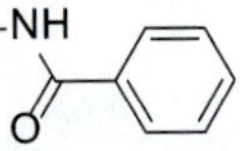	0
2	AAAAAAAA	0
3	TTTTTTTT	0
4	CCCCCCCC	0
5	GGGGGGGG	0

[a] Determined by LCMS

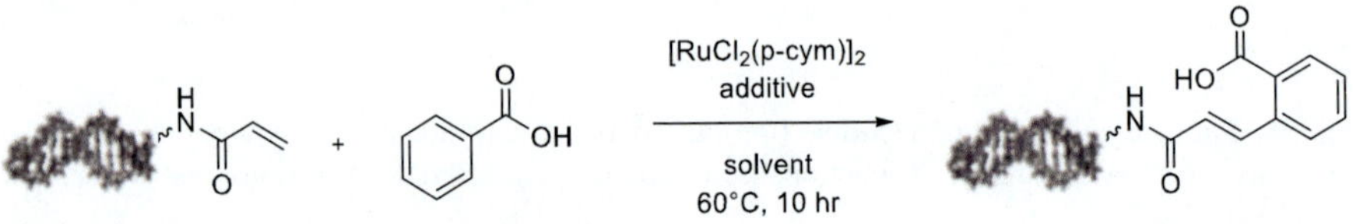

Fig. 39 Optimised conditions for the C-H activation reaction between a DNA-conjugated acrylamide and benzoic acid

acrylate with four single-stranded octa-nucleotides, under the same conditions, showed there is no detectable reaction with the DNA backbone.

Following this, the authors found that the versatility and availability of aromatic carboxylic acids could be better represented in DELs if they are exploited as building blocks rather than being conjugated to the DNA. Therefore, the two reaction components were switched over and the optimised conditions extended to the new reaction setting. Unfortunately, using large excesses of aromatic carboxylic acid led to decomposition of the DNA. After several reaction optimisations and changing the acetate forms, the desired product was obtained with 90% conversion (Fig. 39).

The scope of this $C(sp^2)$-H activation reaction was very limited with few examples >80% conversion and 3 examples between 60–80%; poor yields for aldehyde, vinyl, and ester functionalised aromatics (<30%) were also observed (Fig. 40).

7.4 $C(sp^3)$-H Activation

Despite $C(sp^2)$-H activation having limited building block scope due to the adjacent directing groups, $C(sp^3)$-H functionalisation offers wider opportunities to explore

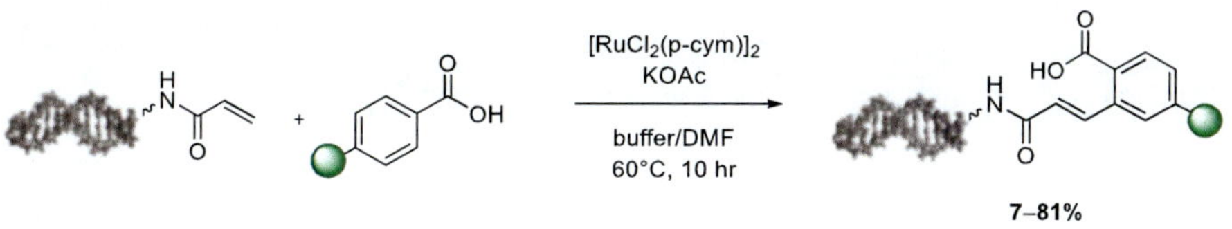

Fig. 40 Application of optimised C-H activation conditions to a variety of *para*-substituted benzoic acids

chemical space. Fan et al. developed the first DEL-compatible palladium-catalysed C(sp^3)-H arylation of aliphatic carboxylic acids, amides, and ketones with aryl iodides (Fig. 41) [61]. Carboxylic acid and (hetero)aryl iodide building blocks are considered cornerstones in DELs. C-H arylation reactions will yield a new axis for chemical space diversity.

The authors investigated the Pd-catalysed β-C-H arylation between pivalic acid and aryl iodide DNA conjugates. The C-H arylation conditions developed previously guided on-DNA reaction condition optimisation, leading to 89% conversion to the desired product without any detection of di- or tri-arylation [62, 63]. It is worth mentioning here that amino acid derived ligands have an impact in improving the yields of this reaction. The scope of the reaction covered different carboxylic acids adjacent to quaternary carbons, with conversions of 50–80%. However, carboxylic acids adjacent to secondary or tertiary carbons gave lower conversion (20–45%). The alkene isosteres, cyclopropane- and cyclobutanoic acids, proved to be competent coupling partners with conversions of 40–50%. Aryl and heteroaryl iodides showed wide scope, reacting with carboxylic acids under the optimised conditions. Furthermore, the obtained arylated products attached to the DNA could undergo second C-H arylation, directed by the carboxylic acid (Fig. 42).

Extending this reaction to amides instead of carboxylic acids opens an opportunity to develop ligand-free Pd-catalysed β-C-H arylation. Therefore, the authors utilised the previously developed reaction conditions, after modification, led to isolation of the desired arylated product with 69% conversion. This reaction modification introduced amides as coupling partners and showed broad tolerability. The amides obtained from amino acids as well as β-amino acids gave the desired products (30–80%) (Fig. 43). The reaction was compatible with different aryl and heteroaryl iodides (Fig. 44).

Ketones could be exploited as building blocks in DELs. Following the strategy of converting monofunctional building blocks to unusual bifunctional ones represents an opportunity for C(sp^3)-H functionalisation. The authors envisioned that the

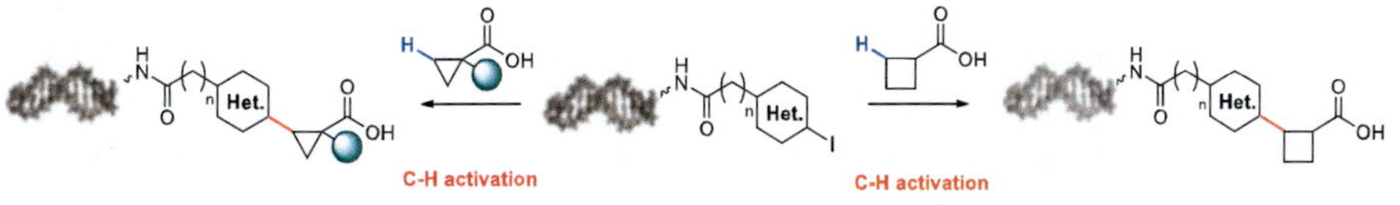

Fig. 41 Pd-catalysed C-H arylation reactions between DNA-conjugated aryl iodides and cyclic aliphatic carboxylic acids

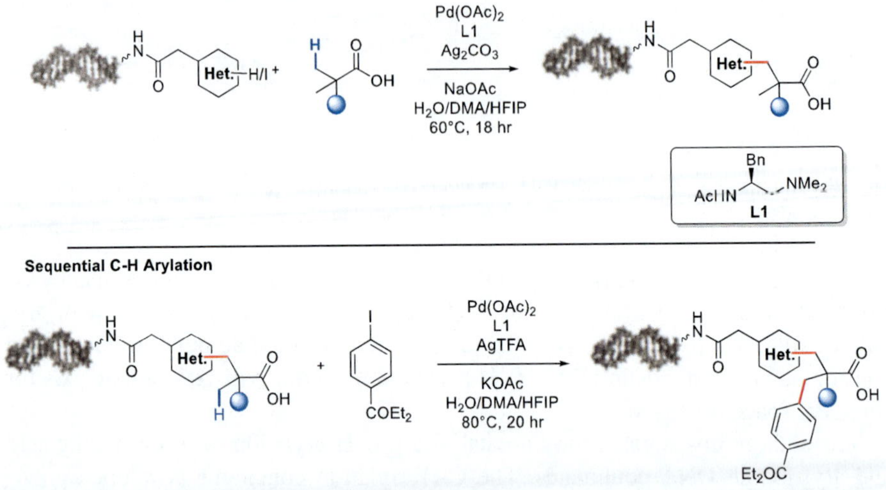

Fig. 42 Pd-catalysed β-C-H arylation reaction between DNA-conjugated aryl iodides and aliphatic carboxylic acids (top). Subsequent C-H arylation of DNA-conjugated acid product with ethyl 4-iodo benzoate (bottom)

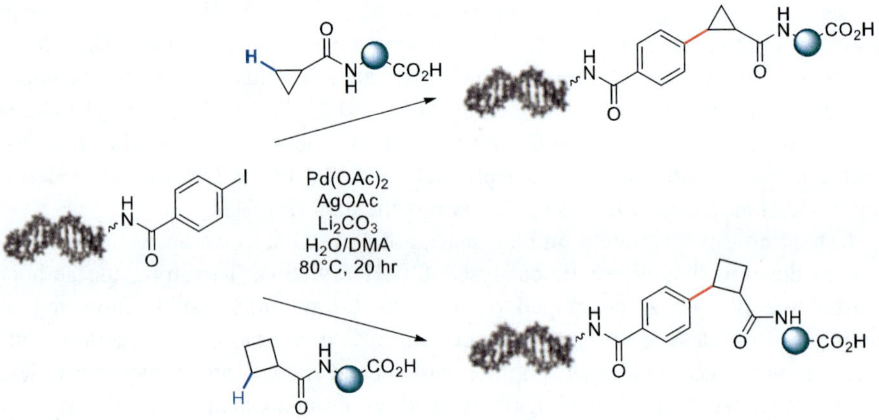

Fig. 43 Expansion of β-C-H arylation conditions to amides (amino acid derived)

aminooxyacetic acid (oxime ether) could be used as a removable directing group for C-H arylation as a masked ketone. The reaction conditions developed in the previous two examples were extended to this masked ketone variant. Optimising the palladium loading, silver salts, temperature, and ligands led to formation of the desired product (62% conversion). The reaction showed wide tolerability for different functional groups and for aryl and heteroaryl iodides with 30–60% conversions (Fig. 45). Most important is the ability to convert the oxime ethers back to ketones. The authors developed DNA-compatible conditions to hydrolyse back to the corresponding ketone using aniline and acetone, which likely works through an equilibrium transamination with aniline and subsequent trapping of the free aminooxyacetic acid with the ketone.

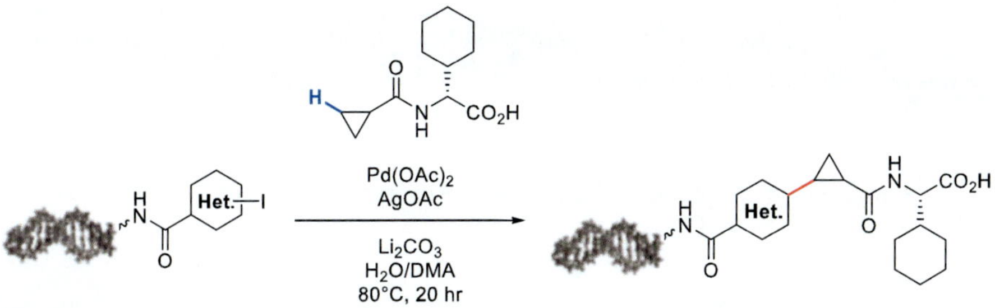

Fig. 44 Utilisation of various (hetero)aryl iodide in β-C-H arylation reactions with an amino acid-derived amide

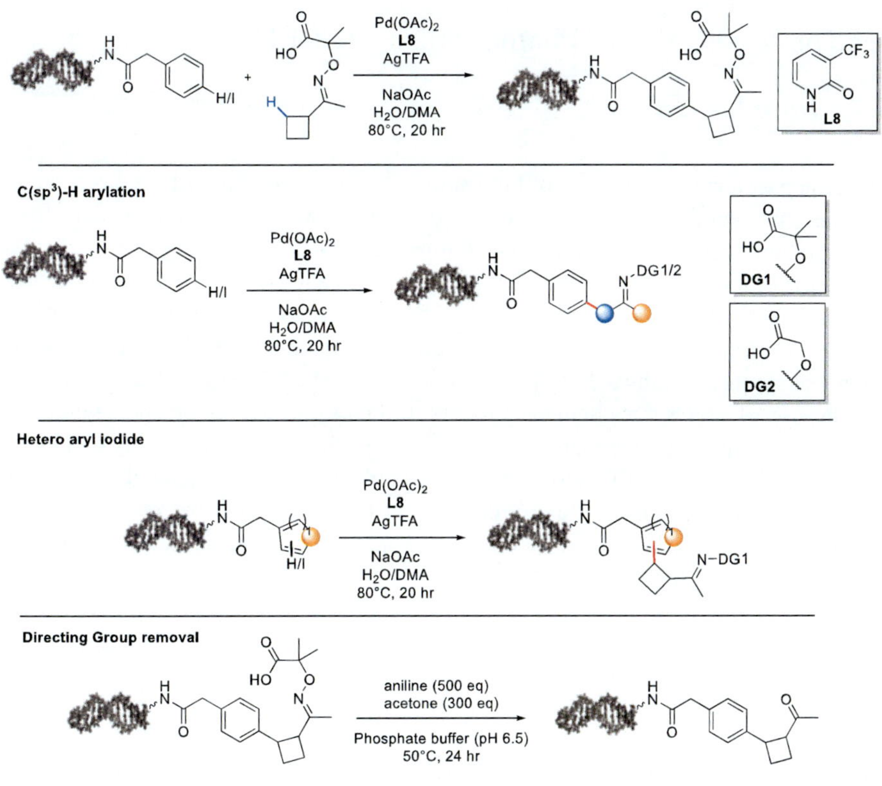

Fig. 45 Optimised β-C-H arylation conditions of oxime ethers, which act as a directing group and as a masked ketone (first). Application of optimised conditions to a variety of aliphatic oxime ethers (second) and heteroaryl systems (third). Directing group removal, hydrolysing the oxime ether back to the corresponding ketone (bottom)

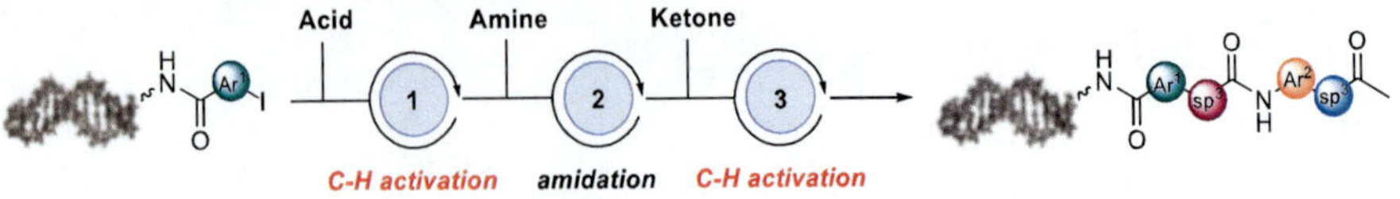

Fig. 46 Schematic of the application of C-H activation to expand the chemical space and number of C(sp3) rich structures in DELs

These new C(sp^3)-H activation reactions are considered a new disconnection, converting current building blocks to unusual bifunctional ones ready to be exploited in DEL syntheses (Fig. 46). These reactions allow access to new chemical space and enhance the number C(sp^3) rich structures, leading to more efficient, drug-like libraries.

8 The Application of Photochemistry to DELs

8.1 Introduction

Over the past few years, the use of light-promoted reactions using LEDs has become an emerging chemical strategy for organic chemists to activate small molecules [64–66]. Among these LED-mediated reactions, photoredox reactions are the most established. Selective excitation of the photoredox catalyst over organic molecules leads to the formation of excited species that can act as either strong oxidants or strong reductants. This, therefore, led to the development of a plethora of diverse chemical synthetic pathways [66].

The use of LEDs for chemical processes allows the access of unconventional modes of reactivity, and in many cases employs readily available building blocks. In the context of DELs, LED-based chemistry could be highly beneficial to increase diversity in libraries, as well as introducing fast, easily accessible reactivity. This section outlines the advancements, considerations, and challenges of LED chemistry for DNA-encoded library synthesis.

8.2 Overview

The need to expand the chemical landscape in DELs has generated a tremendous number of DNA-compatible reactions derived from established LED off-DNA chemistry. One of the most established off-DNA photoredox chemical processes is the decarboxylative coupling. As a consequence of carboxylic acids being widely commercially available, application of this reaction to DELs could allow the generation of very large libraries. Multiple groups have successfully developed on-DNA decarboxylative coupling methodologies [67–69]. Similarly, transformations

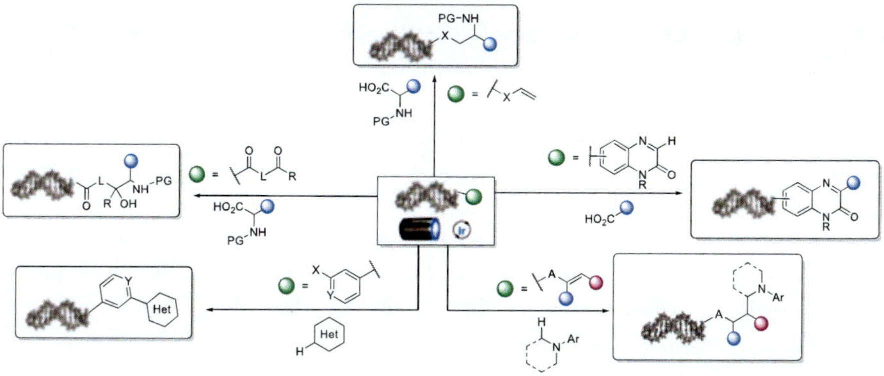

Fig. 47 Selection of available on-DNA photoredox chemistries, including decarboxylative couplings, cross-couplings, and C-H activation reactions

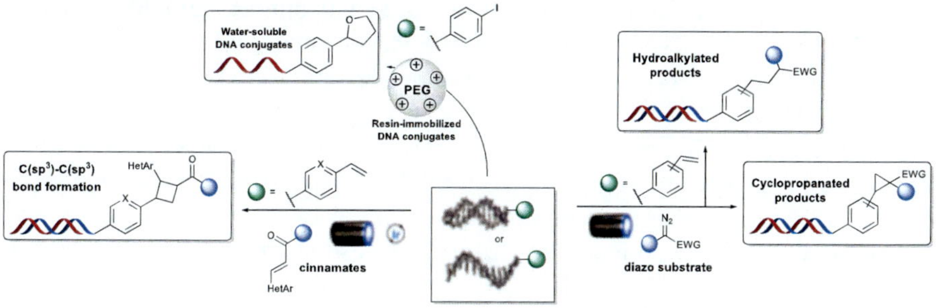

Fig. 48 Current advancements in LED mediated on-DNA reactions: resin-immobilised DNA conjugates (top) and introducing stereodiversity from DNA-conjugated styrenes via a [2 + 2] cycloaddition (left) or diazo reactions (right)

affording C(sp^2)-C(sp^3) bonds, such as C-H activation, have been recently explored by Krumb et al. and Wu et al. [70, 71] (Fig. 47).

Further advancement of on-DNA photoredox has been outlined by Ruff et al. via a polymer-supported strategy to enable photoredox transformations under anhydrous conditions [40]. A metallaphotoredox decarboxylative cross-coupling of carboxylic acids to on-DNA aryl halides in pure DMSO was successfully achieved. Stereodiversity of DNA-encoded substrates is poorly represented within the DEL landscape. Kölmel et al. and Fu et al. investigated the use of LED-mediated reactions with readily available on-DNA styrenes, introducing an expanded stereodiversity (Fig. 48) [72, 73]. Two *trans* diastereomers were formed during the cycloaddition reaction, irrespective of the presence or absence of the DNA tag. To achieve this finding, the authors ran co-injections of the [2 + 2] cycloaddition sample and the acylation with the corresponding acid via LC-MS.

8.3 Impact of Oxygen on DEL Photoredox

Reactive oxygen species (ROS) are highly reactive chemicals formed from O_2 which include peroxides, superoxides, hydroxyl radicals, singlet oxygen, and α-oxygen. While some of these ROS can be beneficial to certain reaction types in common organic synthesis [74], the presence of ROS can be detrimental for DNA. The removal of hydrogen atoms on the sugar-phosphate backbone leads to degradation of the DNA chain. These reactions can occur in the presence of highly reactive radicals, such as a hydroxyl radical (HO•), on the condition that the site is sterically reachable [75, 76].

The presence of oxygen within reaction media has been identified to affect reaction yield on different scales, depending on the reaction type, and a trend is yet to be identified.

Kölmel et al. noted the absence of desired product if the [2 + 2] photocatalytic cycloaddition reaction was not thoroughly degassed with nitrogen prior to irradiation with blue light [72]. The authors presumed the decomposition was due to the formation of reactive oxygen species (ROS); the decomposition product, however, was not disclosed.

In metallaphotoredox reactions, diverse observations were made. Kölmel et al. reported yield maximisation for cross-electrophile coupling in the absence of oxygen (Fig. 49) [77]. With optimised conditions, coupling between the DNA-tagged aryl iodide and *N*-Boc-3-bromo-piperidine led to 92% desired product, while in absence of thorough degassing 27% desired product was observed, accompanied by 7% protodehalogenation (Table 2). This was supported by Kölmel et al. who conducted the coupling between the DNA-tagged aryl iodide and *N*-Boc-morpholine-3-

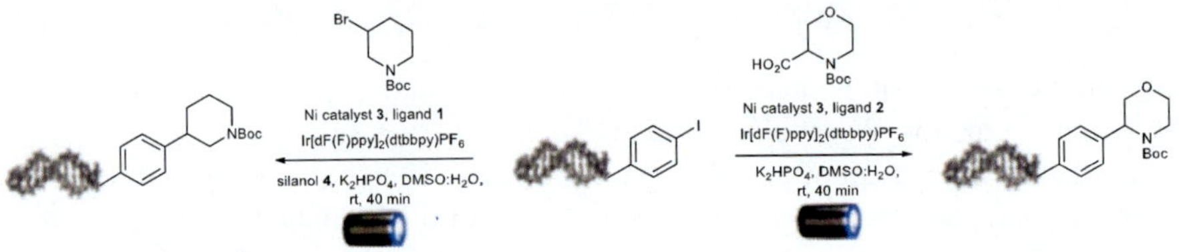

Fig. 49 Cross-electrophile couplings yielding bicyclic compounds form DNA-conjugated aryl iodides and either *N*-Boc-3-bromopiperidine (left), or *N*-Boc-morpholine-3-carboxylic acid (right). Ni catalyst **3** is [(TMEDA)Ni(o-tolyl)Cl], ligand **1** is Pyridine-2,6-bis(carboximidamide) dihydrochloride, ligand **2** is 4-(Trifluoromethyl)pyridine-2-carboxamidine hydrochloride, and silanol **4** is $(Me_3Si)_3SiOH$

Table 2 Selected entries for reactions in Fig. 49, comparing the effect of degassing on conversion. X is the major side product which is the protodehalogenation product from the starting material on-DNA aryl iodide

Entry	Conditions	% Yield (left/X)	Entry	Conditions	% Yield (right/X)
1	As shown	92/0	**1**	As shown	89/4
7	No degassing	27/7	**7**	No degassing	31/7

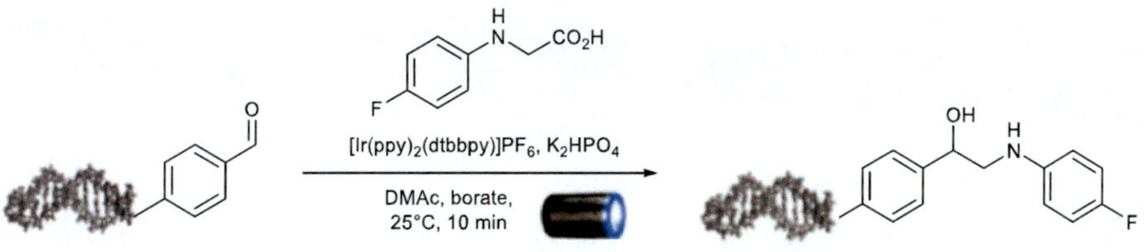

Fig. 50 Decarboxylative coupling with DNA-conjugated aldehyde. Conditions: on-DNA aldehyde starting material (5.0 nmol, 1.0 mM in sodium borate buffer, pH 9.5), (4-fluorophenyl)glycine (200.0 eq, 200.0 mM in DMA), K_2HPO_4 (200.0 eq, 200 mM in H_2O), [Ir(ppy)2(dtbbpy)]PF_6] (2.0 eq, 10.0 mM in DMA), 25°C for 10 min, blue LEDs, protected with N_2

Table 3 Selected entries for reaction in Fig. 50, comparing the effect of degassing on conversion

Entry	Conditions	% Conversion (to right)
2	As shown	91
10	No degassing	87

carboxylic acid, obtaining 89% desired product and 4% protodehalogenation with the optimised conditions [78]. In contrast, the absence of thorough degassing of the reaction afforded 31% desired product and 7% protodehalogenation. In contrast, Phelan et al. and Badir et al. both report metallaphotoredox catalysed cross-couplings easily performed on the benchtop, open to air, and under ambient conditions, requiring minimal effort by the chemist in terms of set-up complexity and timing [78, 79]. These methodologies do not provide comparative data with degassing media but provided yields ranging from 30–84% for C(sp^2)-C(sp^3) cross-couplings on DNA with alkyl bromides (Table 2).

Ruff et al. investigated the potential of polymer-supported catch-and-release decarboxylative metallaphotoredox coupling [40]. Some substrates for photocatalytic decarboxylation do not require the use of an inert atmosphere, yet they observed that these reactions are best performed under inert atmosphere to ensure optimal reproducibility.

Wen et al. attempted to run a decarboxylative coupling under no protection with N_2 prior to irradiation with blue light, the photoredox reaction would generate a slightly decreased conversion leading to 87% desired product in comparison with 91% (Fig. 50, Table 3) [67].

Concordantly, Kölmel et al. identified the presence of oxygen in the reaction media as detrimental to the yield of the reaction [68]. The decarboxylative alkylation between DNA-tagged acrylamide and Boc-Phe-OH yielded 89% desired product with optimised conditions, yet only 27% desired product was observed when the reaction was not thoroughly degassed (Fig. 51, Table 4).

Concerning C-H activation under photoredox conditions, both Wu et al. and Krumb et al. developed methodologies with ambient conditions [70, 71]. Conversions for C-H activation of various on-DNA alkenes with 4-(pyrrolidine-1-yl)pyridine of between 75–94% were achieved by Wu et al. [71], while Krumb et al. disclosed

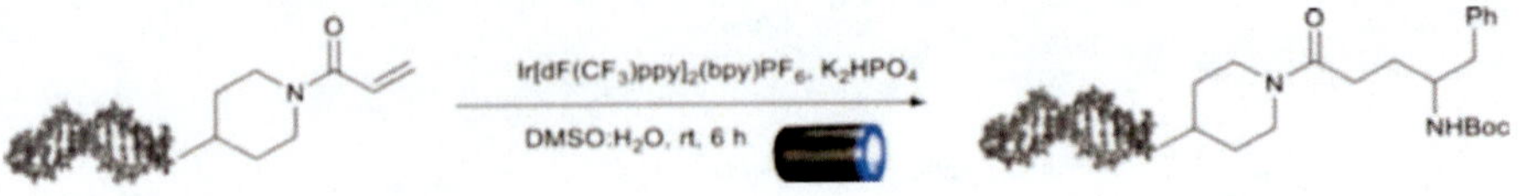

Fig. 51 Decarboxylative alkylation of DNA-conjugated acrylamide with Boc-Phe-OH. Conditions: photocatalyst (1 eq), Boc-Phe-OH (1,000 eq), K_2HPO_4 (1,200 eq), DMSO/H_2O (2:3). $dF(CF_3)ppy$ = 2-(2,4-difluorophenyl)-5-(trifluoromethyl)pyridine, bpy = 2,2′-bipyridine, dtbbpy = 4,4′-di-tert-butyl-2,2′-bipyridine

Table 4 Selected entries for reaction in Fig. 51, comparing the effect of degassing on conversion

Entry	Conditions	% Conversion (to right)
1	As shown	89
4	No degassing	27

yields from 34–93% for C-H activation reactions between DNA-conjugated 4-fluoro-3-iodobenzamide and a range of heteroarenes [70].

Fu et al. identified two cases of photocatalytic hydroalkylation and cyclopronation via diazo compounds, where both methodologies were conducted under atmosphere of N_2 [73]. The authors, however, did not disclose conversions in the presence of oxygen.

To conclude, the presence of oxygen in photoredox chemistry differently affects the conversions obtained, in a reaction type dependent manner. While some substrates do not require the presence of an inert atmosphere, the use of such an atmosphere for developing DNA-compatible photoredox chemistry is recommended to ensure optimal reproducibility.

8.4 *Overcoming the Inconsistency of Photoredox Reactions for DELs*

Excitation of a suitable photocatalyst allows the molecule to undergo intersystem crossing (ISC) to access an excited triplet state. Subsequently, the photocatalyst can transfer its energy to the reaction substrate of interest via single-electron transfer (SET) generating the desired radical that initiates the ensuing mechanism pathways [68, 72]. Similarly, choice of radical precursor will depend on the photocatalytic reaction of interest (utilising carboxylic acid, 1,4 dihydropyridines, silicates, etc.).

Phelan et al. observed that the number of equivalents of photocatalyst used did not prove to play a significant role in photoredox C-H activation reaction yields [79]. This was supported by Krumb et al., whereby 0.25 eq of $Ir(ppy)_3$ generated the product in a 60% yield during a C-H activation reaction between DNA-conjugated 4-fluoro-3-iodobenzamide and a range of heteroarenes [70]. When the authors increased to 1 eq of $Ir(ppy)_3$, the reaction yield slightly decreased to 56%. However,

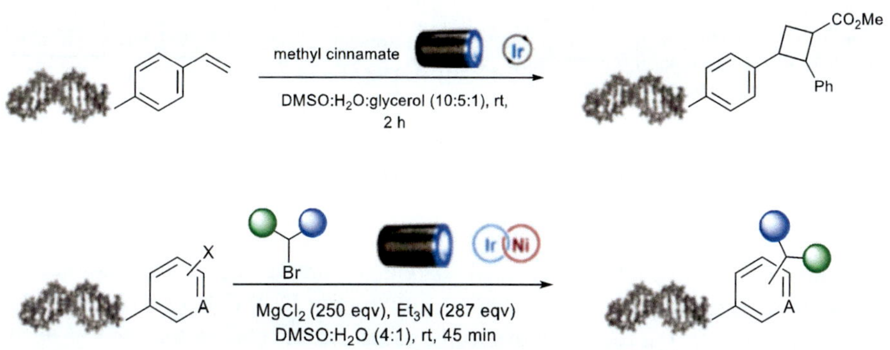

Fig. 52 Examples of the use of additives to enhance stability of radicals for on-DNA photoredox reactions, using glycerol for a [2 + 2] cycloaddition (top) and $MgCl_2$ for a reductive coupling (bottom)

it is important to note that the catalyst is crucial for the reaction to occur; no conversion was observed in the absence of catalyst.

The stability of radicals formed during photoredox chemistry was presented to be crucial to the success of the transformations by Zhang et al. [69] Multiple strategies were resultingly attempted to enhance the stability of the radicals formed.

Secondary or tertiary radicals usually led to better yields than primary radicals, as noted by Kölmel et al., in photocatalytic decarboxylative couplings [68]. Conversely, in the context of metallaphotoredox couplings, the opposite observation was reported; tertiary or sterically encumbered radical precursors systematically generated low yields, as highlighted by Kölmel et al. and Phelan et al. [77–79] Kölmel et al. speculated that this observation was due to the steric congestion at the nickel catalyst [78].

A similar observation was made by Ruff et al. for polymer-supported catalytic metallaphotoredox decarboxylative couplings [40]. Steric hindrance significantly reduced conversion (92% with an unhindered aryl halide, relative to 17% with an *ortho* methyl substituent). The authors also reported destabilisation of radicals in the presence of a significant amount of water.

Kölmel et al. stated that fast radical generation induced by a powerful light source ($\lambda = 450$ nm photoreactor) resulted in both high yields and also shortened the reaction times to 40 min [78]. The authors postulated that a fast reductive elimination step was crucial to outcompete protodehalogenation in aqueous media.

Kölmel et al. added a small amount of glycerol (6% v/v) to enhance their photocatalytic [2 + 2] cycloaddition. They observed an overall beneficial effect on their conversion; the radical scavenging activity of glycerol was proposed as justification (Fig. 52) [72].

Badir et al. used addition of 250 eq $MgCl_2$ to stabilise the DNA backbone during their reductive photoredox coupling (Fig. 52) [80].

Ruff et al. report, for polymer-supported catalytic metallaphotoredox decarboxylative couplings, a destabilisation of radicals in the presence of significant amount of water [40].

Table 5 Selected entries for reaction in Fig. 50, comparing the effect of solvent on conversion

Entry	Solvent	% Conversion of 1
2	DMA	91
5	NMP	52
6	ACN	75
7	DMSO	72

DMSO is the most commonly used solvent for on-DNA catalytic photoredox reactions. As DNA requires solubilisation, a mixture of DMSO:H_2O is traditionally used. DMSO can favour reactivity in specific cases; Krumb et al. witnessed this in an effective heteroaryl radical addition to an electron-deficient isoquinoline [70]. The authors rationalised this observation by DMSO's ability to enhance the nucleophilicity of pyridyl radicals, as reported in the literature [81]. It is noted that the quantity of water can vary significantly depending on reaction type.

DMA can represent an effective replacement of DMSO as highlighted by Wen et al. and Fu et al. [67, 73] Indeed, Wen et al. did an extensive study of cosolvent influence on the synthesis of 1,2-amino alcohols by photoredox decarboxylative coupling of α-amino acids [67]. Interestingly, they identified DMA as the preferred solvent in comparison with NMP, CH_3CN, and DMSO which afforded the desired product in 91%, 52%, 75%, and 72% yield, respectively (Fig. 50, Table 5).

8.5 *DNA Compatibility of LED Chemistry*

Overall, the use of LEDs for on-DNA chemistry has proved to be suitable for the construction of DELs. Zhang et al. and Fu et al. report no obvious DNA decomposition with, respectively, 94% of DNA amplifiable relative to the blank control with $Ir[dF(CF_3)ppy]_2(bpy)PF_6$ as a photocatalyst, and 84–88% amplifiable DNA by qPCR measurement for functionalisation of DNA-alkenes with diazo compounds via photocatalysis (Fig. 53) [69, 73].

Krumb et al. report 1.13×10^{13} molecules per μL of sample utilising developed conditions for on-DNA photocatalytic C-H activation, in comparison with the control sample in the absence of light in which 9.52×10^{12} molecules per μL of sample were observed [70]. Badir et al. observed no significant differences in the ability of their samples post-metallaphotoredox to undergo ligation, PCR amplification, quantification, and sequencing (Fig. 54, Table 6) [80]. Phelan et al. used a 4-cycle tag elongated headpiece and proved that their conditions for the defluorinative alkylation using Si and DHP radical precursors provided 6.10×10^{13} and 4.42×10^{13} molecules per μL of sample, respectively, in comparison with 1.48×10^{14} molecules per μL of sample for the control [79].

However, Zhang et al. noted minor DNA damage with the use of a 4-CzIPN system, leading to 83% amplifiable DNA in comparison with the blank control [69]. The authors hypothesised this difference is observed due to the extended period

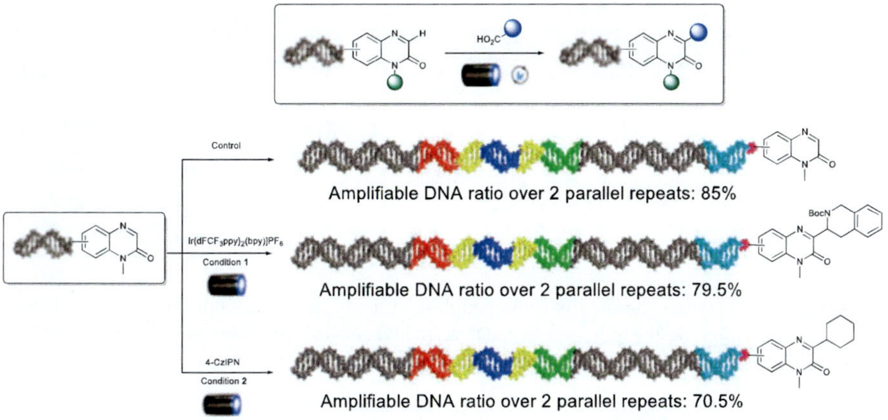

Fig. 53 qPCR measurements and DNA damage evaluation using an SYBR Green Master Mix kit (1,804,053, Life technologies, USA) and a StepOnePlus™ Real-Time PCR System (Thermo Sci., USA). Overall, for the Ir{dFCF$_3$ppy}$_2$(bpy)]PF$_6$ system (middle) 94% amplifiable DNA, relative to the blank control. For the 4-CzIPN system, 83% amplifiable DNA relative to the blank control

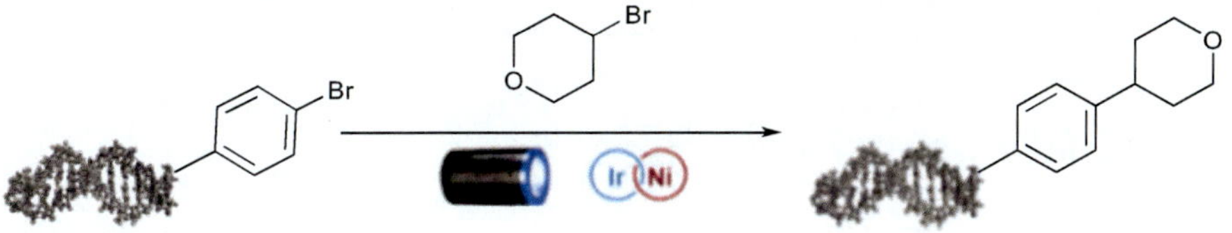

Fig. 54 Debrominative metallaphotoredox coupling between DNA-conjugated aryl bromide and 4-bromotetrahydropyran

Table 6 qPCR and sequencing results following alkylation reaction in Fig. 54, with various deviation from standard conditions

Sample	Deviations	Molecules/uL sample (qPCR)	% Mutated sequences (Illumina sequencing)
70-1	No deviations	2.59×10^{13}	4.01
70-2	No nickel	2.10×10^{13}	6.80
70-3	No photocatalyst	3.97×10^{13}	5.29
70-4	No triethylamine	3.10×10^{13}	5.18
70-5	No MgCl2	7.27×10^{12}	9.80
70-6	No light (control)	3.76×10^{13}	5.06

of irradiation time or strong reactivity of radicals generated from the aliphatic carboxylic acids (Fig. 53).

Ruff et al. reported 48% of amplifiable DNA recovered after release from resin following photoredox decarboxylative coupling (Fig. 55) [40]. While the authors

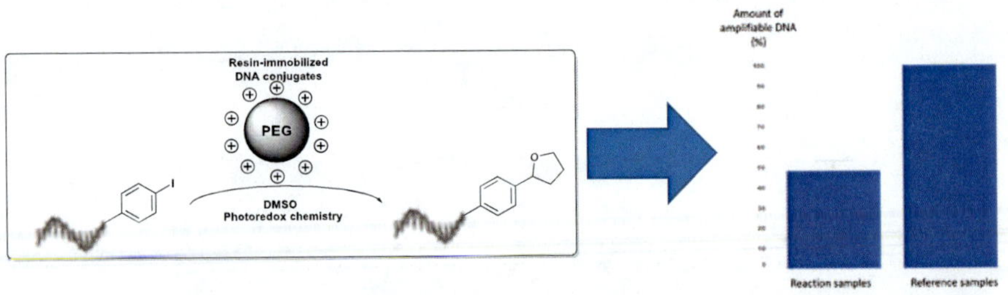

Fig. 55 Decarboxylative photoredox cross-coupling utilising resin-immobilised DNA conjugates (left) and evaluation of the post-reaction DNA recovery via quantification by qPCR (right). Reprinted with permission from ACS Comb. Sci. 2020, 22, 3, 120–128. Copyright 2020 American Chemical Society

report loss of DNA after the catch-react-release cycles, they concluded that these recoveries were comparable with those observed for other well-established DNA-compatible chemical transformations. Most importantly, the authors proved that the recovered material is fully competent in subsequent elongations by ligation.

8.6 Metallaphotoredox and DELs

Merging photoredox catalysis with transition metal catalysis, in other words metallaphotoredox, has become the main area of focus to facilitate coupling sp^3-hybridised fragments [82, 83]. Expanding the application of C–C bond formation through nickel catalysis with the use of simple, cheap, and versatile starting materials is outlined.

Kölmel et al. report a photoredox cross-electrophile coupling of alkyl bromides with on-DNA aryl iodides [15, 77]. This reaction identified a new ligand: pyridyl bis (carboxamidine) for obtention of the desired product. The authors produced a metallaphotoredox decarboxylative arylation of α-amino acids with on-DNA aryl halides. Badir et al. identified a photocatalytic $C(sp^2)$-$C(sp^3)$ cross-coupling between on-DNA aryl halides and alkyl bromides [80]. The authors report a further photocatalytic cross-coupling between aryl halides and silylamine radical precursors. Phelan et al. report a photoredox-catalysed alkylation reaction as DNA compatible [79]. Their methodology involves various radical precursors: dihydropyridines (DHP), carboxylic acids, and bis(catecholato)-silicates (Fig. 56).

The choice of cosolvent can be critical as the nickel catalyst has varying solubility in different solvent types. As reported by Kölmel et al. some nickel catalysts frequently employed in metallaphotoredox procedures, such as $NiCl_2$ or $Ni(COD)_2$, display suboptimal performance due to limited solubility in aqueous media [78]. In order to overcome catalyst decomposition, the authors suggest boosting the reaction by further addition of fresh stocks of reagents.

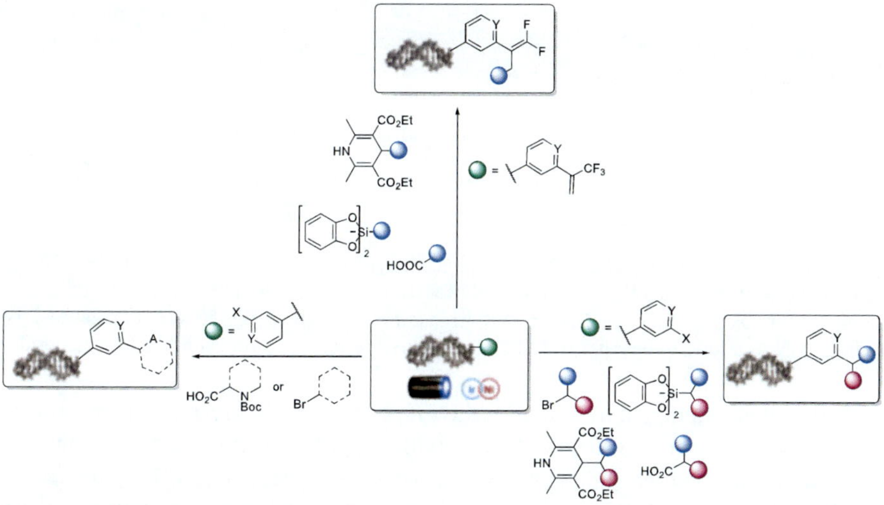

Fig. 56 Summary of the available metallaphotoredox reactions emploing Ni/Ir catalyst systems: alkylation (top), decarboxylative arylation of α-amino acids (left), and photocatalytic cross-couplings (right)

Table 7 Selected entries for reactions in Fig. 57, comparing the necessity of various factors. X is the major side product which is the protodehalogenation product from the starting material on-DNA aryl iodide

Entry	Conditions	% Yield (left)	Entry	Conditions	% Yield (right/X)
1	As shown	63	1	As shown	89/4
7	No Ni catalyst	0	8	No photocatalyst	0/0
8	No 4CzIPN catalyst	0	9	No light	0/0
			10	No nickel catalyst	0/12

Metallaphotoredox reactions typically involve two catalytic cycles, thus product distribution proves to be significantly impacted by the choices of photocatalyst/ catalyst. Phelan et al. and Badir et al. determined that 250 eq of radical precursor and a 4:1 nickel precatalyst to photocatalyst ratio were suitable for reactivity, while concurrently synchronising the nickel and photoredox catalytic cycles [79, 80].

Kölmel et al. investigated the metallaphotoredox decarboxylative arylation of DNA-tagged aryl iodide and N-Boc-morpholine-3-carboxylic acid [78]. Both photocatalyst and metal catalyst were indispensable for the reaction to occur (Table 7). The authors identified a loss in yield obtaining 87% and 77% of desired product with photocatalysts Ir[dF(CF_3)ppy]$_2$(bpy)PF_6 and Ir[dF(CF_3)ppy]$_2$(dtbbpy)-PF_6, respectively (Fig. 57).

Both Phelan et al. and Badir et al. reported the importance of at least 20% H_2O in the reaction media, an essential requirement to solubilise the DNA sufficiently to permit the reaction to proceed [79, 80]. However, the former found out that increasing the fraction of water significantly slowed the rate of the reaction of DHP cross-couplings between on-DNA aryl bromides and DHP coupling partners.

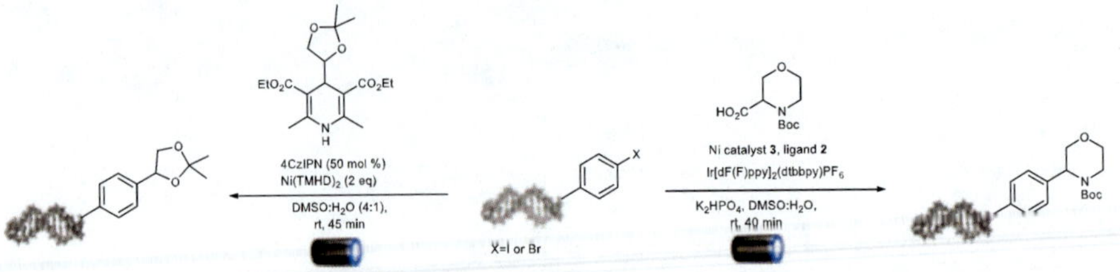

Fig. 57 Photoredox couplings from DNA-conjugated aryl halides. Conditions – left: DHP (250 eq, 6.25 μmol), 4CzIPN (50 mol%, 12.5 nmol), $Ni(TMHD)_2$ (2.0 eq, 50 nmol), aryl halide (25 nmol, 1.0 eq), 80:20 DMSO:H2O (1 mM), 45 min, irradiating with blue LED (30 W). Ni catalyst 3 is [(TMEDA)Ni(o-tolyl)Cl], ligand 2 is 4-(Trifluoromethyl)pyridine-2-carboxamidine hydrochloride

Phelan et al. attempted to modulate the pK_a of the reaction media, either with an aqueous buffer, such as 100 mM MOPS buffer (pH 8), or other additives such as magnesium salts, yet the authors report a detrimental impact on reactivity for cross-couplings between on-DNA aryl bromides and the corresponding DHP coupling partner [79]. Nonetheless, they suggest that MOPS buffer proved optimal for minimising the formation of protodehalogenation and phenol side products.

9 Conclusion

The ability to carry out a wide range of synthetic transformations on DNA-conjugated substrates is critical for the advancement of the DEL technology. A DEL-compatible reaction must be DNA-compatible and high-yielding with limited numbers of side products. A DEL-compatible reaction must also be split-and-pool compatible, and run robustly on many substrates which are (ideally) commercially available. An appreciation for *all* of these factors is often overlooked in the literature, and complete DEL compatibility is often unmet. Whilst this is still an emerging field, the developments described in this chapter demonstrate that new chemical approaches to DEL synthesis can make a significant impact on the scope and efficiency of the reactions employed. Modifications to the DNA tag and to the reaction media have been demonstrated to both improve existing reactions and to enable new chemistry. Together with the continued development of new reactions, these advances will continue to expand the chemical space accessible in DELs. Much of the new chemistry has not yet been shown in the context of a true DEL synthesis, and in this regard it remains to be seen whether these chemistries can make an impact on DEL technology.

MJA, IFSFC, JSG, HH, JO, CLAS, CT, MJW declare no conflict of interest.

Compliance and Ethical Standards

Funding: We thanks Cancer Research UK (Programme Grant funding to the group, ref. DRCDDRPGMApr2020\100002), EPSRC Centre for Doctoral Training in Molecular Sciences for Medicine (funding for MA, JSG, JO, ref. EP/S022791/1), EPSRC and AstraZeneca (funding for IFSFC), The Wellcome Trust (funding for HH and CT, Ref. 221302/Z/20/Z), Pharmaron (funding for JO), Genentech (funding for CLAS) for financial support.

Informed Consent: No patients were studied in this chapter.

Ethical Approval: This Chapter does not contain any studies with human participants or animals performed by any of the authors.

References

1. Malone ML, Paegel BM (2016) What is a "DNA-Compatible" reaction? ACS Comb Sci 18(4): 182–187. https://doi.org/10.1021/acscombsci.5b00198
2. Kunig V, Potowski M, Gohla A, Brunschweiger A (2018) DNA-encoded libraries—an efficient small molecule discovery technology for the biomedical sciences. Biol Chem 399(7):691–710. https://doi.org/10.1515/hsz-2018-0119
3. Shi Y, Wu YR, Yu JQ, Zhang WN, Zhuang CL (2021) DNA-encoded libraries (DELs): a review of on-DNA chemistries and their output. RSC Adv 11(4):2359–2376. https://doi.org/10.1039/d0ra09889b
4. Castan IFSF, Graham JS, Salvini CLA, Stanway-Gordon HA, Waring MJ (2021) On the design of lead-like DNA-encoded chemical libraries. Bioorg Med Chem 43:116273. https://doi.org/10.1016/j.bmc.2021.116273
5. Fair RJ, Walsh RT, Hupp CD (2021) The expanding reaction toolkit for DNA-encoded libraries. Bioorg Med Chem Lett 51:128339. https://doi.org/10.1016/j.bmcl.2021.128339
6. Franzini RM et al (2015) Identification of structure-activity relationships from screening a structurally compact DNA-encoded chemical library. Angew Chem 127(13):3999–4003. https://doi.org/10.1002/anie.201410736
7. Kollmann CS et al (2014) Application of encoded library technology (ELT) to a protein-protein interaction target: discovery of a potent class of integrin lymphocyte function-associated antigen 1 (LFA-1) antagonists. Bioorg Med Chem 22(7):2353–2365. https://doi.org/10.1016/j.bmc.2014.01.050
8. Deng H et al (2012) Discovery of highly potent and selective small molecule ADAMTS-5 inhibitors that inhibit human cartilage degradation via encoded library technology (ELT). J Med Chem 55(16):7061–7079. https://doi.org/10.1021/jm300449x
9. Satz AL (2015) DNA encoded library selections and insights provided by computational simulations. ACS Chem Biol 10(10):2237–2245. https://doi.org/10.1021/acschembio.5b00378
10. An R et al (2014) Non-enzymatic depurination of nucleic acids: factors and mechanisms. PLoS One 9(12):e115950. https://doi.org/10.1371/journal.pone.0115950
11. Duncan BK, Miller JH (1980) Mutagenic deamination of cytosine residues in DNA. Nature 287(5782):560–561. https://doi.org/10.1038/287560a0
12. Sikorsky JA, Primerano DA, Fenger TW, Denvir J (2007) DNA damage reduces Taq DNA polymerase fidelity and PCR amplification efficiency. Biochem Biophys Res Commun 355(2): 431–437. https://doi.org/10.1016/j.bbrc.2007.01.169
13. Epe B, Pflaum M, Boiteux S (1993) DNA damage induced by photosensitizers in cellular and cell-free systems. Mutat Res Toxicol 299(3–4):135–145. https://doi.org/10.1016/0165-1218(93)90091-Q

14. Gates KS (2009) An overview of chemical processes that damage cellular DNA: spontaneous hydrolysis, alkylation, and reactions with radicals. Chem Res Toxicol 22(11):1747–1760. https://doi.org/10.1021/tx900242k
15. Wang X et al (2018) Ruthenium-promoted C−H activation reactions between DNA-conjugated acrylamide and aromatic acids scheme 1. Different applications of aromatic acids in DEL synthesis. Org Lett 20(16):4764–4768. https://doi.org/10.1021/acs.orglett.8b01837
16. Li J-Y, Huang H (2018) Development of DNA-compatible Suzuki-Miyaura reaction in aqueous media. Bioconjug Chem 29(11):3841–3846. https://doi.org/10.1021/acs.bioconjchem.8b00676
17. Gerry CJ, Yang Z, Stasi M, Schreiber SL (2019) DNA-compatible [3 + 2] nitrone−olefin cycloaddition suitable for DEL syntheses. Org Lett 21(5):1325–1330. https://doi.org/10.1021/acs.orglett.9b00017
18. Qu Y et al (2020) Copper-mediated DNA-compatible one-pot click reactions of alkynes with aryl borates and TMS-N 3. Org Lett 22(11):4146–4150. https://doi.org/10.1021/acs.orglett.0c01219
19. Abel GR, Calabrese ZA, Ayco J, Hein JE, Ye T (2016) Measuring and suppressing the oxidative damage to DNA during Cu(I)-catalyzed Azide−alkyne cycloaddition. Bioconjug Chem 27(3):698–704. https://doi.org/10.1021/acs.bioconjchem.5b00665
20. Paciaroni NG, Ndungu JM, Kodadek T (2020) Solid-phase synthesis of DNA-encoded libraries via an "Aldehyde Explosion" strategy. Chem Commun 56(34):4656. https://doi.org/10.1039/d0cc01474e
21. Lee Satz A et al (2015) DNA compatible multistep synthesis and applications to DNA encoded libraries. Bioconjug Chem 26(8):1623–1632. https://doi.org/10.1021/acs.bioconjchem.5b00239
22. Priego JJ et al (2020) On-DNA palladium-catalyzed hydrogenation-like reaction suitable for DNA-encoded library synthesis. Bioconjug Chem 32(1):88–93. https://doi.org/10.1021/acs.bioconjchem.0c00566
23. Stanway-Gordon HA, Graham JS, Waring MJ (2022) On-DNA transfer hydrogenolysis and hydrogenation for the synthesis of DNA-encoded chemical libraries. Angew Chem 61(3): e202111927. https://doi.org/10.1002/anie.202111927
24. Ratnayake AS et al (2019) A solution phase platform to characterize chemical reaction compatibility with DNA-encoded chemical library synthesis. ACS Comb Sci 21(10): 650–655. https://doi.org/10.1021/acscombsci.9b00113
25. Sauter B, Schneider L, Stress C, Gillingham D (2021) An assessment of the mutational load caused by various reactions used in DNA encoded libraries. Bioorg Med Chem 52:116508. https://doi.org/10.1016/j.bmc.2021.116508
26. Škopić MK, Salamon H, Bugain O, Jung K, Gohla A, Doetsch LJ, Dos Santos D, Bhat A, Wagner B, Brunschweiger A (2017) Acid-and Au (I)-mediated synthesis of hexathymidine-DNA-heterocycle chimeras, an efficient entry to DNA-encoded libraries inspired by drug structures. Chem Sci 8(5):3356–3361. https://doi.org/10.1039/C7SC00455A
27. Škopić MK, Willems S, Wagner B, Schieven J, Krause N, Brunschweiger A (2017) Exploration of a Au (I)-mediated three-component reaction for the synthesis of DNA-tagged highly substituted spiroheterocycles. Org Biomol Chem 15(40):8648–8654. https://doi.org/10.1039/C7OB02347B
28. Potowski M, Kunig VB, Losch F, Brunschweiger A (2019) Synthesis of DNA-coupled isoquinolones and pyrrolidines by solid phase ytterbium-and silver-mediated imine chemistry. Med Chem Commun 10(7):1082–1093. https://doi.org/10.1039/C9MD00042A
29. Kunig VB, Ehrt C, Dömling A, Brunschweiger A (2019) Isocyanide multicomponent reactions on solid-phase-coupled DNA oligonucleotides for encoded library synthesis. Org Lett 21(18): 7238–7243. https://doi.org/10.1021/acs.orglett.9b02448
30. Potowski M, Kunig VB, Eberlein L, Vakalopoulos A, Kast SM, Brunschweiger A (2021) Chemically stabilized DNA barcodes for DNA-encoded chemistry. Angew Chem 60(36): 19744–19749. https://doi.org/10.1002/anie.202104348

31. Needels MC, Jones DG, Tate EH, Heinkel GL, Kochersperger LM, Dower WJ, Barrett RW, Gallop MA (1993) Proc Natl Acad Sci U S A 90:10700–10704
32. Winssinger N, Harris JL, Backes BJ, Schultz PG (2001) From split-pool libraries to spatially addressable microarrays and its application to functional proteomic profiling. Angew Chem 40(17):3152–3155. https://doi.org/10.1002/1521-3773(20010903)40:17<3152::AID-ANIE3152>3.0.CO;2-P
33. Pels K, Dickson P, An H, Kodadek T (2018) DNA-compatible solid-phase combinatorial synthesis of β-cyanoacrylamides and related electrophiles. ACS Comb Sci 20(2):61–69. https://doi.org/10.1021/acscombsci.7b00169
34. MacConnell AB, McEnaney PJ, Cavett VJ, Paegel BM (2015) DNA-encoded solid-phase synthesis: encoding language design and complex oligomer library synthesis. ACS Comb Sci 17(9):518–534. https://doi.org/10.1021/acscombsci.5b00106
35. Cistrone PA, Dawson PE (2016) Click-based libraries of SFTI-1 peptides: new methods using reversed-phase silica. ACS Comb Sci 18(3):139–143. https://doi.org/10.1021/acscombsci.5b00195
36. Flood DT, Yan NL, Dawson PE (2018) Post-translational backbone engineering through selenomethionine-mediated incorporation of Freidinger lactams. Angew Chem 57(28): 8697–8701. https://doi.org/10.1002/anie.201804885
37. Halpin DR, Lee JA, Wrenn SJ, Harbury PB, Joyce G (2004) DNA display III. Solid-phase organic synthesis on unprotected DNA. PLoS Biol 2(7):e175. https://doi.org/10.1371/journal.pbio.0020175
38. Flood DT, Asai S, Zhang X, Wang J, Yoon L, Adams ZC, Dillingham BC, Sanchez BB, Vantourout JC, Flanagan ME, Piotrowski DW (2019) Expanding reactivity in DNA-encoded library synthesis via reversible binding of DNA to an inert quaternary ammonium support. J Am Chem Soc 141(25):9998–10006. https://doi.org/10.1021/jacs.9b03774
39. Flood DT, Zhang X, Fu X, Zhao Z, Asai S, Sanchez BB, Sturgell EJ, Vantourout JC, Richardson P, Flanagan ME, Piotrowski DW (2020) RASS-enabled S/P− C and S−N bond formation for DEL synthesis. Angew Chem 132(19):7447–7453. https://doi.org/10.1002/anie.201915493
40. Ruff Y, Martinez R, Pellé X, Nimsgern P, Fille P, Ratnikov M, Berst F (2020) An amphiphilic polymer-supported strategy enables chemical transformations under anhydrous conditions for DNA-encoded library synthesis. ACS Comb Sci 22(3):120–128. https://doi.org/10.1021/acscombsci.9b00164
41. Götte K, Chines S, Brunschweiger A (2020) Reaction development for DNA-encoded library technology: from evolution to revolution? Tetrahedron Lett 61(22):151889. https://doi.org/10.1016/j.tetlet.2020.151889
42. Dwars T, Paetzold E, Oehme G (2005) Reactions in micellar systems. Angew Chem 44(44): 7174–7199. https://doi.org/10.1002/anie.200501365
43. Lipshutz BH, Aguinaldo GT, Ghorai S, Voigtritter K (2008) Olefin cross-metathesis reactions at room temperature using the nonionic amphiphile "PTS": just add water. Org Lett 10(7): 1325–1328. https://doi.org/10.1021/ol800028x
44. Lipshutz BH, Taft BR (2008) Heck couplings at room temperature in nanometer aqueous micelles. Org Lett 10(7):1329–1332. https://doi.org/10.1021/ol702755g
45. Lipshutz BH, Petersen TB, Abela AR (2008) Room-temperature Suzuki−Miyaura couplings in water facilitated by nonionic amphiphiles. Org Lett 10(7):1333–1336. https://doi.org/10.1021/ol801712e
46. Škopić MK, Gotte K, Gramse C, Dieter M, Pospich S, Raunser S, Weberskirch R, Brunschweiger A (2019) Micellar Brønsted acid mediated synthesis of DNA-tagged heterocycles. J Am Chem Soc 141(26):10546–10555. https://doi.org/10.1021/jacs.9b05696
47. Hunter JH, Prendergast L, Valente LF, Madin A, Pairaudeau G, Waring MJ (2019) High fidelity Suzuki–Miyaura coupling for the synthesis of DNA encoded libraries enabled by Micelle forming surfactants. Bioconjug Chem 31(1):149–155. https://doi.org/10.1021/acs.bioconjchem.9b00838

48. Graham JS, Hunter JH, Waring MJ (2021) Micellar Buchwald–Hartwig coupling of aryl and heteroarylamines for the synthesis of DNA-encoded libraries. J Org Chem 86(23): 17257–17264. https://doi.org/10.1021/acs.joc.1c02325
49. Hunter JH, Anderson MJ, Castan IF, Graham JS, Salvini CL, Stanway-Gordon HA, Crawford JJ, Madin A, Pairaudeau G, Waring MJ (2021) Highly efficient on-DNA amide couplings promoted by micelle forming surfactants for the synthesis of DNA encoded libraries. Chem Sci 12(27):9475–9484. https://doi.org/10.1039/D1SC03007H
50. Steinreiber GA, Pichler H, May O, Gruber-Khadjawi M (2011) Application of designed enzymes in organic synthesis. Chem Rev 111(7):4141–4164. https://doi.org/10.1021/cr100386u
51. Sheldon RA, Brady D, Bode ML (2020) The Hitchhiker's guide to biocatalysis: recent advances in the use of enzymes in organic synthesis. Chem Sci 11(10):2587–2605. https://doi.org/10.1039/C9SC05746C
52. Satz AL (2018) What do you get from DNA-encoded libraries? ACS Med Chem Lett 9(5): 408–410. https://doi.org/10.1021/acsmedchemlett.8b00128
53. Thomas B et al (2017) Application of biocatalysis to on-DNA carbohydrate library synthesis. Chembiochem 18(9):858–863. https://doi.org/10.1002/cbic.201600678
54. Chai J, Lu X, Arico-Muendel CC, Ding Y, Pollastri MP (2021) Application of l-threonine aldolase to on-DNA reactions. Bioconjug Chem 32(9):1973–1978. https://doi.org/10.1021/acs.bioconjchem.1c00363
55. Fesko K (2016) Threonine aldolases: perspectives in engineering and screening the enzymes with enhanced substrate and stereo specificities. Appl Microbiol Biotechnol 100(6):2579–2590. https://doi.org/10.1007/s00253-015-7218-5
56. Wang J, Lundberg H, Asai S, Martín-Acosta P, Chen JS, Brown S, Farrell W, Dushin RG, O'Donnell CJ, Ratnayake AS, Richardson P, Liu Z, Tian Q, Blackmond DG, Baran PS (2018) Kinetically guided radical-based synthesis of $C(sp^3)$–$C(sp^3)$ linkages on DNA. Proc Natl Acad Sci 115(28):E6404–E6410. https://doi.org/10.1073/pnas.1806900115
57. Zhu C, Ang NW, Meyer TH, Qiu Y, Ackermann L (2021) Organic electrochemistry: molecular syntheses with potential. ACS Cent Sci 7(3):415–431. https://doi.org/10.1021/acscentsci.0c01532
58. Crabtree RH, Lei A (2017) Introduction: CH activation. Chem Rev 117(13):8481–8482. https://doi.org/10.1021/acs.chemrev.7b00307
59. Gerry CJ, Wawer MJ, Clemons PA, Schreiber SL (2019) DNA barcoding a complete matrix of stereoisomeric small molecules. J Am Chem Soc 141(26):10225–10235. https://doi.org/10.1021/jacs.9b01203
60. Maetani M, Zoller J, Melillo B, Verho O, Kato N, Pu J, Comer E, Schreiber SL (2017) Synthesis of a bicyclic azetidine with in vivo antimalarial activity enabled by stereospecific, directed C(sp3)–H arylation. J Am Chem Soc 139(32):11300–11306. https://doi.org/10.1021/jacs.7b06994
61. Fan Z, Zhao S, Liu T, Shen PX, Cui ZN, Zhuang Z, Shao Q, Chen JS, Ratnayake AS, Flanagan ME, Kölmel DK, Piotrowski DW, Richardson P (2020) Yu, J-Q; merging $C(sp^3)$–H activation with DNA-encoding. Chem Sci 11(45):12282–12288. https://doi.org/10.1039/d0sc03935g
62. Shen P-X, Hu L, Shao Q, Hong K, Yu J-Q (2018) Pd (II)-catalyzed enantioselective $C(sp^3)$–H arylation of free carboxylic acids. J Am Chem Soc 140(21):6545–6549. https://doi.org/10.1021/jacs.8b03509
63. Chen G, Zhuang Z, Li GC, Saint-Denis TG, Hsiao Y, Joe CL, Yu J-Q (2017) Ligand-enabled β-C–H arylation of α-amino acids without installing exogenous directing groups. Angew Chem 56(6):1506–1509. https://doi.org/10.1002/anie.201610580
64. Graetzel M (1981) Artificial photosynthesis: water cleavage into hydrogen and oxygen by visible light. Acc Chem Res 14(12):376–384. https://doi.org/10.1021/ar00072a003
65. Kalyanasundaram K, Grätzel M (1998) Applications of functionalized transition metal complexes in photonic and optoelectronic dDvices. Coord Chem Rev 177(1):347–414. https://doi.org/10.1016/S0010-8545(98)00189-1

66. Skubi KL, Blum TR, Yoon TP (2016) Dual catalysis strategies in photochemical synthesis. Chem Rev 116(17):10035–10074. https://doi.org/10.1021/acs.joc.6b01449
67. Wen H, Ge R, Qu Y, Sun J, Shi X, Cui W, Yan H, Zhang Q, An Y, Su W, Yang H (2020) Synthesis of 1, 2-amino alcohols by photoredox-mediated decarboxylative coupling of α-amino acids and DNA-conjugated carbonyls. Org Lett 22(24):9484–9489. https://doi.org/10.1021/acs.orglett.0c03461
68. Kölmel DK, Loach RP, Knauber T, Flanagan ME (2020) Employing photoredox catalysis for DNA-encoded chemistry: decarboxylative alkylation of α-amino acids. ChemMedChem 13(20):2159–2165. https://doi.org/10.1002/cmdc.201800492
69. Zhang Y, Luo H, Ma H, Wan J, Ji Y, Shaginian A, Li J, Deng Y, Liu G (2021) On-DNA derivatization of quinoxalin-2-ones by visible-light-triggered alkylation with carboxylic acids. Bioconjug Chem 32(8):1576–1580. https://doi.org/10.1021/acs.bioconjchem.1c00346
70. Krumb M, Kammer LM, Badir SO, Cabrera-Afonso MJ, Wu VE, Huang M, Csakai A, Marcaurelle LA, Molander GA (2022) Photochemical C–H arylation of heteroarenes for DNA-encoded library synthesis. Chem Sci 13(4):1023–1029. https://doi.org/10.1039/D1SC05683B
71. Wu R, Du T, Sun W, Shaginian A, Gao S, Li J, Wan J, Liu G (2021) Functionalization of DNA-tagged alkenes enabled by visible-light-induced C–H activation of N-Aryl tertiary amines. Org Lett 23(9):3486–3490. https://doi.org/10.1021/acs.orglett.1c00924
72. Kölmel DK, Ratnayake AS, Flanagan ME, Tsai MH, Duan C, Song C (2020) Photocatalytic [2+2] cycloaddition in DNA-encoded chemistry. Org Lett 22(8):2908–2913. https://doi.org/10.1021/acs.orglett.0c00574
73. Fu X, Tang J, Hua R, Li X, Kang Z, Qiu H, Hu W (2022) Functionalization of DNA-tagged alkenes with diazo compounds via photocatalysis. Org Lett 24(11):2208–2213. https://doi.org/10.1021/acs.orglett.2c00516
74. Rettig ID, McCormick TM (2021) Enrolling reactive oxygen species in photon-to-chemical energy conversion: fundamentals, technological advances, and applications. Adv Phys X 6(1):1950049. https://doi.org/10.1080/23746149.2021.1950049
75. Gates KS (2007) Chemical reactions of DNA damage and degradation. In: Reviews of reactive intermediates. Wiley, pp 333–378. https://doi.org/10.1002/9780470120828.ch8
76. Moss RA, Platz MS, Jones Jr M (eds) (2007) Reviews of reactive intermediate chemistry. Wiley
77. Kölmel DK, Ratnayake AS, Flanagan ME (2020) Photoredox cross-electrophile coupling in DNA-encoded chemistry. Biochem Biophys Res Commun 533(2):201–208. https://doi.org/10.1016/j.bbrc.2020.04.028
78. Kölmel DK, Meng J, Tsai MH, Que J, Loach RP, Knauber T, Wan J, Flanagan ME (2019) On-DNA decarboxylative arylation: merging photoredox with nickel catalysis in water. ACS Comb Sci 21(8):588–597. https://doi.org/10.1021/acscombsci.9b00076
79. Phelan JP, Lang SB, Sim J, Berritt S, Peat AJ, Billings K, Fan L, Molander GA (2019) Open-air alkylation reactions in photoredox-catalyzed DNA-encoded library synthesis. J Am Chem Soc 141(8):3723–3732. https://doi.org/10.1021/jacs.9b00669
80. Badir SO, Sim J, Billings K, Csakai A, Zhang X, Dong W, Molander GA (2020) Multifunctional building blocks compatible with photoredox-mediated alkylation for DNA-encoded library synthesis. Org Lett 22(3):1046–1051. https://doi.org/10.1021/acs.orglett.9b04568
81. Boyington AJ, Riu MLY, Jui NT (2017) Anti-Markovnikov hydroarylation of unactivated olefins via pyridyl radical intermediates. J Am Chem Soc 139(19):6582–6585. https://doi.org/10.1021/jacs.7b03262
82. Tasker SZ, Standley EA, Jamison TF (2014) Recent advances in homogeneous nickel catalysis. Nature 509(7500):299–309. https://doi.org/10.1038/nature13274
83. Chan AY, Perry IB, Bissonnette NB, Buksh BF, Edwards GA, Frye LI, Garry OL, Lavagnino MN, Li BX, Liang Y, Mao E (2021) Metallaphotoredox: the merger of photoredox and transition metal catalysis. Chem Rev 122(2):1485–1542. https://doi.org/10.1021/acs.chemrev.1c00383

Top Med Chem (2022) 40: 123–144
https://doi.org/10.1007/7355_2022_147

Published online: 27 October 2022

Design Considerations in Constructing and Screening DNA-Encoded Libraries

Ying Zhang and Raphael M. Franzini

Contents

Abstract Preparing a DNA-encoded chemical library platform is a major undertaking, which requires careful planning. Here we outline general design principles for DNA-encoded libraries on the levels of library topology, chemical reactions, and selection of building blocks. The effects of design parameters on the coverage of the chemical space by a DNA-encoded library and on the properties of encoded compounds are discussed.

Keywords Chemical diversity, DEL design, DNA-encoded libraries, Library synthesis, Library topology, On-DNA chemistry

Y. Zhang (✉)
X-Chem, Inc., Waltham, MA, USA
e-mail: YZhang@X-Chemrx.com

R. M. Franzini (✉)
Department of Medicinal Chemistry, University of Utah, Salt Lake City, UT, USA
e-mail: Raphael.franzini@utah.edu

Abbreviations

AI	Artificial intelligence
AS-MS	Affinity selection mass spectrometry
BALI-MS	Bead-assisted mass spectrometry
BB	Building blocks
bRO5	Beyond Lipinski's rule of five
cLogP	Calculated octanol/water partition coefficient
DEL	DNA-encoded chemical library
ECFP	Extended connectivity fingerprint
HTS	High-throughput screening
LC-MS	Liquid chromatography mass spectrometry
ML	Machine learning
MW	Molecular weight
PAINS	Pan-assay interference compounds
PROTACs	Proteolysis targeting chimeras
TMAP	Tree MAP

1 Introduction

DNA-encoded library (DEL) screening technology has provided new options for small-molecule hit finding and lead identification in pharmaceutical research, complementing high-throughput screening (HTS) of repository small-molecule compound collections [1]. DEL hits are starting to make a significant contribution to lead molecules (Fig. 1).

As Dragovich et al. [1] point out, the selection of a lead chemical series to progress is one of the most important decisions made by research teams. The progressivity of lead series to clinical candidates is influenced by the nature and quality of the chosen chemical matter.

DELs make it possible to identify drug leads in a process that is simpler, faster, and cheaper than conventional methods such as HTS [2, 3]. However, establishing a state-of-the-art DEL platform is a major undertaking, which requires considerable investments. Chemical building blocks, DNA coding sequences, extensive reaction validation studies, and reliable logistics are all essential when constructing DELs [4]. It is therefore advisable to carefully design a DEL before engaging in such endeavors to ensure a satisfactory return on investment. Thorough understanding of how design parameters affect the productivity of a DEL is necessary. The objective of this chapter is to examine central considerations in DEL design and to propose guidelines for best practices in DEL construction.

The value of a DEL is determined by its capacity to deliver molecules suitable for lead identification and subsequent development. It is obviously impossible to predict what library will lead to a new blockbuster drug. Practically, one can assess the

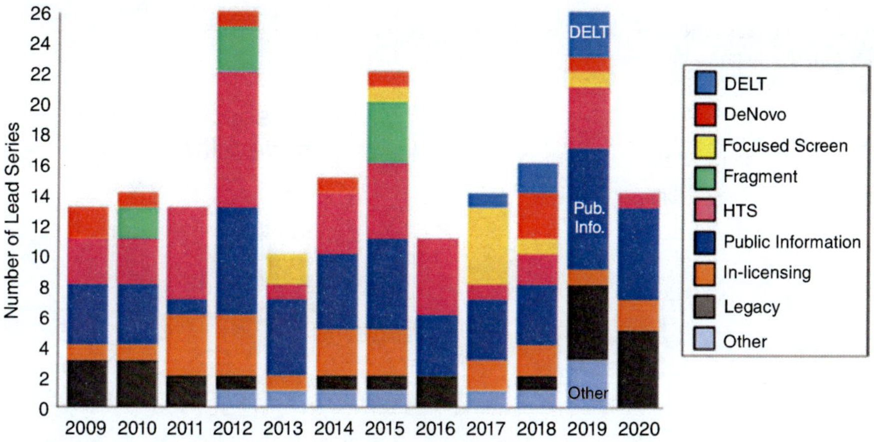

Fig. 1 Distribution of lead series according to methods of origin. In recent years, DEL-derived hits have begun to contribute a growing share of lead series. (The shown statistics originates from Roche/Genentech and was adopted from Dragovich et al. [1])

utility of a DEL by evaluating the frequency with which it delivers appropriate hit molecules for targets of interests, which can be a broad range of biomacromolecules, members of a protein family, or a specific target. While the value of a DEL only becomes apparent retrospectively, knowledge from previous library construction and hit discovery efforts provides insights on what parameters influence the rate with which a library delivers molecules of interest.

To be interesting for drug development, molecules must meet several criteria. The properties of DEL hits tend to align with those of the molecules present in a library, and it is important to ensure at the design stage that a large fraction of encoded molecules meets these criteria. Foremost, a molecule must be active and be preferentially as potent as possible. The complementarity of library molecules and targeted binding sites is decisive for success and calls for structurally diverse compounds [5]. Additionally, hit molecules should have properties that are favorable from a lead development standpoint (i.e., meet Lipinski's rule of five criteria) [6, 7]. With few exceptions, DELs are combinatorial libraries and unless a library is carefully designed, hit molecules from a number of DELs in the past tend to be unacceptably large, highly hydrophobic (or hydrophilic), and overly flexible [8]. Molecules with low solubility or having toxicity liabilities are also problematic. Ideally, a DEL selection yields several structurally diverse clusters of hits providing a medicinal chemist with options. Molecules in a DEL should therefore cover a chemical space that is relevant for a target of interest and cover a lead-like property space as much as possible (Fig. 2). Practical considerations are further relevant when designing DELs. The design of a DEL directly affects the effort it takes to identify a hit molecule. Having to find active molecules from thousands of related structures without a reliable way of triaging them is laborious especially if false positives are common. The difficulty of synthesis of DEL-derived molecules further determines how many

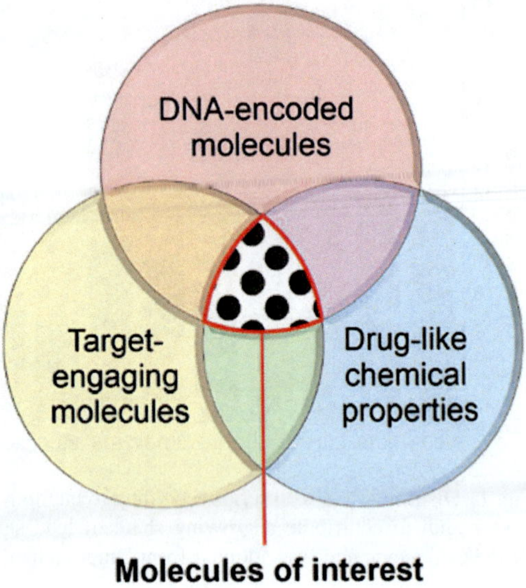

Fig. 2 The effectiveness of a DEL in discovering hit compounds is represented as the intersection of target-engaging molecules possessing drug-like chemical properties, rather than the library's numeric size

hit molecules one can test. The simpler the synthesis of a DEL is, the easier it is to prepare the hit molecules for evaluation.

Three levels of design parameters determine the properties of a DEL. These parameters are library topology, assembly chemistry, and selection of building blocks [8]. Library topology defines the number of building block sets and how they are arranged. Most DELs are prepared in a split-and-pool protocol in which building block sets are introduced in sequential reaction-encoding cycles [9]. The number of such cycles affects the properties of encoded molecules. The connectivity of these sets of building blocks varies for different libraries resulting in linear, branched, and cyclic DEL compounds. The reactions used to synthesize a DEL influence structural properties of DEL molecules and also dictate which set of building block classes being used. Ultimately, the selection of the building blocks profoundly affects the complementarity of molecules and a target of interest. Best practices for DEL design depend on the specific application and the philosophy of the medicinal chemists. In the following paragraphs, we critically evaluate the connections of each of these aspects to library properties and hit discovery rates.

2 Synthesis Cycles, Library Topology, and Numeric Library Size

The topology of a DEL reflects the connectivity of the chemical constituents. In particular, DEL topologies define the number of building block sets (i.e., the number of synthesis cycles) and their geometric arrangements. The number of synthesis cycles critically impacts both the numeric size of a DEL and the properties of the

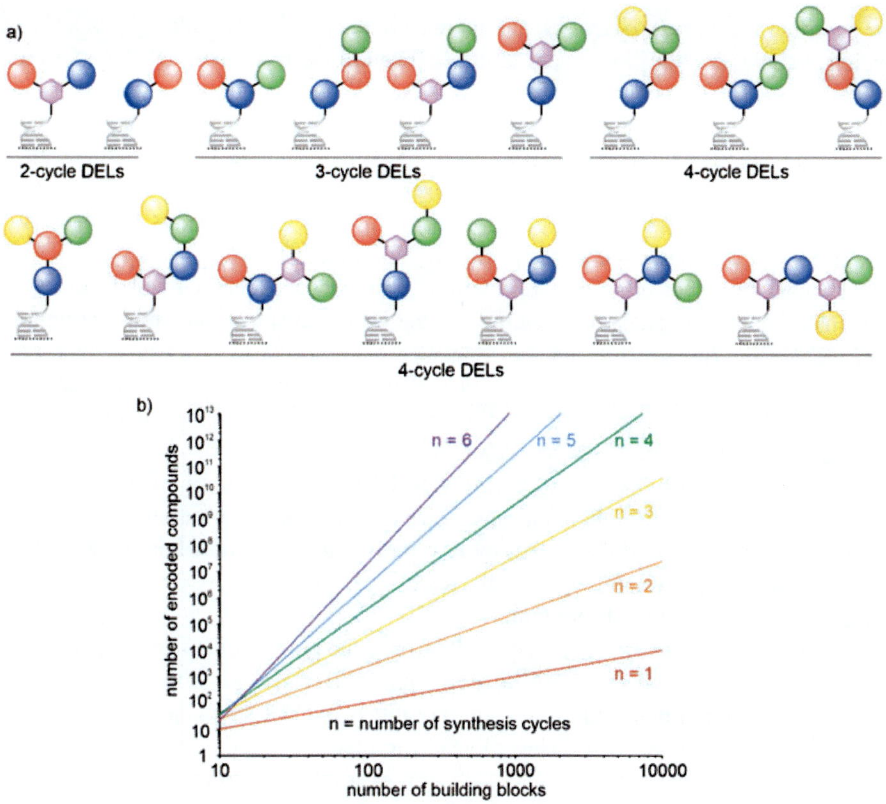

Fig. 3 Library topologies for acyclic DELs. (**a**) Examples of library topologies with two to four sets of building blocks (circles). Hexagons are conserved branching scaffolds. (**b**) Relationship between number of building block and numeric library size as a function of the number of synthesis cycles

encoded molecules. Most DELs contain between two and four sets of building blocks, which are arranged in either a linear, branched or cyclic geometry. Although the number of reported DELs is growing rapidly [10], the number of DEL topologies is limited (Fig. 3). Only two ways of arranging building blocks are available for 2-cycle DELs. There are four 3-cycle acyclic topologies, and the number of topologies increases considerably for DELs with four and more sets of building blocks. There are many DEL topologies with ring structures, but despite increasing efforts to generate macrocyclic DELs [11] most reported DELs are acyclic.

The numeric size of a DEL is an attractive yet deceptive parameter for appraising the prospect of a DEL. The size of a DEL depends on the number of synthesis cycles and the number of building blocks utilized in each. Because the numeric size of a DEL grows exponentially with the number of synthesis cycles, using multiple sets of building blocks is a straightforward way of enlarging DELs at limited expenses and effort. Library sizes of billions (and even trillions) of compounds are well within reach for 4-cycle DELs [2]. In contrast, the numeric size increases with cycle size to the power of the number of cycles, and increasing the number of building blocks in a cycle is expensive. For illustration, at least 10^4 building blocks and DNA codes are needed to reach 10^8 encoded molecules for a 2-cycle DEL but only 100 for a 4-cycle DEL. However, multi-cycle libraries intrinsically suffer from structural redundancy among the encoded molecules, which means that many molecules are structurally similar to each other. Therefore, the numeric size poorly mirrors the chemical diversity of a DEL. There is no compelling empirical evidence that the numeric size of a DEL correlates with its capacity to yield hits [12, 13].

The enthusiasm for ultra-large DELs may stem from the idea that with billions or even trillions of molecules the coverage of the chemical space must be so comprehensive that the discovery of active molecules is highly likely. However, even the largest DEL only represents a tiny fraction of the synthetically accessible chemical space [14]. If the chemical space of a library is incongruent with the chemical space of target binders, no hits will be found independent of how many molecules are present within a DEL. Actually, multi-cycle DELs provide a significantly worse coverage of the respective chemical space because the number of possible molecules increases with compound size much more rapidly than the exponential relationship between DEL size and number of synthesis cycles. This effect is the same concept underlying fragment-based screening.

The number of synthesis cycles directly affects the physicochemical properties of encoded molecules and therefore hit compounds. Because of their combinatorial nature, DEL compounds tend to have unfavorable physicochemical properties unless carefully designed [8, 36]. The physicochemical properties of the X-Chem ReadiDEL library are shown in Fig. 4b as an example. By limiting the number of cycles, the resulting compounds in the library (Fig. 4b) exhibit desirable physicochemical properties. Increasing the number of synthesis cycles typically worsens compliance with criteria of oral bioavailability and lead-likeness such as molecular weight, number of hydrogen bond acceptors and donors, polar surface area, and rotatable bonds. As a result, hits from DELs with four or more synthesis cycle are largely outside the property range that most medicinal chemists consider suitable for lead development [6, 7]. It has been argued that the large size of DEL compounds is unproblematic because often truncated molecules are used for lead development [13]. The experience of the authors is that full-length DEL hits are generally most useful for lead development. Furthermore, it is far more economical to construct several DELs with fewer synthesis cycles rather than searching for truncates in ultra-large DELs. Figuratively speaking, searching for truncates is like looking for the same number of needles in much more hay.

Increasing the number of synthesis cycles also negatively affects the reliability of DEL screen output data. DEL compounds are not individually purified and increasing the number of synthetic steps inadvertently leads to a greater fraction of incomplete molecules and side products. The heterogeneity of libraries therefore is exacerbated by an increase in the number of synthetic steps and so does the frequency of false positives caused by side reactions. While there are reports suggesting that for small 2-cycle DELs sequence counts can predict relative activity at least within clusters of closely related molecules [15–17], such correlations

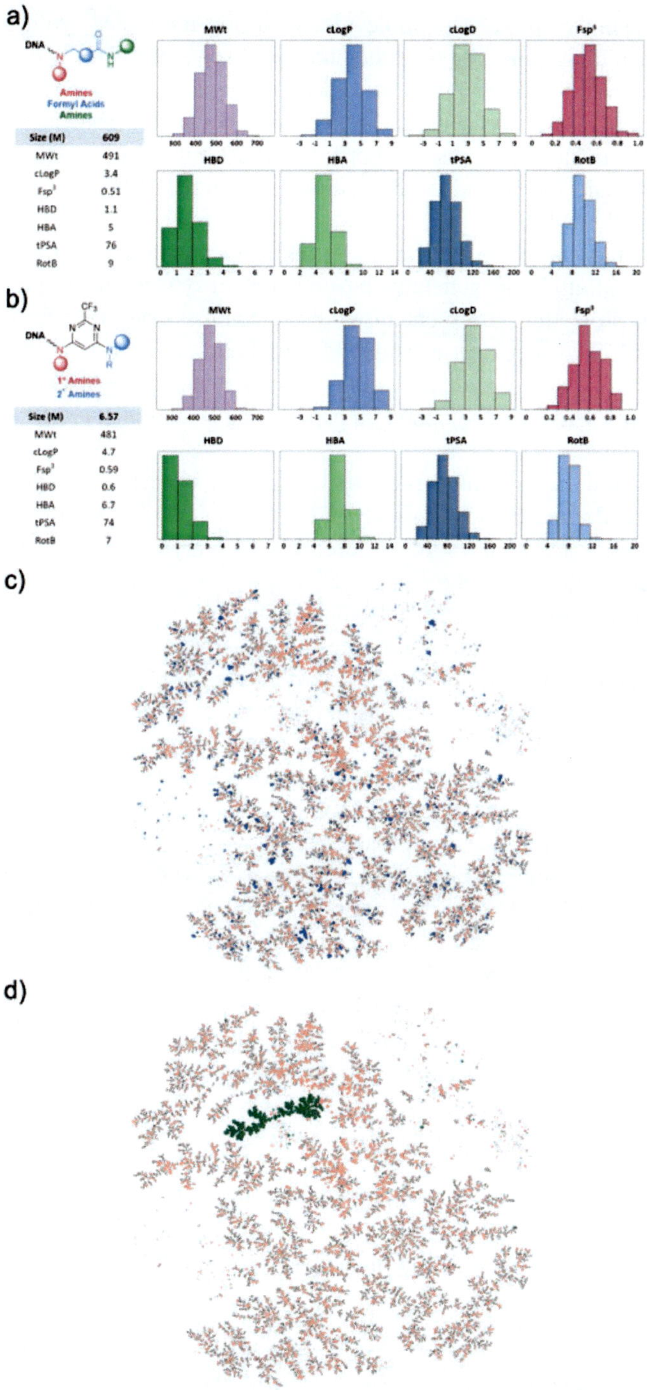

Fig. 4 Comparison of two DNA-encoded libraries. (**a**) Structure and physicochemical properties of the X-Chem ReadiDEL library, which is a 3-cycle linear library. (**b**) Structure and physicochemical properties of a 2-cycle linear library with a conserved 4,6-diamino-2-trifluoromethylpyrimidine scaffold. (**c**) TMAP plot of X-Chem ReadiDEL (dark blue) overlaps with EnamineREAL (gray) and

become increasingly unlikely for molecules requiring multi-step synthesis and for which statistical undersampling (i.e., the number of sequence reads is insufficient to establish meaningful statistics) in the screening output is likely [18, 19]. For numerically very large DELs the available sequencing power leads to undersampling, which can lead to statistical noise except for the best binders [19]. These issues are especially relevant for scientists that aim to combine DEL data with artificial intelligence methods [20].

Restricting the number of synthesis cycles further facilitates hit evaluation. A 2-cycle hit compound can generally be prepared in one to three synthetic steps depending on protecting group chemistry used. Synthesizing a hit from a 4-cycle DEL conversely can require a considerable synthetic effort. Low-cycle DELs are therefore preferred for lead identification. Similarly, using sophisticated reactions when preparing DELs may look good on paper, but also complicates the synthesis of hits to be validated.

Besides the number of synthesis cycles, the geometric arrangement of the sets of building blocks defines the topology of a DEL compound. The geometry of protein binding site varies and so does the preference for library topology. A deep narrow binding pocket will be best served by DELs encoding compounds with elongated molecules typical for linear DELs. Branched structures are preferred for defined pockets with several surface accessible sites. An illustrative example of such pockets are NAD^+-binding sites that have two adjacent sites binding aromatic molecules [5, 12]. Extended binding sites such as those found in protein–protein interactions are best explored with DELs containing molecules with a large surface area. If information on the target site is available, one can select the DEL topology accordingly when designing target-focused DELs. Otherwise, panels of libraries with different geometries are a good approach.

In addition to building blocks, DELs can also contain conserved elements that are not part of building blocks (Fig. 3). These structural elements can be formed as part of a reaction such as in the case of an example of X-Chem ReadiDEL in Fig. 4a. The secondary amines generated in the first cycle of the reaction scheme can be further reacted with amine-reactive building blocks (the aminomethylene moiety in the X-Chem ReadiDEL is an example, Fig. 4a). Alternatively, a scaffold (or core) moiety can be added as a separate step to the structure. In this situation, the scaffold can be considered as a synthesis cycle with a single building block. For illustration, a branched two-cycle library with a scaffold corresponds to a three-cycle library with one member at cycle 1 (i.e., top left structure in Fig. 3 is a special case of the third library from the left with one building block at cycle 1). Branch points have the advantage that they allow the assembly of two sets of monovalent building blocks that are structurally more diverse than bivalent ones [7]. However, conserved

Fig. 4 (continued) ChEMBL (pink) in chemical space. (**d**) TMAP plot of library shown in Fig. 4b (dark green) clusters in limited chemical space and is distinct from EnamineREAL and ChEMBL chemical space. (TMAP plots are based on ECFP-6 fingerprint, courtesy of Enko)

scaffolds increase the size of the molecules and can have other detrimental effects on compounds physicochemical properties without contributing to the structural variability of the compounds. Conserved linear elements are especially problematic in this regard. This effect is visible in the comparison of the property space of X-Chem ReadiDEL with a 2-cycle DEL that has a conserved linear 4,6-diamino-2-trifluoropyrimidine scaffold. The two libraries have similar average molecular weights even though the structural diversity of the ReadiDEL is clearly superior to the library with the pyrimidine scaffold (Fig. 4). Because of the conserved trifluoropyrimidine moiety, this library clusters in a comparably small chemical space because molecules are similar to each other (Fig. 4d), which contrasts the much broader chemical space of ReadiDEL (Fig. 4c) that covers a similar chemical space as the EnamineREAL library [21] or compounds in ChEMBL. The effect is illustrated using TMAP, which generates and distributes intuitive visualizations of large data sets of chemical space [22]. It is noteworthy though that algorithms of molecular similarity tend to emphasize central parts of the molecule and even the pyrimidine-scaffold library covers molecules of impressive chemical diversity. As a rule of thumb, it is advisable to minimize conserved structural elements relative to variable parts of molecules, and such scaffolds are prone to replacement during hit-to-lead development [23].

3 Choosing the Right Chemistry

The synthesis and screening of non-templated, non-bead-based DNA-encoded libraries was first reported in detail over a decade ago [24, 25]. Given the wide embrace and enthusiastic uptake of DEL screening for hit finding in drug discovery, chemists have optimized many reactions that are routinely used in medicinal chemistry for DEL synthesis. As outlined in the mini-perspective by Franzini and Randolph [8], reactions such as amide formation, reductive amination, reductive alkylation, nucleophilic substitution, Suzuki-Miyaura and benzimidazole formation have been widely reported and rendered highly applicable for encoded library synthesis (Fig. 5).

Inherent challenges in the preparation of DEL compound libraries are the restraints introduced by the requisite DNA tag. The requirement of a DNA tag on each library compound necessitates the use of fully aqueous or aqueous-organic solvent mixtures and prevents the use of reagents such as strong acids or oxidants that would hydrolyze or oxidize the oligonucleotide. Rising to these challenges, numerous groups in both industry and academia have sought to expand the synthetic scope of DEL accessibility by increasing the number of chemical reactions that can be conducted on DNA. Much success has followed and been reported such as: metal-catalyzed C-C bond formation [26], photoredox chemical approaches [27–29], enzymatic catalysis [30], solid-phase immobilization [31], and multi-component condensation reactions [32]. The number of publications reporting DNA-compatible chemical reactions has increased greatly in recent years as more

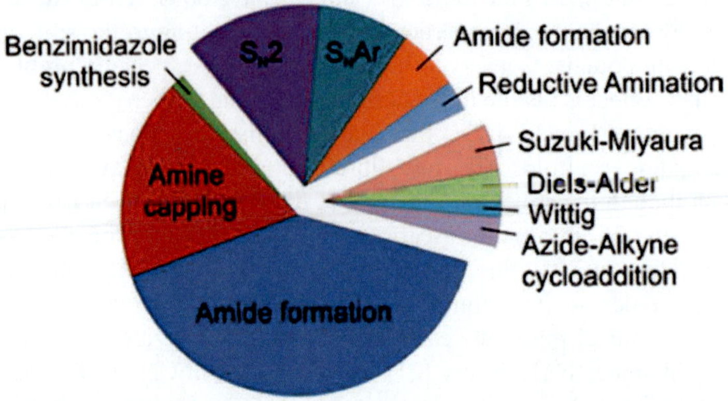

Fig. 5 Summary of widely used chemical reactions in DEL preparation (circa 2016; adopted from Franzini and Randolph [8])

chemists enter the field and now far exceed the number of most used chemical reactions in medicinal chemistry such as amide bond formation, Suzuki-Miyaura coupling, SNAr reactions, and protection/deprotection of functional groups [33]. These newly adopted DNA-compatible reactions include coupling reactions which form new carbon–carbon, carbon–heteroatom, or heteroatom–heteroatom bonds. In addition, photo-catalysis, C-H activation, and metal-mediated cross coupling reactions are among the fast-growing list of coupling chemistries demonstrated to occur on-DNA. Another body of work focuses on cyclization reactions, including aromatic and non-aromatic heterocyclization reactions. For a comprehensive review of on-DNA reactions and practical utilization of these reactions, please refer to chapter "Advancements in DEL-Compatible Chemical Reactions" of this book and "The expanding reaction toolkit for DNA-encoded libraries" [34].

While many reported DNA-compatible chemical reactions have greatly expanded the synthetic chemistry toolbox, the design and synthesis of highly diverse and numerically large encoded libraries remain one of the challenges in establishing a robust DEL platform. Planning and execution of a high-quality library product is a multi-faceted problem that requires mechanistic understanding of organic chemistry (how to make), knowledge of medicinal chemistry (what to make), and cheminformatics support (see chapter "Cheminformatics Approaches Aiding the Design and Selection of DNA-Encoded Libraries" of this book) for assessing the lead-likeness and drug-likeliness of chemical space covered by a given library design. All these elements work together to ensure that the compound library is suitable for small-molecule drug discovery. Discerning "what to make" takes on greater importance when so many reported synthetic methodologies are available; successful library design demands more than simply applying available tools in the toolbox.

3.1 *Library Design: Non-scaffold Based*

One design approach for diversity-oriented DELs is the catenation of mono-, bi-, and tri-functional building blocks, which creates ultra-diverse structures with no fixed motif. In the example shown in Fig. 4a, for a 3-cycle library, reductive alkylation of bi-functional formyl acids to the on-DNA secondary amino moieties, followed by an amide formation reaction with the resulting amino acids gives rise to a library of over 600 million drug-like compounds with excellent predicted physicochemical properties.

3.2 *Scaffold-Based Design*

A common strategy in DEL synthesis is to start with a core or template scaffold appended to the DNA tag and to introduce diversity elements by the addition of building blocks to the core scaffold. This approach requires cores with functional groups that can combine with building blocks (BBs) in an orthogonal manner and in addition to a reaction handle for DNA linkage; depending on the used chemistry orthogonal protecting groups are required. This is a similar strategy to the combinatorial chemistry approach for the preparation of library compounds for High-throughput Screening (HTS) campaigns. As shown in Fig. 4b, a library which shares a common pyrimidine core, even when functionalized with the most diverse set of amine BBs, results in a library with limited molecular diversity in chemical space when compared to libraries without a common motif. The diversity of the core/template approach is hampered by the repetition of the core motif in each compound. Additionally, core scaffolds, especially those with a reaction handle for linking to DNA, are not widely available from commercial sources and often require custom synthesis; therefore, a significant upfront investment results for the on-DNA library synthesis.

4 Build Block (BB) Inclusivity

The most important factor influencing DEL design is the utilization of a large number of BBs in a given synthetic operation. Given the vast encodable sequence space of even a short segment of DNA, there is no practical limit for the inclusion of building blocks. All available chemical fragments can be encoded, although it is worth noting that the number of available building blocks varies considerably for different reaction classes; tens of thousands of carboxylic acids and primary/secondary amines are available, whereas, for example, the number of commercially available azides or terminal alkynes is much smaller [35]. On the other hand, because of the inclusive nature of BBs for DEL synthesis, a DEL often bears the burden of high MW from the appending BBs, resulting in the production of compounds with undesirable properties (high MW and high cLogP). To achieve a split size (the number of BBs used in a synthetic step for "split and pool" library synthesis method) of 5,000 BBs at a given chemistry cycle, the average residue MW added by the incorporation of the building blocks is ~150 [36]. Scaffolded library designs tend to result in compounds with high MW, as the starting point, as the average residue MW of bi-functional cores and templates is around 220. The starting point MW can be even higher for tri-functional scaffolds with increased molecular complexity and a DNA attachment linkage point. Recognizing the importance of DEL design to produce compounds within a desired property space, DEL practitioners opt to apply property filters at the building block level to confine the overall library physicochemical properties to reside in the more desirable Lipinski Ro5 parameter space [37]. However, this filtering exercise to achieve desirable physicochemical property space may be at the cost of excluding structurally complex and attractive building blocks, thereby undermining the inclusive power of encoded libraries [35].

5 Atom Efficient Library Design

To keep the library properties in a useful range for hit generation, one must minimize the molecular weight burden of the non-BB portions. One way of achieving this is by designing a scheme for an "atom efficient" library. An example of such a library design resulting in a 1,2,4-oxadiazole library was reported by Zhang and Clark [36]. Constructing an oxadiazole core within the DEL synthesis scheme by utilizing the most prevalent and commercially available building block sets of amines and carboxylic acids offers the unparalleled advantage of building block inclusivity and diversity, resulting in library compounds with superior predicted physicochemical properties (Fig. 6).

6 Escape from Flatland

Molecular "flatland" and the challenges it introduces in the optimization of hits into drug candidates have been a topic of discussion in small-molecule drug discovery [38]. Although synthetically challenging to access such compounds, efforts have been exerted in the design and synthesis of encoded libraries that construct saturated cores. These libraries present compounds with a high sp^3 content and leverage the "atom efficient" nature of DEL design to produce these libraries. One example, also reported by Zhang and Clark [36] is a piperidine library. Instead of pre-formed core, the piperidine moiety is created as part of the synthetic scheme. These libraries increase the three-dimensionality of the library compounds by using highly diverse, widely available building blocks in concert with atom efficient chemistry to produce molecules with desirable physicochemical properties (Fig. 7).

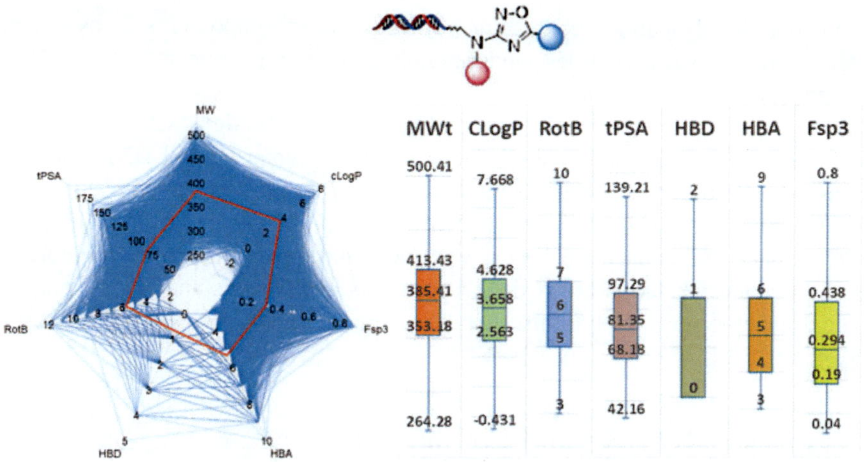

Fig. 6 Physicochemical profile of two-cycle 1,2,4-triazole library

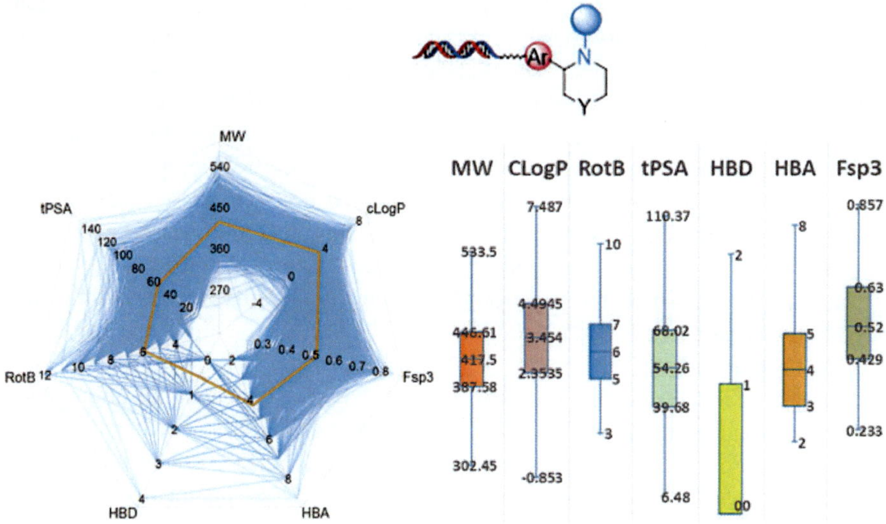

Fig. 7 Physicochemical profile of aliphatic azacycle library

7 Chemical Space Beyond Rule of Five

In recent years, there has been an increasing occurrence of drug discovery programs that require the development of orally available and cell permeable classes of drugs derived from the chemical space beyond Lipinski's rule of 5 (bRO5). The chameleonic physicochemical properties and differentiated binding modes of bRO5 compounds make them suitable starting points especially for the difficult-to-drug targets beyond traditional small-molecule drug discovery. bRO5 compounds such as natural products, macrocyclic peptides, and proteolysis targeting chimeras (PROTACs) have attracted much attention in the field [39–41]. DEL synthesis of natural product inspired libraries and macrocyclic peptide libraries [42, 43] aligns with these efforts and have been an important addition to the rapidly growing field of DNA-encoded small-molecule library screening. DEL screening of de novo designed bRO5 compounds will aid the exploration of bRO5 chemical space. The bRO5 DEL design is in its nascent stage. As it evolves it will merit further discussion as chemistries beyond peptide formation are widely reported and bRO5 hits and leads from the platform emerge prevalently in the literature.

8 Practical Remarks and Considerations

1. Chemical integrity of encoded libraries.
   Significant issues with incomplete reaction conversion and side product formation should be addressed during the library development stage when reaction scope and reproducibility of high reaction yields should be thoroughly studied. Ambiguity in the final product should be carefully avoided in the execution of the synthesis. Exclusion of building blocks with ambiguous reactive handles and removal of unintended reactive intermediates should be meticulously planned [44]. Tag sequences should reliably represent the library compounds of the original intentions. Although methods such as BALI-MS (Bead-Assisted Ligand Isolation Mass Spectrometry) and AS-MS (Affinity Selection Mass Spectrometry) of on-DNA hit resynthesis have been reportedly applied to the post-selection triage of false positive binders from DEL screens, identifying off-scheme hits costs unnecessary time and resources.
2. Fidelity of encoding.
   Reaction conditions that may affect the integrity of the encoding oligonucleotides should be avoided. This may include metal-catalyzed reactions, radical reactions, prolonged heating at high temperatures and reactions that require high acidity. DNA damage is best observed during the development of a chemical reaction when reaction mixtures were carefully characterized using LC-MS. One example is an on-DNA reaction of the Cu(I)-catalyzed azide-alkyne cycloaddition, commonly known as click chemistry. Increasing baseline and decreasing intensity of the MS peak of the desired reaction product over time in the presence of Copper (I) is readily observable in LC-MS. Therefore, the active catalyst should be quenched and scavenged as soon as the reaction is complete to preserve the integrity of the encoding DNA. DNA compatibility with chemical reactions has been frequently evaluated [29, 45]. DNA damage induced by chemical reactions has been recently reported [46].
3. Best practices for choosing building blocks.
   The importance of selecting the right building blocks (BBs) cannot be overstated.

Table 1 Feature prevalence table showing structures of the top five building blocks at cycle B that co-enrich with the indicated cycle A and C building block pairs and define tolerated variability within the active chemotype as indicated in the selection output data at cycle B. The relative prevalence of each building block B, in terms of the percent representation in the total number of co-enriched compounds, is indicated below each building block. No tolerated variation was observed at cycle A presumably because of the relative uniqueness of this building block [47]

Cycle A	Cycle B	Cycle C
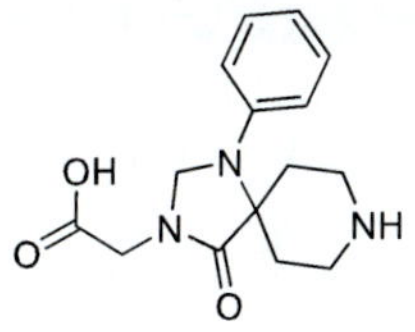	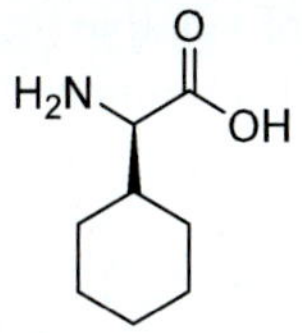 34%	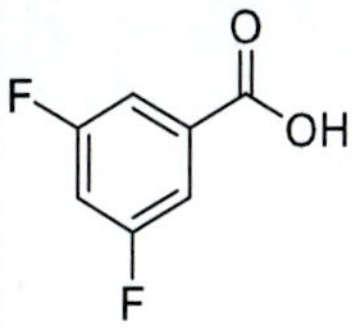
	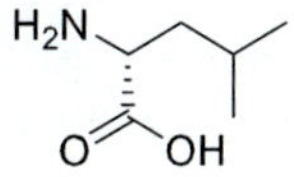 24%	
	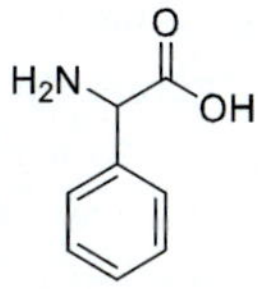 22%	
	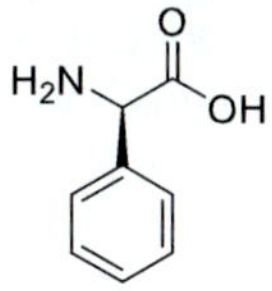 17%	
	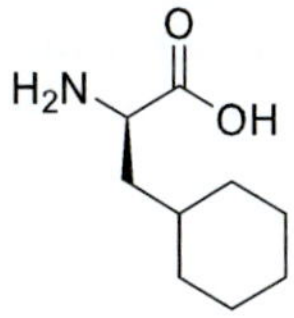 2%	

When selection output contains a large number of strongly enriched features, BB SAR/SER (*s*tructure *a*ffinity/*e*nrichment *r*elationship) often emerge [47] (Table 1).

The observation of clear BB SAR within enriched features increases confidence of the identification of putative binders and offers rich information during the hit to lead phase when the hit compounds are confirmed. This further illustrates the power of the inclusive nature of BBs for encoded libraries. The choice of BBs should cover broad structural diversity, and equally importantly, include sufficiently similarly related structures to permit the exploration of SAR. Physicochemical property filters, such as MW, cLogP and number of rotatable bonds, should be applied at BB purchase. However, med chem filters other than PAINS (pan-assay interference compounds) filter can be applied when choosing which compounds to resynthesize post DEL screen. There is no significant downside to the inclusion of an unattractive BB in a library although applying PAINS filters [48–50] is general practice. An undesirable affinity binder among a vast number of enriched features is merely a data point and can be simply discarded without being resynthesized.

4. Design of library scheme with feasibility of off-DNA synthesis in mind.

   Schemes that generate multiple undefined stereocenters should be avoided because they give rise to a mixture of stereoisomers with the same encoding sequence. This complicates off-DNA hit confirmation as each stereoisomer is required to be synthesized off-DNA. On the other hand, enantiopure BBs and the corresponding racemic BBs are advised to be included in the library synthesis and individually encoded. These BBs are found to be prudent in providing SAR information during hit confirmation.

   An upfront investment to utilize specialty linkable scaffolds is merited when such core scaffolds lead to library compounds with better drug-like properties or alleviate the burden of complex off-DNA synthesis for hit confirmation. In the example shown below, library compounds with a pyrimidine core are fitted with a long fatty acid chain (A) for DNA attachment which eliminates the possibility of binding interactions with the carboxamide moiety and target which could occur if a commercially available pyrimidine core (B) were used for this library. The off-DNA compounds can be easily synthesized from the readily available di-chloromethylpyrimidine (or di-chloropyrimidine), truncating the alkyl chain to a methyl group (or des-methyl) to represent the aliphatic DNA attachment linker of the library compounds (Figs. 8 and 9).

5. BB validation for a library scheme.

   There is a need for all BBs in a library to react with a reasonably high yield (different scientists have different standards on what an acceptable reaction yield is, but typically at least 50%); therefore, their reactivity should be assessed thoroughly during chemistry development. Although not a critical requirement, in theory, even a low yielding library member can be enriched and discoverable as long as the binding event occurred in affinity selection, it is desirable to assure the quality of library with high yielding library members.

   BB validation has been a routine process for DEL practitioners. The reactivity of the proposed set of BBs is evaluated using a model reaction on short DNA-oligonucleotide, and only those BBs that afford a yield above a threshold

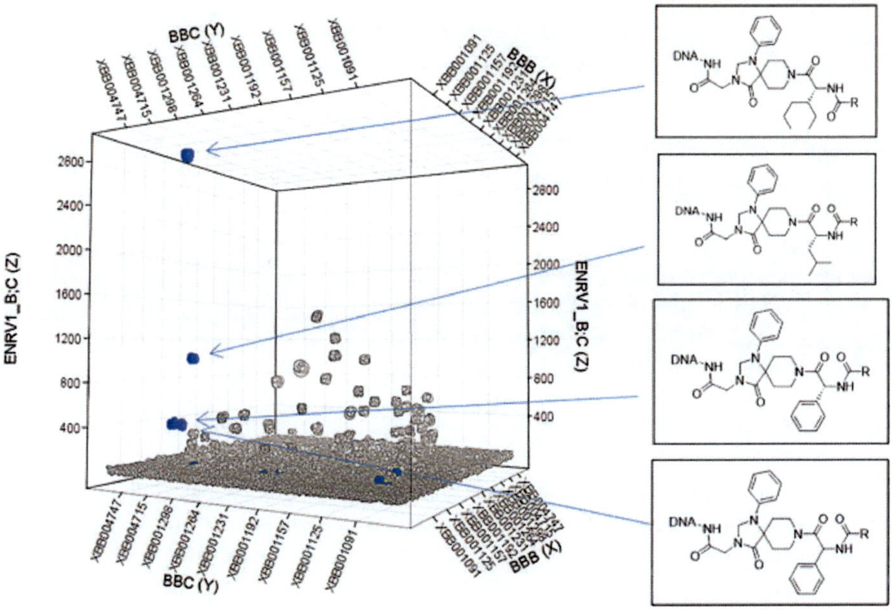

Fig. 8 3D scatterplots showing individual enriched features against Autotaxin. The x- and y-axes represent individual building blocks at the indicated cycles of synthesis. The z axis represents a statistical metric of enrichment significance (ENRv1, negative Log10 of the asymptotic significance value) for each disynthon [47]

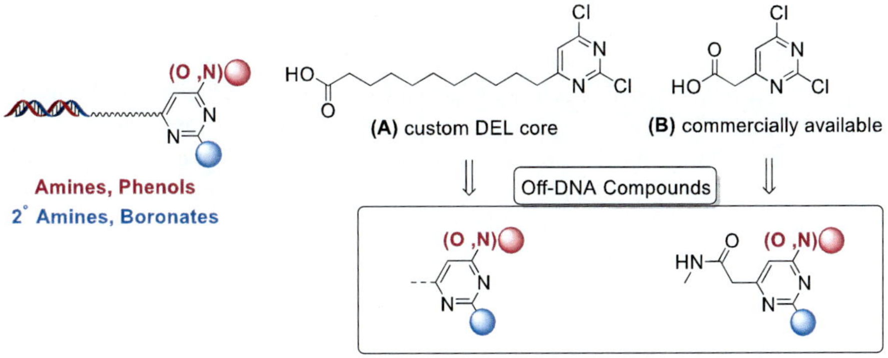

Fig. 9 Utilizing specialty linkable scaffolds in DEL synthesis

(e.g., 50%) of product pass validation and are included in the library synthesis (Fig. 10).

Artificial intelligence (AI) has provided the ability to unlock the full potential of vastly rich and high-quality datasets generated by DEL technology [20]. In the future, machine learning (ML) models based on rich datasets generated in DEL BB validation for a wide variety of chemical reactions may allow for virtual BB

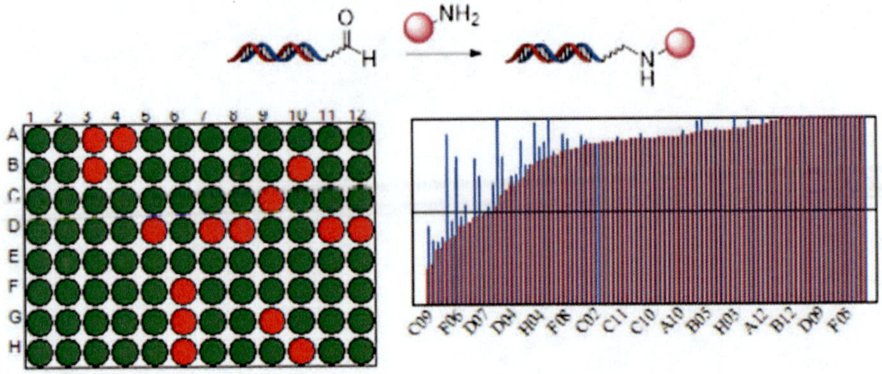

Fig. 10 BB with reaction yields below the threshold are excluded from the library

validation. ML for BB validation can provide prediction of pass/fail with high precision and ultimately provide guidance in BB collection to maximize the diversity and balanced similarity.

9 Conclusions

DEL design parameters define how effective the resultant libraries will be at providing attractive advanceable compounds. Design parameters include library topology, reaction scheme, and building block identities. Library diversity principally flows from the extent to which large numbers of building blocks can be accessed and utilized, which is in turn dependent upon the scheme and the availability of building blocks of the relevant classes. Atom-efficient schemes with monovalent building blocks are best able to leverage available building block diversity by incorporating them into compounds with few invariant atoms. Lastly, limitations of known DNA-compatible reactions have historically constrained DEL designs, however the reaction toolbox is much more extensive today.

Acknowledgments The authors would like to acknowledge Matthew Clark and Anthony Keefe for their helpful comments.

Compliance with Ethical Standards

Funding: Ying Zhang received no financial support to assist with the preparation of this manuscript. Raphael Franzini is supported by the National Institutes of Health (R35GM138335).

Declaration of Conflicting Interests: Ying Zhang is employed by X-Chem, Inc.; the research and authorship of this chapter were completed within the scope of the employment with X-Chem, Inc. Raphael Franzini is a scientific advisor for Leash Laboratories and ConfometRx.

Ethical approval: This Chapter does not contain any studies with human participants or animals performed by any of the authors.

References

1. Dragovich PS, Haap W, Mulvihill MM, Plancher J, Stepan AF (2022) Small-molecule Lead-finding trends across the Roche and Genentech research organizations. J Med Chem 65(4): 3606–3615
2. Gironda-Martinez A, Donckele EJ, Samain F, Neri D (2021) DNA-encoded chemical libraries: a comprehensive review with successful stories and future challenges. ACS Pharmacol Transl Sci 4(4):1265–1279
3. Goodnow Jr RA, Dumelin CE, Keefe AD (2017) DNA-encoded chemistry: enabling the deeper sampling of chemical space. Nat Rev Drug Discov 16(2):131–147
4. Yuen LH, Franzini RM (2017) Achievements, challenges, and opportunities in DNA-encoded library research: an academic point of view. Chembiochem 18(9):829–836
5. Yuen LH, Dana S, Liu Y, Bloom SI, Thorsell AG, Neri D, Donato AJ, Kireev D, Schuler H, Franzini RM (2019) A focused DNA-encoded chemical library for the discovery of inhibitors of NAD(+)-dependent enzymes. J Am Chem Soc 141(13):5169–5181
6. Bickerton GR, Paolini GV, Besnard J, Muresan S, Hopkins AL (2012) Quantifying the chemical beauty of drugs. Nat Chem 4(2):90–98
7. Hefti FF (2008) Requirements for a lead compound to become a clinical candidate. BMC Neurosci 9(Suppl 3):S7
8. Franzini RM, Randolph C (2016) Chemical space of DNA-encoded libraries. J Med Chem 59(14):6629–6644
9. Shi B, Zhou Y, Huang Y, Zhang J, Li X (2017) Recent advances on the encoding and selection methods of DNA-encoded chemical library. Bioorg Med Chem Lett 27(3):361–369
10. Halford B (2017) How DNA-encoded libraries are revolutionizing drug discovery. Chem Eng News 95:25
11. Plais L, Scheuermann J (2022) Macrocyclic DNA-encoded chemical libraries: a historical perspective. RSC Chem Biol 3(1):7–17
12. Montoya AL, Glavatskikh M, Halverson BJ, Yuen LH, Schuler H, Kireev D, Franzini RM (manuscript submitted) Predicting lead compounds from DNA-encoded chemical libraries: understanding limitations of selection data, integrating pharmacophore building with docking, and discovery of Tankyrase 1 inhibitors
13. Satz AL (2018) What do you get from DNA-encoded libraries? ACS Med Chem Lett 9(5): 408–410
14. Martin A, Nicolaou CA, Toledo MA (2020) Navigating the DNA encoded libraries chemical space. Commun Chem 3:127
15. Franzini RM, Ekblad T, Zhong N, Wichert M, Decurtins W, Nauer A, Zimmermann M, Samain F, Scheuermann J, Brown PJ, Hall J, Graslund S, Schuler H, Neri D (2015) Identification of structure-activity relationships from screening a structurally compact DNA-encoded chemical library. Angew Chem Int Ed Engl 54(13):3927–3931
16. Samain F, Ekblad T, Mikutis G, Zhong N, Zimmermann M, Nauer A, Bajic D, Decurtins W, Scheuermann J, Brown PJ, Hall J, Graslund S, Schuler H, Neri D, Franzini RM (2015) Tankyrase 1 inhibitors with drug-like properties identified by screening a DNA-encoded chemical library. J Med Chem 58(12):5143–5149
17. Wang S, Denton KE, Hobbs KF, Weaver T, McFarlane JMB, Connelly KE, Gignac MC, Milosevich N, Hof F, Paci I, Musselman CA, Dykhuizen EC, Krusemark CJ (2020) Optimization of ligands using focused DNA-encoded libraries to develop a selective cell-permeable CBX8 Chromodomain inhibitor. ACS Chem Biol 15(1):112–131
18. Satz AL (2016) Simulated screens of DNA encoded libraries: the potential influence of chemical synthesis Fidelity on interpretation of structure-activity relationships. ACS Comb Sci 18(7):415–424
19. Satz AL, Hochstrasser R, Petersen AC (2017) Analysis of current DNA encoded library screening data indicates higher false negative rates for numerically larger libraries. ACS Comb Sci 19(4):234–238

20. McCloskey K, Sigel EA, Kearnes S, Xue L, Tian X, Moccia D, Gikunju D, Bazzaz S, Chan B, Clark MA, Cuozzo JW, Guie MA, Guilinger JP, Huguet C, Hupp CD, Keefe AD, Mulhern CJ, Zhang Y, Riley P (2020) Machine learning on DNA-encoded libraries: a new paradigm for hit finding. J Med Chem 63(16):8857–8866
21. Gorgulla C, Boeszoermenyi A, Wang ZF, Fischer PD, Coote PW, Padmanabha Das KM, Malets YS, Radchenko DS, Moroz YS, Scott DA, Fackeldey K, Hoffmann M, Iavniuk I, Wagner G, Arthanari H (2020) An open-source drug discovery platform enables ultra-large virtual screens. Nature 580(7805):663–668
22. Probst D, Reymond J-L (2020) Visualization of very large high-dimensional data sets as minimum spanning trees. J Cheminform 12:12
23. Lemke M, Ravenscroft H, Rueb NJ, Kireev D, Ferraris D, Franzini RM (2020) Integrating DNA-encoded chemical libraries with virtual combinatorial library screening: optimizing a PARP10 inhibitor. Bioorg Med Chem Lett 30(19):127464
24. Clark MA, Acharya RA, Arico-Muendel CC, Belyanskaya SL, Benjamin DR, Carlson NR, Centrella PA, Chiu CH, Creaser SP, Cuozzo JW, Davie CP, Ding Y, Franklin GJ, Franzen KD, Gefter ML, Hale SP, Hansen NJ, Israel DI, Jiang J, Kavarana MJ, Kelley MS, Kollmann CS, Li F, Lind K, Mataruse S, Medeiros PF, Messer JA, Myers P, O'Keefe H, Oliff MC, Rise CE, Satz AL, Skinner SR, Svendsen JL, Tang L, van Vloten K, Wagner RW, Yao G, Zhao B, Morgan BA (2009) Design synthesis and selection of DNA-encoded small-molecule libraries. Nat Chem Biol 5(9):647–654
25. Mannoccia L, Zhang Y, Scheuermann J, Leimbacher M, De Bellis G, Rizzi E, Dumelin C, Melkko S, Neri D (2008) High-throughput sequencing allows the identification of binding molecules isolated from DNA-encoded chemical libraries. Proc Natl Acad Sci 105(46): 17670–17675
26. Wang X, Sun H, Liu J, Dai D, Zhang M, Zhou H, Zhong W, Lu X (2018) Ruthenium-Promoted C–H Activation Reactions between DNA-Conjugated Acrylamide and Aromatic Acids. Org Lett 20(16):4764–4768
27. Kölmel DK, Loach RP, Knauber T, Flanagan ME (2018) Employing photoredox catalysis for DNA-encoded chemistry: decarboxylative alkylation of α-amino acids. Chem Med Chem 13(20):2159–2165
28. Badir SO, Sim J, Zhang X, Dong W, Molander GA, Billings K, Csakai A (2020) Multifunctional building blocks compatible with photoredox-mediated alkylation for DNA-encoded library synthesis. Org Lett 22(3):1046–1051
29. Kölmel DK, Ratnayake AS, Flanagan ME, Tsai M-H, Duan C, Song C (2020) Photocatalytic [2 + 2] cycloaddition in DNA-encoded chemistry. Org Lett 22(8):2908–2913
30. Thomas B, Lu X, Birmingham WR, Huang K, Both P, Reyes Martinez JE, Young RJ, Davie CP, Flitsch SL (2017) Application of biocatalysis to on-DNA carbohydrate library synthesis. Chem Bio Chem 18:858
31. MacConnell AB, McEnaney PJ, Cavett VJ, Paegel BM (2015) DNA-encoded solid-phase synthesis: encoding language design and complex oligomer library synthesis. ACS Comb Sci 17(9):518–534
32. Kunig VBK, Ehrt C, Dömling A, Brunschweiger A (2019) Isocyanide multicomponent reactions on solid-phase-coupled DNA oligonucleotides for encoded library synthesis. Org Lett 21(18):7238–7243
33. Brown DG, Boström J (2016) Analysis of past and present synthetic methodologies on medicinal chemistry: where have all the new reactions gone? J Med Chem 59(10):4443–4458
34. Fair R, Walsh RT, Hupp CD (2021) The expanding reaction toolkit for DNA-encoded libraries. Bioorg Med Chem Lett 51:128339
35. Kalliokoski T (2015) Price-focused analysis of commercially available building blocks for combinatorial library synthesis. ACS Comb Sci 17(10):600–607
36. Zhang Y, Clark MA (2021) Design concepts for DNA-encoded library synthesis. Bioorg Med Chem 51:116189

37. Zhu H, Flanagan ME, Stanton RV (2019) Designing DNA encoded libraries of diverse products in a focused property space. J Chem Inf Model 59:4645–4653
38. Lovering F, Bikker J, Humblet C (2009) Escape from flatland: increasing saturation as an approach to improving clinical success. J Med Chem 52(21):6752–6756
39. Caron G, Kihlberg J, Goetz G, Ratkova E, Poongavanam V, Ermondi G (2021) Steering new drug discovery campaigns: permeability, solubility, and physicochemical properties in the bRo5 chemical space. Med Chem Lett 12(1):13–23
40. Doak BC, Over B, Giordanetto F, Kihlberg J (2014) Oral druggable space beyond the rule of 5: insights from drugs and clinical candidates. Chem Biol 21(9):1115–1142
41. DeGoey DA, Chen H, Cox PB, Wendt MD (2018) Beyond the rule of 5: lessons Learned from AbbVie's drugs and compound collection. J Med Chem 61(7):2636–2651
42. Onda Y, Bassi G, Elsayed A, Ulrich F, Oehler S, Plais L, Scheuermann J, Neri D (2021) A DNA-encoded chemical library based on peptide macrocycles. Chem A Eur J 27:7160
43. Koesema E, Roy A, Paciaroni NG, Coito C, Tokmina-Roszyk M, Kodadek T (2022) Synthesis and screening of a DNA-encoded library of non-peptidic macrocycles. Angew Chem Int Ed 61: e202116999
44. Foley TL, Burchett W, Chen Q, Flanagan ME, Kapinos B, Li X, Montgomery JI, Ratnayake AS, Zhu H, Peakman M (2021) Selecting approaches for hit identification and increasing options by building the efficient discovery of actionable chemical matter from DNA-encoded libraries. SLAS Discov 26(2):263–280
45. Phelan JP, Lang SB, Sim J, Berritt S, Peat AJ, Billings K, Fan L, Molander GA (2019) Open-air alkylation reactions in photoredox-catalyzed DNA-encoded library synthesis. J Am Chem Soc 141(8):3723–3732
46. Ratnayake AS, Flanagan ME, Foley TL, Smith JD, Johnson JG, Bellenger J, Montgomery JI, Paegel BM (2019) A solution phase platform to characterize chemical reaction compatibility with DNA-encoded chemical library synthesis. ACS Comb Sci 21(10):650–655
47. Cuozzo JW, Clark MA, Keefe AD, Kohlmann A, Mulvihill M, Ni H, Renzetti LM, Resnicow DI, Ruebsam F, Sigel EA, Thomson HA, Wang C, Xie Z, Zhang Y (2020) Novel autotaxin inhibitor for the treatment of idiopathic pulmonary fibrosis: a clinical candidate discovered using DNA-encoded chemistry. J Med Chem 63(14):7840–7856
48. Dahlin JL, Nissink JWM, Strasser JM, Francis S, Higgins L, Zhou H, Zhang Z, Walters MA (2015) PAINS in the assay: chemical mechanisms of assay interference and promiscuous enzymatic inhibition observed during a sulfhydryl-scavenging HTS. J Med Chem 58(5): 2091–2113
49. Baell JB, Holloway GA (2010) New substructure filters for removal of pan assay interference compounds (PAINS) from screening libraries and for their exclusion in bioassays. J Med Chem 53(7):2719–2740
50. Bolz SN, Adasme MF, Schroeder M (2021) Toward an understanding of pan-assay interference compounds and promiscuity: a structural perspective on binding modes. J Chem Inf Model 61(5):2248–2262

Top Med Chem (2022) 40: 145–182
https://doi.org/10.1007/7355_2022_148

Published online: 1 November 2022

Cheminformatics Approaches Aiding the Design and Selection of DNA-Encoded Libraries

Hongyao Zhu, Justin I. Montgomery, and Robert V. Stanton

Contents

H. Zhu (✉)
Machine Learning and Computational Sciences, Pfizer Worldwide Research Development and Medical, Groton, CT, USA
e-mail: hongyao.zhu@pfizer.com

J. I. Montgomery
Discovery Sciences, Pfizer Worldwide Research Development and Medical, Groton, CT, USA
e-mail: justin.montgomery@pfizer.com

R. V. Stanton
Machine Learning and Computational Sciences, Pfizer Worldwide Research Development and Medical, Cambridge, MA, USA
e-mail: robert.stanton@pfizer.com

Abstract The combinatorial nature of DNA-Encoded Libraries (DEL) creates a vast number of compounds covering a wide range of chemistry space, enabling DEL target selections for possible binders that generate enormous amounts of data. This makes advanced cheminformatics and data analysis approaches indispensable for advancing DEL technology for practical applications in the pharmaceutical industry. When designed DEL libraries can be biased to contain compounds with more drug like features through the enumeration of the potential virtual products, based on their combinatorial assembly from available building blocks, and the calculation of molecular descriptors. Computational approaches are also critical in analyzing selection data. In a DEL target selection, it is essential to apply stringent analysis methods to the raw sequencing data to separate real hits from noise and to mitigate the amount of resource required for off-DNA synthesis for hit confirmation. In this chapter we focus on the important role of cheminformatics and computational analysis at all phases of the DEL hit identification workflow by discussing in detail DEL product enumeration, designing DELs of diverse products with focused in specific property spaces, estimating data noise levels in DEL selections, and the process of DEL selection data triage and analysis that eventually identifies hits for on- and off-DNA synthesis. We also outline perspectives of emerging cheminformatics methods for DELs that could further strengthen the technology to facilitate hit identification processes and improve the hit confirmation rate. Emerging methods such as efficient structure searching, visualizing DEL chemistry space, and potential machine learning applications for DEL selection analysis are discussed.

Keywords Building-block selection, DEL, DNA-encoded library, DEL design, DEL hit identification, Library screening, Noise level estimate, Selection data triage

1 Cheminformatics in DEL Technology Platform

The utility of DELs [1, 2] for the identification of small-molecule lead compounds in drug discovery projects has been widely published for the last decade [3–7]. The pace of DEL technology advances has only increased in recent years with: (1) expansion of the chemical space accessible to DNA-compatible synthesis through the development of novel synthetic routes, [8, 9] (2) improvements to the selection protocols used, allowing access to novel target types [10–12], and (3) innovations in the hit conformation technologies employed which have translated to improved confirmation rates for compounds synthesized off-DNA [13–15]. These advances have resulted in a significant number of reports of DEL screens producing chemical lead material which has been optimized into clinical candidates [16–18]. The increased use of DELs is a significant paradigm shift from the High Throughput Screening (HTS) campaigns which have dominated lead identification for several decades and now competes with other emerging technologies such as ASMS [19, 20], fragment-based screening [21, 22], and virtual screening [23], all of which can be done at a fraction of the cost needed for an HTS of a large pharmaceutical company's corporate compound collection.

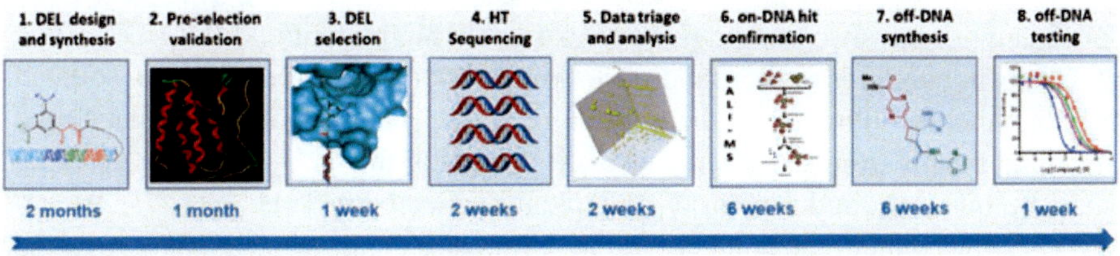

Fig. 1 Key steps of the DEL technology platform for lead discovery at Pfizer

Figure 1 outlines the key steps of the DEL technology platform implemented at Pfizer for lead discovery with steps 1, 6, and 7 being chemistry efforts, steps 2, 3, 4, 5, and 8 biology-related, and steps 1 and 5 involving cheminformatics. The time frames displayed are averages from actual projects. Steps 6 and 7, the on-DNA hit confirmation and off-DNA synthesis, can be done either sequentially or in parallel to speed up the hit confirmation process.

One critical element of DEL screening that is less frequently discussed is the data analysis and informatics steps involved. These start well before any selection experiment with the design of the DEL library itself. Several authors have discussed the importance of chemical diversity in the library design and have demonstrated library design algorithms focused either on building block (the chemical monomers used in the split and pool synthesis) or synthon (combinations of these monomers) based approaches to create diversity in the final DEL library [24–28]. The effect of off-target products (by-products of the intended chemical reaction which are the result of incomplete reactions and the lack of purification during synthesis) has also been debated with both computational and experimental methods to help control for these effects published [29, 30]. All of these methods depend on an underlying set of cheminformatics tools which allow for the enumeration of structures (the computational reaction of building blocks to create the intended product structure), property calculations (molecular weight (MW), number of rotatable bonds, cLogP, etc.) as well as database and storage capabilities to allow these results to be quickly searched. While the general cheminformatics toolsets necessary have been available since the advent of combinatorial chemistry, the scale of DEL libraries let alone their design space introduces unique challenges.

Once the DEL library is (1) synthesized, (2) pooled with other libraries, (3) used in a selection experiment, and (4) the result sequenced, a new set of bioinformatics challenges is created. The tools used for Next Generation Sequencing (NGS) analysis are well suited for the primary image to sequence conversion (bcl2fastq: v2.20; RRID: SCR_015058). However, the decoding of the resulting sequences tags into their corresponding unique compound identifiers is a task that can become computationally intensive. This has become particularly true as the capacity of sequencers (number of reads) has increased and the cost for sequencing has decreased. These factors can easily result in hundreds of millions of sequence reads which need to be decoded. For large DEL libraries sequenced at reasonable depths this can require either a cloud enabled or high-performance computing environment to allow for parallelization. While short read (100–300 bases) sequencing has advanced to the point of extremely high fidelity (99%+ / Q20), this still results in hundreds of thousands of base calling errors that need to be accounted for during the decoding across the millions of reads. One approach to insulating the compound counts from sequencing errors is designing the DNA tags such that they remain unique even in the event of multiple miscalled bases [31].

With a raw count of the number of times each compound was detected in hand the results can be easily plotted into a classic "cube" representation with the size of each ordinal point representing the read count. While this can be a useful starting point, there are several additional data normalization and statistical steps which can greatly help with the interpretation of the data [32]. These steps can include removal of the signal associated with a blank control (a selection done using only the resin to which the protein is normally affixed detecting compounds that might have affinity for the resin), noise estimation [32], normalization across replicates [33], and statistical significance estimation at the synthon level. Methods have been described outlining the use of Z-scores [31], normalized Z-scores [34], population fractions [35], and p-values [36] to help identify compounds with a strong signal as compared to background and noise. The interpretation of copy counts is further complicated by the distribution of input copy numbers for each specific chemical species as outlined by Satz [30, 37].

Using the normalized and corrected data, the selection of compounds for synthetic follow-up still requires an expert who understands the data processing steps, the experimental selection details, and the underlying chemistry of the DEL libraries. Visualization and data access tools are essential in this process to identify trends such as structure–activity relationships across chemically similar building blocks/synthons, or from different screening conditions (variations of protein concentration, addition of an active site blocker, etc.) which can be used as either inclusion or exclusion criteria for the selections. For example, the identification of compounds which show a dose response effect across selections done at multiple protein concentrations can be used to help build confidence that specific compounds are true positives [38]. In contrast, the addition of an active site blocker in a selection might be used to help differentiate between active site and allosteric binders. As a final step in compound selection, the accumulated results from previous screening campaigns should be used to exclude "frequent hitter" compounds from consideration. Given the hundreds of thousands of compounds which can still be part of the analysis at this late stage, visualization and data organization tools are essential to aid in compound selection.

All the data analysis steps outlined here (cheminformatics, bioinformatics, statistical, chemistry) are critical to the proverbial identification of the "needle in the haystack." For DEL screening the haystack can consist of hundreds of billions of compounds, meaning it is necessary to distinguish a hit rate of a millionth of a percent. While it is possible for large sets of compounds to be resynthesized, increasing the probability of hit identification, this adds significant expense to the overall experiment. It is therefore typical for tens of compounds to be made for conformation rather than hundreds making each selected compound critical to the screen's success. It is interesting to note the hit conformation rate needed for DEL screening is many orders of magnitudes greater than what is expected in a traditional HTS, where only a few million compounds are screened but hundreds are retested. One recent advance that has greatly increased the chances of identifying true hits from DEL selections is the use of ASMS technologies to screen on-DNA mixtures that have been synthesized using the same unpurified route as the original library [13, 14]. This allows all the possible side products of the initial reaction to be tested simultaneously and the actual binding entity to be identified.

In summary, the advances in data analysis techniques have paced the improvements and innovations in DEL chemistry and biology enabling the broad applicability and success rates seen today. In this chapter we will describe in depth some of the practical cheminformatics and bioinformatics concepts necessary for successful DEL screening. While many times overlooked, the most trivial indexing or statistical mistake in the data analysis can result in meaningless data. In contrast good data management practices can enable the wave of machine learning (ML) and artificial intelligence (AI) analysis techniques which are currently being introduced [39].

2 Molecular Property Space Focused DEL Design

DEL libraries should be designed to cover as much chemical space as possible, but in the bounds of reasonable property space. To achieve this goal, one would need to develop a wide range of DEL reaction schemes from available DEL compatible chemical reactions as reviewed in Chap. 3 and to include a structurally diverse set of building blocks (BBs). Recently, reaction schemes were explored by combinatorially coupling functional groups along with their corresponding BB selections [40]. We have opted to have reaction schemes designed by experienced synthetic chemists incorporate their insights of reaction compatibility among multiple cycles, chemistry fidelity in DEL synthesis, and drug likeness of DEL products, etc. Even though DNA-compatible reactions continue to emerge, the limited availability of reaction schemes leaves the BB selection a vital component for DEL design. In this section, we outline the strategy of BB selection for a given reaction scheme to yield diverse DEL products.

2.1 BB Collections and Initial Structural Filtering

For a given DEL, BB collections were first created by searching internal solid stores and vendor's BB inventories. BB access can be further increased through collaboration [41]. General structural filters are applied to help to improve the quality of hits from DEL selections. Some of the common filters are: (1) reactive groups, for example, Michael acceptors, acetals/aldehydes, and a-hetero N-F as shown in Fig. 2. (2) Undesirable functional groups: nitro, iodide, silicate, long alkyl chains, terminal butyl, and extended PEG as shown in Fig. 3. (3) Undefined chiral centers –

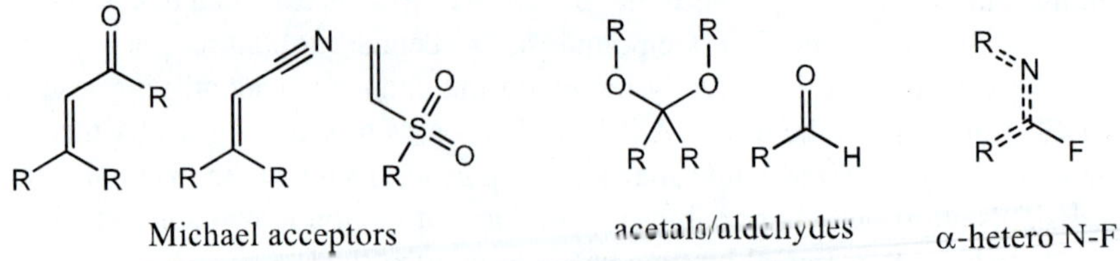

Fig. 2 Examples of reactive functional groups to be excluded in BBs

Fig. 3 Undesirable functional groups

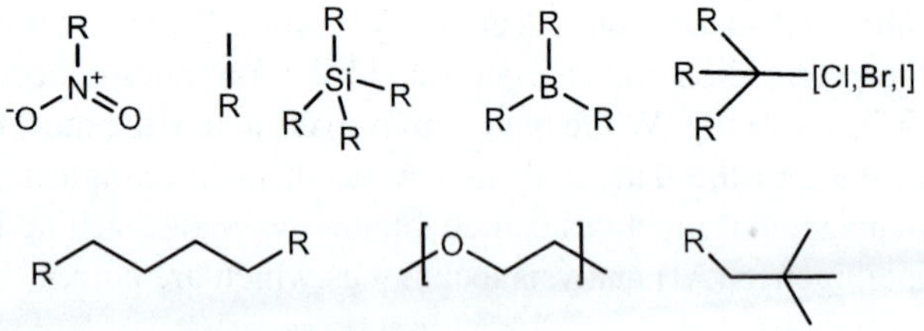

maximum of one allowed for a 3-cycle DEL or two for 2-cycle DELs. As an example, we collected 2,864 fluorenylmethoxycarbonyl (Fmoc) protected amino-acids, BBs used in cycle 1 of PF-DEL-0001 as discussed in 5.2.3, from both internal (~53%) and vendors' (~47%) collections.

While many DEL library designs are done simply at the monomer level, this can result in many compounds with undesirable properties. It is therefore important to consider fully enumerated products as part of the library design process.

2.2 *Robust and Efficient Enumeration of DEL Products*

In this section, an approach for efficent DEL enumeration is outlined. A DEL enumeration engine was implemented in C++ using OEChem toolkit [42]. The workflow along with requird input data and operations is depicted in Fig. 4. PF-DEL-0026 [43] is used as an example to demonstrate the approach in Fig. 5.

Enumeration input data consists of a user-sketched reaction (RXN) and lists of monomer structures along with deprotection and exclusion rules. Following a substructure search (SSS) via a subgraph isomorphism algorithm, the monomers are clipped to generate R-groups. Product structures are then combinatorially assembled via a cursive algorithm. In Fig. 5, the enumeration of PF-DEL-0026 is illustrated by (a) Markush structure for the 3-cycle DEL which uses optimized reaction conditions for the photocatalytic [2 + 2] cycloaddition [43] in the second chemistry cycle, (b) the defined reaction scheme showing the disconnected core being used for the first-cycle chemistry, (c) a representative product of PF-DEL-0026 from corresponding BBs with substructures color-highlighted.

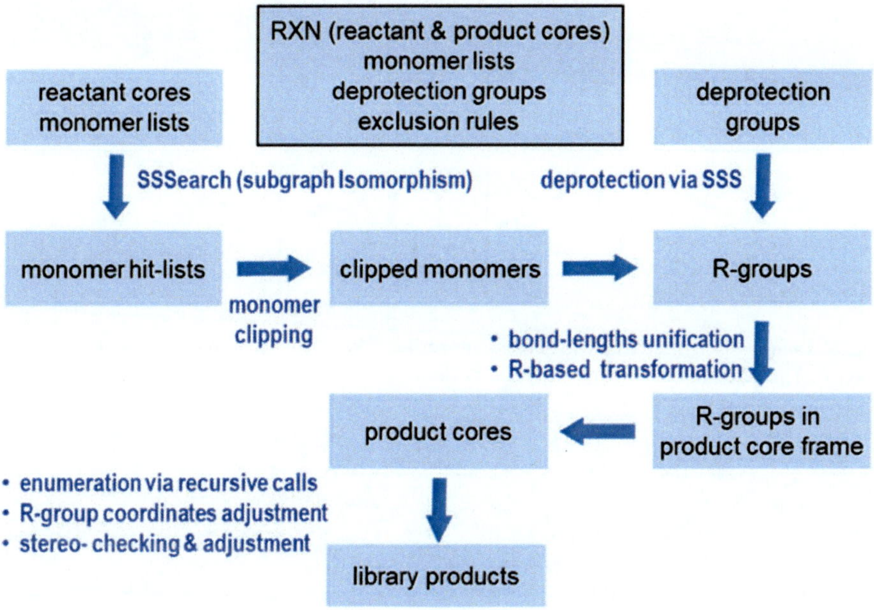

Fig. 4 Library enumeration workflow

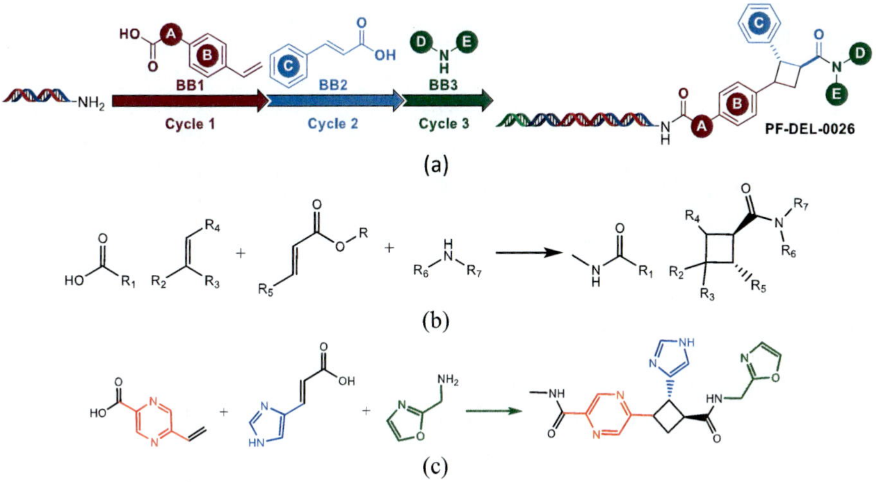

Fig. 5 (**a**) Markush structure of PF-DEL-0026, a 3-cycle DEL [43] (**b**). The reaction scheme used for enumeration (**c**). An example product

Attaching to the DNA headpiece is essential for the first cycle of chemistry. It is worthwhile pointing out that using disconnected cores to handle DEL reaction schemes prevents ambiguity for attachment points, as depicted in Fig. 6.

Often, a reactant-core is loosely defined in an RXN to cover a structurally diverse set of monomers, and a particular functional group can be included in broader

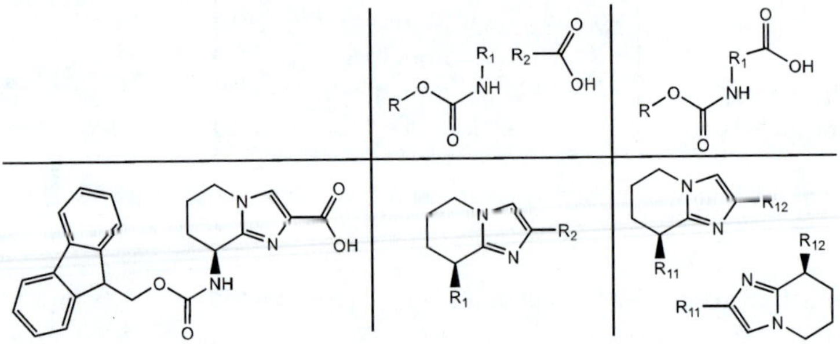

Fig. 6 Using a disconnected core to define a "linker-like" fragment versus multi-connected R-label "–R_1–". The latter introduces ambiguity for attachment points

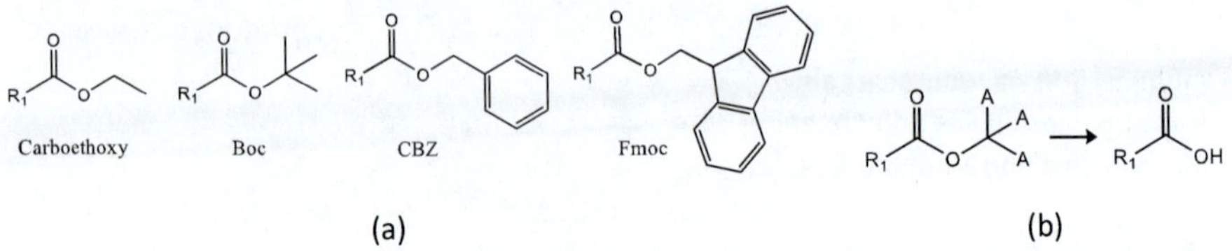

Fig. 7 (**a**) commonly used protection groups for amines; (**b**) a general saponification transformation to remove the protection groups for carbonic acids

Table 1 Enumeration performance of two DELs

DEL	# Compounds (million)	Single process (minute)	# Process	Multi-process (minute)
PF-DEL-0001	47.2	159	45	9
PF-DEL-0026	52.9	141	49	5

functional group definitions. For example, if amine is the intended chemotype in a monomer set, all amides, sulfonamides, or amidines must be filtered out. These exclusion rules are necessary to optimize enumeration performance.

Protection groups in monomers can be replaced either in the RXN as shown in Fig. 5b with -R to represent the Fmoc as a leaving group or through a monomer-based approach with the following steps: protection groups are tagged; reactant-core hits are identified; the overlaps are examined; protection groups are replaced; and monomers are clipped. The latter option performs general transformation from a protection group in Fig. 7a to hydrogen or last step saponification transformation as depicted in Fig. 7b.

Using a recursive function in assembling products enables support of an unlimited number of reaction components and large library enumerations. With an Intel Xeon Gold 6148@2.40GHz processor, enumerating one million DEL products takes an average time of ~3 min. Table 1 summarizes the enumeration performance for two DELs and shows that for a large library, the enumeration can be completed in minutes through parallelization of the enumeration processes.

In summary, the use of these methods makes enumeration of billions of products practical, which in turn facilitates further DEL design, structural similarity searching, data analysis, and triage.

2.3 *Designing DELs of Diverse Products in a Focused Property Space*

Diverse DEL libraries can be designed by focusing on individual monomer sets or by considering the enumerated products of the full-matrix combination of all monomer sets. Benefits and problems come with either choice. Monomer-based selections are computationally efficient but suffer from poor representation of the final products due to rearrangement of functional groups in chemical reactions. Another approach is to fully enumerate target product properties. However, when working in the DEL chemistry space, where the full enumeration of a virtual library in the potential design space can easily reach into the tens of trillions of products, full product-based property calculations for design are not practical. For the reaction scheme of PF-DEL-0001, the triazine library as shown in Fig. 8, after an initial structural filtering, we had 2,864 Fmoc-amino acids for cycle-1, 2,691 amines for cycle-2, and 2,075 amines for cycle-3. Combining all of these would result in ~16 billion library products to consider when selecting monomers. To overcome hurdles from monomer-based and product-based designs, we developed techniques that allow for the selection of monomers for a DEL library that result in products that are not only chemically diverse but also fall within attractive medicinal chemistry space [28].

Through a series of rational and systematic sub-library enumerations and property calculations (1) a set of monomer representatives are identified, and then (2) representative sub-library are defined and used to determine product property violations for associated monomers against six different property attributes. Minimization of product property violatioins is then used to drive the monomer selection for the full library. This technique helps minimize the computational expense while allowing the design to be done in product space instead of monomer space. The compounds in the resultant library can be biased to fall within a lead-like or drug-like property space that is attractive for medicinal chemisty follow-up.

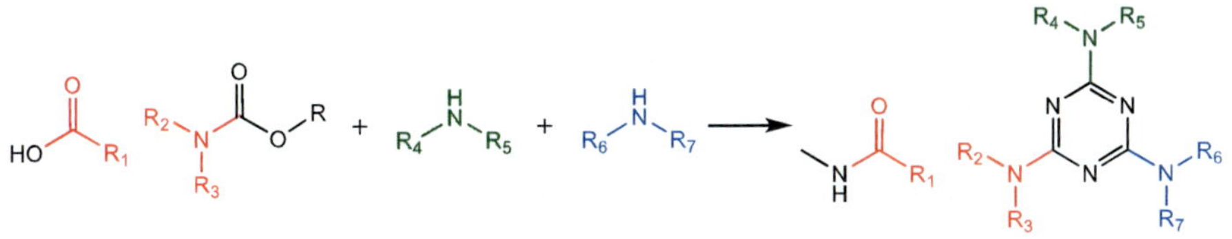

Fig. 8 RXN for PF-DEL-0001, the triazine library by Clark et al. [2]

Identifying representative monomers: For a 3-cycle DEL, individual monomers in a set are a_i, $i = 1, 2, \ldots, L$; b_j, $j = 1, 2, \ldots, M$; c_k, $k = 1, 2, \ldots, N$. The full potential design space for the DEL library is every potential combination of $a_ib_jc_k$. To define a subset of compounds which can be used to embody the full library a set of representative monomers are selected (a_{rep}, b_{rep}, c_{rep}), one from each of the monomer sets. The representative monomers are chosen to best fit the median of the property profiles for the monomers in that set. The properties considered here include molecular weight (MW), the number of hydrogen bond donors (nDonors), the count of nitrogens and oxygens (NOcount), the number of rotatable bonds (nROT), the polar surface area (PSA), and lipophilicity, the calculated partition coefficient between water and octanol (cLogP). cLogP was calculated using bioByte v4.3 [44] while the remaining properties were generated using in-house tools built on OEChem [42]. For each monomer in the DEL sub-library a score S is calculated using Eq. 1.

$$\text{S} = \sum_{i=1}^{n} \sum_{m} \frac{|P_i^m - \overline{P_i}|}{(P_i^{\max} - P_i^{\min})} \tag{1}$$

Here i is a loop over the number of properties being optimized, and m is a list of all basis products with that specific monomer. The basis products are defined as the products formed by combining all the reactants for a given reaction component with the simplest set of complementary reactant partners [45, 46]. For example, for the simplest set of ($a_{\min}$, $b_{\min}$) this would be a summation over all c_k in $a_{\min}b_{\min}c_k$. P_i^m is the property value with $\overline{P_i}$ being the average value for that property in the set of compounds m. This functional form for S allows each of the properties to be normalized to its corresponding range. The monomer with the minimum value of S is used as the representative for that monomer set.

Enumerating a representative sub-library: With the identified monomers (a_{rep}, b_{rep}, c_{rep}), compounds in the three sets ($a_{\text{rep}}b_jc_k$, $a_ib_{\text{rep}}c_k$, and $a_ib_jc_{\text{rep}}$) are defined as the representative sub-libraries, which are then enumerated, and the properties are calculated. The computational savings of only enumerating the sub-libraries instead of the full library is dependent on the size of the monomer lists being considered. In the case of PF-DEL-0001, the triazine library in Fig. 8, the reduction is ~831-fold given its dimension of 2,864 × 2,691 × 2,075. As shown in Fig. 9, in a cube representation of a three-component DEL library, the representative subsets can be visualized as planes in the cube.

Ranking monomers based on product property violations: Design criteria are defined according to desired property ranges. A property out of the range will contribute a violation count. Figure 10 shows the heatmap view of property violation counts for each product in the representative sub-libraries. The coloring in this figure highlights the number of property violations for each compound using cutoffs of: (1) ≤5 hydrogen bond donors; (2) ≤12 hydrogen bond acceptors; (3) 400–625 molecular weight (violations counted for either too low or high); (4) ≤6 cLogP; (5) ≤12 rotatable bonds; and (6) a polar surface area of ≤140Å^2. The unsorted

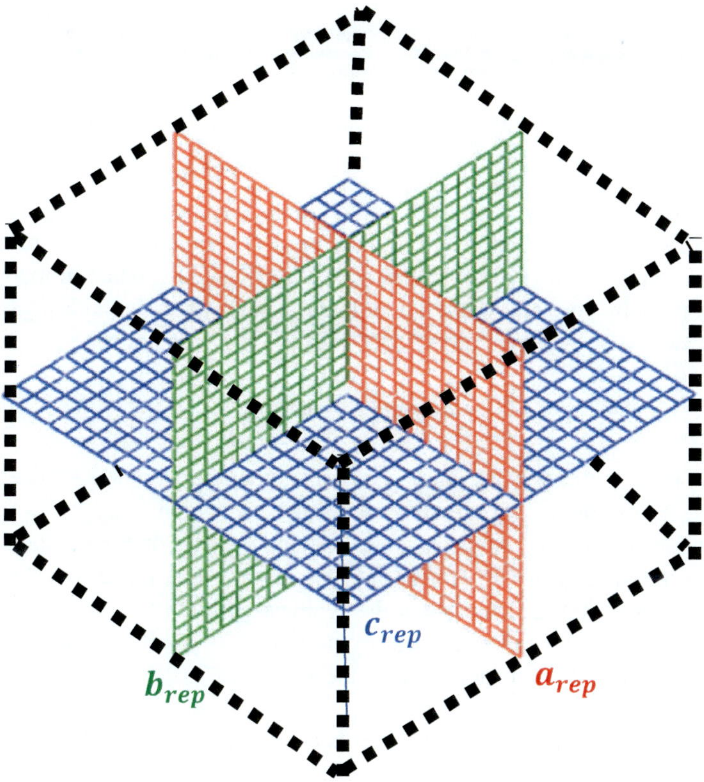

Fig. 9 Cube representation of a DEL where each coordinate (a_i, b_j, c_k) represents a DEL product. Representative sub-libraries form a subset of the full array library as shown by the three colored planes. The $(a_{\text{rep}}, b_{\text{rep}}, c_{\text{rep}})$ are representative monomers identified by minimizing S. The grids in three colored planes represent compounds in the DEL subsets $(a_{\text{rep}}, b_j, c_k)$, $(a_i, b_{\text{rep}}, c_k)$, and $(a_i, b_j, c_{\text{rep}})$. Adapted with permission from [28]. *Copyright © 2019, American Chemical Society*

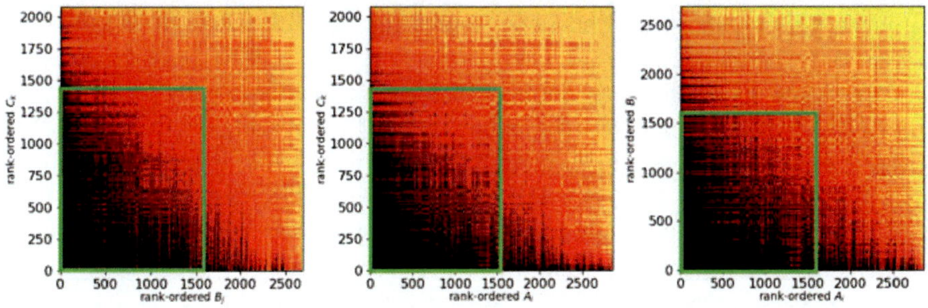

Fig. 10 Heatmap view of property violation counts for DEL products in the representative sub-libraries of PF-DEL-0001, where black indicates minimal property violations and yellow a greater number

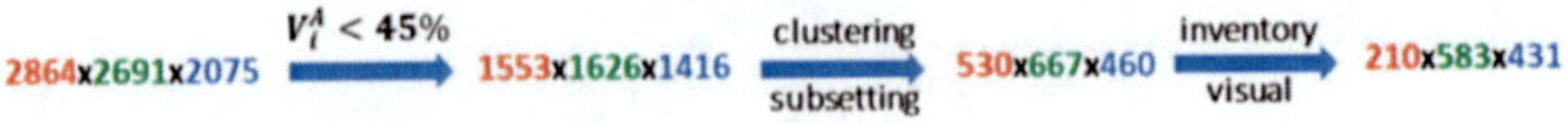

Fig. 11 Monomer selection for PF-DEL-0001: from virtual library to designed library

heatmaps are colored by monomer contribution to product property attributes, wherein black indicates minimal products property violation of 0, and yellow suggests a greater number of violations. The percentage of product property violations is defined by normalized overall violation counts, as shown in Eq. 2.

$$V_i^A = \sum_f \left(\frac{100}{n_{\text{prop}} \times n_f} \sum_j v_{ij}^f \right) \tag{2}$$

where f is the building block index, n_f is the number of monomers for the building block, n_{prop} is the number of properties used in the calculation, and j is a loop over all the compounds from the other specific monomers that form the representative sub-library set. The scores V_i^A for each monomer are then ranked with the lower scoring monomers selected for use in the synthesized library. Figure 10 shows the heatmap with rank-ordered monomers by normalized overall violation counts V_i^A monomers selected within green boxes are predicted to form products in better property space.

Removing highly similar monomers to increase DEL diversity: When selecting a property focused set of monomers, it remains desirable to choose as structurally diverse compounds as possible. To do this monomer redundancy is eliminated based on the 2D similarity of each monomer's nearest neighbors. The nearest neighbors are identified based on pairwise fingerprint similarity of >0.85 [47, 48]. Since large protection groups can lower the resolution of fingerprint similarity (for example, Fmoc), these groups are removed computationally from all monomers prior to the 2D similarity calculations. To help illustrate the number of monomers removed in different filtering steps, the monomer counts for PF-DEL-0001, from virtual to designed, is shown in Fig. 11.

The scores V_i^A, defined as normalized overall violation counts in Eq. 2, are directly used to rank each set of monomers. For the sub-libraries in Fig. 10, the lower left portion of the heatmaps represents the compounds with desired properties. Based on V_i^A, 1,553 × 1,626 × 1,416 monomers were selected from the monomer sets 2,864 × 2,691 × 2,075 followed with clustering, subsetting, and final inventory check and visual selection to yield a final set of 210 × 583 × 431 as the designed DEL. The mean properties for both virtual and designed DELs are summarized in Table 2. As indicated by values of μ and σ, all six properties are simultaneously improved from the virtual to the designed DELs, as illustrated earlier.

Through this dimensionality reduction strategy, the representative sub-library design approach provides a computationally efficient way to bias the design of DELs to fall within a specified property profile. In general, full enumeration and property calculations for products of a virtual DEL are computationally prohibitive

Table 2 Properties for full products

	Virtual		Designed	
Property	μ	σ	μ	σ
MW	645.0	66.6	560.2	44.8
HD	3.1	1.0	2.9	0.9
HA	13.2	1.7	12.6	1.1
cLogP	3.0	1.7	1.6	1.4
RB	12.5	2.2	10.3	1.9
PSA	148.7	21.7	146.4	17.3

let alone the computational cost of the algorithms necessary for design optimization. Simple methods are instead necessary to make the problem of monomer selection for a property focused library tractable [28]. Extremely large DELs can also create practical limitations on the selection. As the library size increases, the copy number of each individual molecule must decrease to maintain a workable molarity in the solubilized library. While focusing a DEL to fall within a specified property space can improve the lead-like nature of any resulting hits, it is also likely that this will remove significant chemical diversity from the library compounds. The chemical space being removed from the library needs to be understood to be one of the limited interests for medicinal chemist hit follow-up. Given the huge number of monomers that can be used even in a property focused DEL, some level of similarity or redundancy across monomers is helpful as it can create a level of internal validation or structure–activity relationships (SAR) within the DEL hits.

In summary, a simple algorithm is described here to select a property focused DEL from a larger design space, allowing the selection to be done at the enumerated compound level as compared to the monomer level. The algorithm greatly reduces the computational expense necessary for selecting a property focused library by only requiring the enumeration and property evaluation of a small subset of compounds from the design space. Smaller DELs in good property space allow for increased copy counts during selection, reduced signal to noise, and chemically attractive hits for medicinal chemistry follow-up.

3 Rapid Decoding of DEL Selection Data

After a DEL selection, high-throughput sequencing yields a large set of sequences in fastq format. While the encoding scheme of DEL DNA tags, which are constructed during library synthesis, varies among DELT practitioners, an example can be found in Chen et al. [11] Details in dealing with encoding strategies can be found in Chap. 6. Decoding refers to mapping DNA tags to a DEL and its corresponding BBs to identify a specific DEL compound. The accumulated number of appearances of a compound is reported as reads. This process is illustrated in Fig. 12.

In this section, an efficient decoding workflow is outlined that allows the processing of routine selections.

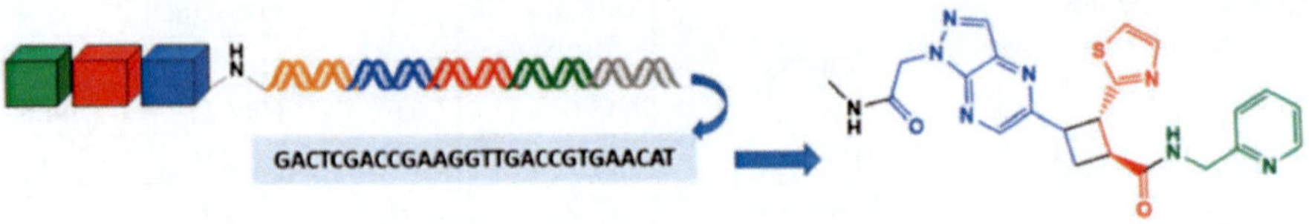

Fig. 12 Mapping library ID and codons in a DNA tag to its corresponding chemical structure

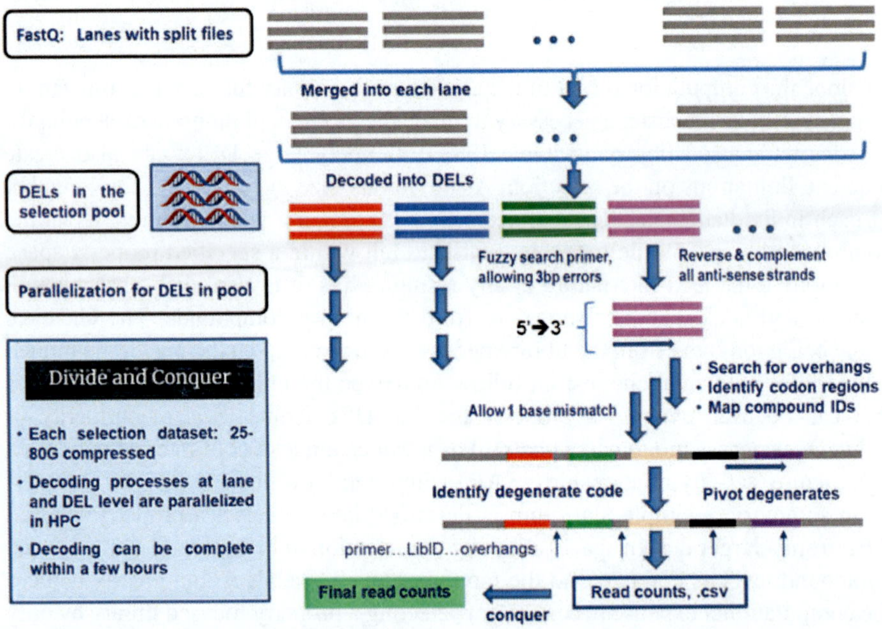

Fig. 13 A decoding workflow in which a divide-and-conquer approach is used for efficiency

3.1 The Decoding Workflow

The decoding workflow, detailed in Fig. 13, consists of a series of steps among which 4 key steps are addressed below.

Sequences assignment to specific DELs is done so that decoding can be parallelized on each of the pooled DELs. The DEL libraries were pooled together proportionally so that the dose can be determined with 1×10^6 molecules per compound in the selection. To reduce the misclassification rate, the 5' PCR primer is used as a reference to identify library ID with an error-tolerance. We chose a tolerance of ±3 bases in our sequence-DEL assignment workflow so that each of the valid sequences is assigned to a specific DEL in the pool.

Cleaning up sequences is a combination of (1) cutting off the 5' PCR primer; (2) converting to the reverse complement for sequences where the 5' PCR primer is not found; (3) cutting off the 5' PCR primer for the converted sequences; and (4) removing the 3' PCR primer. This data cleaning is conducted with Cutadapt, a tool for sequence trimming tasks that finds the adapter or primer sequences in an error-tolerant way [49]. Biopython is used for sequence I/O and anti-sense conversion tasks [50].

Mapping codons to BBs can be done using overhanging bases as references with an error-tolerance of ±2 bases for matching. The base mismatching is possible as the DNA codons for each monomer are designed such that each monomer code remains unique even with several sequencing mismatches. The sequence counts are aggregated, and the degenerate codes which were randomly generated as part of the tag on each compound in PCR amplification are pivoted.

Efficient tag decoding through parallelization: In general, multi-rounds of selection are carried out with consistent recovery of molecules across all samples for each round as determined by qPCR. Selection outputs can range from 10^8 to 10^9 total molecules per sample proceeding PCR amplification. For efficiency, the decoding process can be parallelized using a divide-and-conquer approach. Divide-and-conquer is used to separate both samples and DEL libraries levels into separate calculations which can be split across processors using SLURM, Simple Linux Utility for Resource Management, system [51]. With a parallelized workflow, decoding for a typical DEL selection can be completed in 2–4 h. For exploratory selections with high sequencing depths with ~3 billion reads, the decoding can be complete in ~38 h.

3.2 Data Noise and Uncertainty

Quantitative interpretation of DEL selection data remains an open challenge with at least two key questions that need to be resolved: (1) how many times a compound needs to be seen to separate it from noise (2) how can data across different selections be normalized? Given the nature of sequence sampling in DEL selection, selection replicates can be a useful tool to help understand data noise and uncertainty. While typical in RNAseq and proteomics experiments, replicates are typically not included in DEL selections due to the expense. In our recent work [33], we used replicate selections to estimate noise levels. The method allows the identification of an appropriate cutoff for the number of times a compound needs to be seen to be considered true signal and illustrates the effect of sequencing depth on separating signal and noise.

As an example data set to illustrate noise calculations, DEL selection results for TPL2 (Tumor progression locus 2), a serine/threonine kinase, [52] were used. DEL hits for this TPL2 had been successfully identified from these selections and confirmed as true binders via binding assays of the corresponding off-DNA compounds. DEL affinity selection was carried out with steps very similar to those described in Chen et al. using a pool of 10 three-cycle DELs [53]. All DELs were prepared using a process of design, logistics, and synthesis like that outlined by Kolmel et al. [43]. The libraries were pooled targeting 1×10^6 copies of each individual molecule. In total, the pool contained ~1.45 billion unique compounds.

As part of the overall experimental design, both the target concentration and the presence of a supplement were explored in different selection conditions. The team was interested in identifying both orthosteric and allosteric inhibitors, which could be accomplished through inclusion of a known active site inhibitor in some of the selection conditions. Several of the samples were run with two replicates providing data that can be used to estimate technical variability in DEL selections. TPL2 datasets were also generated using different sequencers, PhiX loading, demultiplexing scripts. Samples were sequenced at a 15% sequencing depth, the ratio of the count of sequences that were perfectly decodable as described below over the number of released molecules as determined by qPCR. After decoding, reads which were duplicates due to PCR amplification were excluded through a pivot on a degenerate code which was randomly generated as part of the tag on each compound.

DEL selections are run at or near a thermodynamic equilibrium. Based on binding thermodynamics and typical selection conditions using an excess of the target protein, the binding free energy can be written:

$$\Delta G_i = c \cdot \log r_i + d_i \tag{3}$$

where r_i is the sequence read counts seen for library compound i, c is the constant, and d_i is the compound-dependent constant. The detailed rational can be found in our recent work [33].

For a pair of replicate selections x and y, the binding affinity of DEL member i, $\Delta G_i^x = \Delta G_i^y$, is therefore

$$\log r_i^x = \log r_i^y + \theta_i \tag{4}$$

where θ_i denotes the compound-dependent constants. Equations 3 and 4 show that normalization between replicates can be achieved by a simple weighting factor or a shift to the log scale $\log r_i$.

In a DEL selection it is possible for a compound to bind to the support matrix (bead) as well as the protein being investigated. Due to this possibility the effective binding affinity $\Delta G_{i,\text{eff}}^x$ for DEL member i is defined as Eq. 5. Based on Eq. 4, read counts r_i^x must be adjusted for r_i^{blank},the corresponding read counts observed in the blank sample, in which the target protein was not present.

$$\Delta G_{i,\text{eff}}^x = \left(\Delta G_i^x - \Delta G_i^{\text{blank}}\right) \propto \log\left(\frac{r_i^x}{r_i^{\text{blank}}}\right) \tag{5}$$

The correction factor $(r_i^x / r_i^{\text{blank}})$ has been used to adjust for bead binding in our DEL selection data triage process for hit identification, as detailed in Sect. 4.2 – Identifying bead binders and read count adjustment using blank controls.

One accepted approach to normalization across samples is to use the total number of observed compounds as a simple ratio [29, 32, 34, 54]. However, this method can

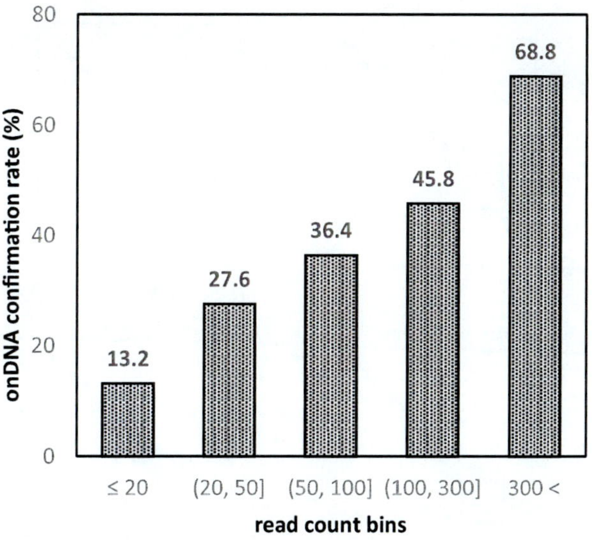

Fig. 14 Average on-DNA confirmation rates by read counts bins. Binding confirmation was conducted by on-DNA resynthesis followed by BALI-MS. In total, 564 on-DNA compounds for 20 targets are included

yield poor correlation for compounds with high signal as the ratio can be dominated by the many compounds which are only seen a few times (noise). While using the total number of compounds aligns well with the intended outcome of the DEL selection experiment where the inputs are carefully controlled by quantitative polymerase chain reaction (qPCR) [16, 55], it does not account for the technical challenges in running the experiment and inherent sources of noise. A simple but effective method is to use the compound with maximum read counts for the selection as the scaling factor ($r^x_{\max}/r^y_{\max}$) for normalization [33]. To eliminate the potential error introduced by using only one compound, a set of compounds with the top counts was instead used:

$$r_i^x = \left(\frac{1}{N_h}\sum\nolimits_h \frac{r_h^x}{r_h^y}\right) \cdot r_i^y, h \in H \tag{6}$$

where H denotes a subset of compounds with the highest counts and N_h is the number of compounds in the set. The value of N_h was empirically set by choosing the top 50 hits for normalization in our study. This approach is independent of the size of the DEL library in contrast to other published works [32]. Note that normalization using the compounds with the highest read counts was motivated by the observation of on-DNA confirmation rates. Figure 14 illustrates the average on-DNA confirmation rate for identifying target-specific binders by bead-assisted ligand isolation mass spectrometry (BALI-MS) [13], a technique used to identify true binders found within mixtures of on-DNA synthesis. A total of ~300 on-DNA compounds for 16 targets are included in the statistics shown. The confirmation rate increases from 11.8% for compounds with less than 20 read counts to 78% for compounds with greater than 300 read counts. This illustrates that compounds with

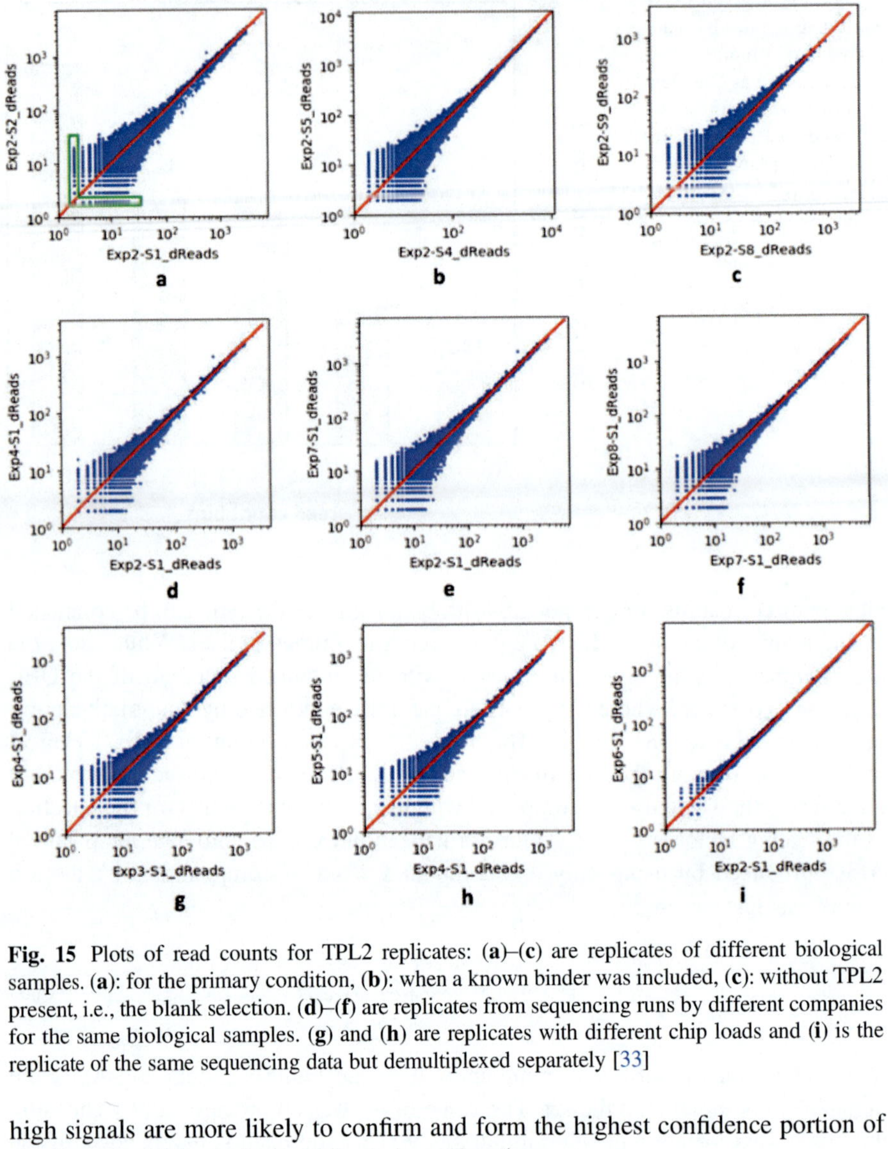

Fig. 15 Plots of read counts for TPL2 replicates: (**a**)–(**c**) are replicates of different biological samples. (**a**): for the primary condition, (**b**): when a known binder was included, (**c**): without TPL2 present, i.e., the blank selection. (**d**)–(**f**) are replicates from sequencing runs by different companies for the same biological samples. (**g**) and (**h**) are replicates with different chip loads and (**i**) is the replicate of the same sequencing data but demultiplexed separately [33]

high signals are more likely to confirm and form the highest confidence portion of the dataset which can be used for noise estimation.

It is possible to visually evaluate normalization methods across replicates by looking at a correlation plot of the adjusted values. A well-functioning method will show: (1) symmetrical patterns along the unity line; (2) must be applicable to all replicate pairs with a high degree of consistency.

Figure 15 shows comparisons of read counts for TPL2 replicates between different samples, which demonstrates consistency for the normalization by Eq. 6. Figure 15a–c are replicates of different biological samples. Data dispersions between replicates are similar among sample conditions, primary samples, samples with a known binder, and blank samples. Plots comparing read counts for replicates from sequencing for the same sample done by different companies are shown in Fig. 15d–f. The dispersions arise from finite sampling at given sequencing depths. This set of comparisons demonstrates consistent outcomes of reproducibility, as emphasized previously, from both the technology maturity and high quality of operations at those companies. Figure 15g, h are replicates of primary condition with different chip loading, contrasting a 100% load with 20% PhiX (Exp3-S1), 80% load and 1% PhiX (Exp4-S1), and 100% load and 1% PhiX (Exp5-S1), respectively. Figure 15i shows replicates of the same sequencing data but demultiplexed separately. Though the dispersion with maximum deviation of 0.05 is much smaller than other condition changes discussed above with maximum deviations greater than 0.15, the difference exists due to software versions and variable parameters. This dispersion is negligible to our downstream data triage, but it demonstrates that any changes along the selection workflow can contribute to the overall dispersion and bring uncertainty to the read counts. The dispersion is quantified by the deviation of variable to expected read counts, defined as

$$\sigma_i^L = \sqrt{\sum_L \left(\log r_i^L - \log r_i^{\exp} \right)^2 / N_L} \tag{7}$$

where r_i^L is the variable read counts and $r_i^{\exp}$ the expected read counts with $r_i^L > r_i^{\exp}$. The set of data points covers all variable read counts in one sample for a given read counts in another; therefore, all data points fall in the L-shape boxed in green color in Fig. 6a for a given level of sampling.

Estimating noise in DEL selections. As a logarithm transformation of counts is used in Eq. 4, one estimation of error for the log function, $f = \log r$, can be derived from standard deviation $\sigma_f = \frac{0.434}{r} \cdot \sigma_r$ as

$$\delta f = \frac{0.434}{r} \cdot \delta r \tag{8}$$

All read counts would fall on the unity line if there was no error variation for a pair of technical replicates that were completely sampled. Thus, the dispersion, a measure of error variation from the unity line between a replicate pair is defined as

$$\log r_y = \log r_x \pm \left(\frac{0.434}{r_x} \cdot \delta r_x \right) \tag{9}$$

where δr_x denotes the possible intrinsic error of counts in a DEL selection.

In Fig. 16, the normalized read counts can be divided into a set with signal **S** and random noise **R**. Set **S** contains read counts that fall in the area between the two green lines which illustrate the deviation range defined by Eq. 9 with $\delta r_x = 1$. The boundaries of the magenta box are empirically defined as 85% agreement between the read counts in the two dimensions. Set **R** falls in the uncertain area as defined by the magenta box. This empirically derived square-shaped dispersion is a consequence of random events during the DEL selection and sequencing process. This

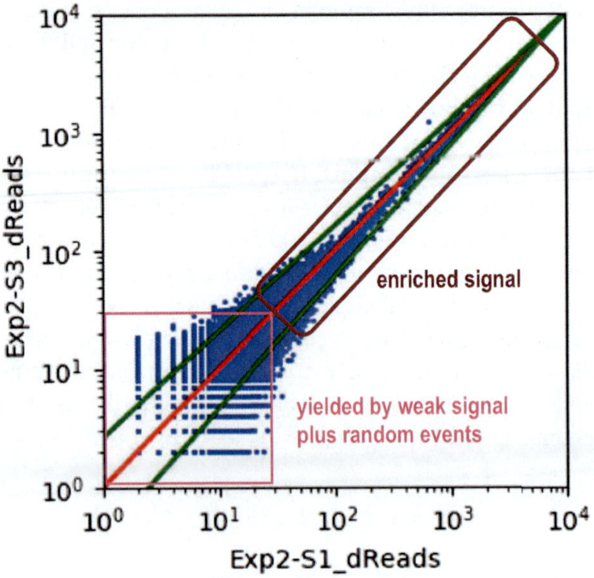

Fig. 16 Normalized read counts for a replicate pair can be classified into a set with enriched signal (**S**) in comparing to random noise and those that may be due to random noise (**R**) along with weak signal are outlined here in orange and magenta boxes, respectively. The green lines are deviation ranges at each of read counts as defined in Eq. 9. Adapted with permission from [33]. *Copyright © 2022, American Chemical Society*

concept will be reinforced in the next section using "naïve" sequencing results from DEL libraries that have not been selected against a target. The schematic orange box should contain compounds with enriched signal. Hits from the union set may still be identified through disynthon features [32, 54] where structure–activity relationships (SAR) are present. The low hit confirmation rate of 11.8% for $r_i < 20$ is a consequence of the signal and noise mixture in the low counts range.

Figure 17a shows the distribution of read counts across compounds for a "naïve sequencing" of PF-DEL-0020, a two-cycle library with 1,009,510 DNA tags from a combination of 314 × 3,215 building blocks. Naïve sequencing refers to the results obtained when a DEL library is sequenced before being selected against a target. The sequencing resulted in 6.88 million reads. With an average of 6+ reads assigned to each compound, it provides a sufficient sampling rate to estimate the product distribution of PF-DEL-0020. The finite random sampling yields a Poisson distribution as fitted by a two-parameter Poisson function with $\lambda = 7.18$ and a scaling factor $s = 0.98$. A plot comparing the number of counts for each compound for replicates of PF-DEL-0020 naïve sequencing is shown as Fig. 17b, where compounds with different read counts are randomly dispersed with the maximum read counts for any specific compound being 25.

To further understand the data pattern of replicate plots and the impact of sequencing depth, PF-DEL-0020 naïve samples and TPL2 selections were evaluated

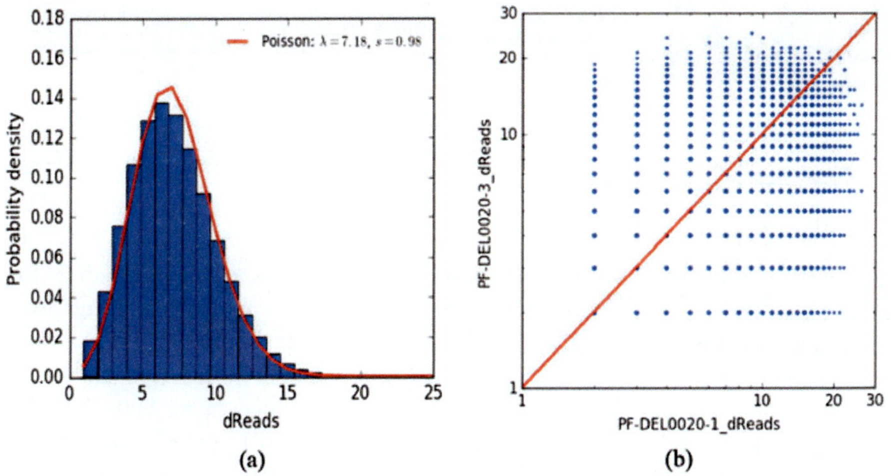

Fig. 17 (**a**) Read counts distribution of naïve sequencing of PF-DEL-0020. Red line is the fitting of two-parameter Poisson distribution; (**b**) read counts comparative plot for replicates of PF-DEL-0020 naïve sequencing. Adapted with permission from [33]. *Copyright © 2022, American Chemical Society*

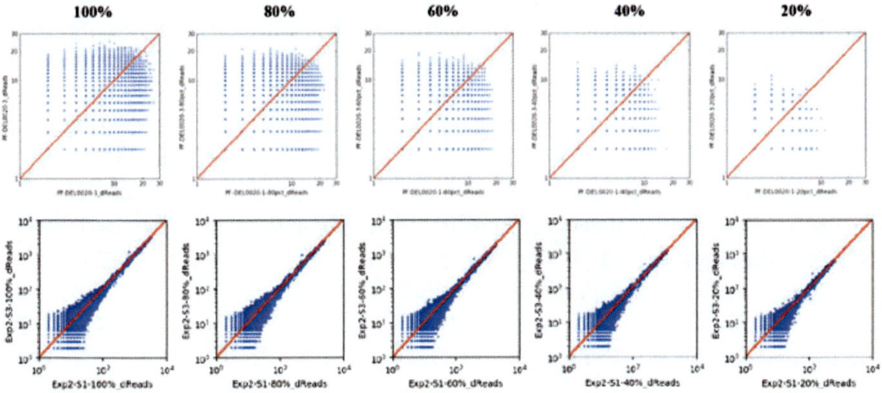

Fig. 18 Comparative read count plots for replicates: top, PF-DEL-0020 naïve sequencing; bottom: TPL2 primary replicate samples. From left to right are in order of all sequences (100%), 80%, 60%, 40%, and 20% which are randomly sampled sequences followed by our decoding process. The corresponding maximum counts of TPL2 $\left(r_{max}^{S1}, r_{max}^{S3}\right)$ are (6,673; 6,806), (5,050; 5,242), (3,768; 3,932), (2,526; 2,658), and (1,241; 1,356). Adapted with permission from [33]. *Copyright © 2022, American Chemical Society*

using 80%, 60%, 40%, and 20% of all available sequences. Comparative plots of replicates are shown in Fig. 18. For PF-DEL-0020 naïve samples, counts are dispersed in two dimensions with the maximum counts decreasing from 25 for 100% of the data to 11 when 20% of the data is used. In comparison the maximum counts for TPL2 replicate samples $\left(r_{max}^{S1}; r_{max}^{S3}\right)$ are decreased from (6,673; 6,806) for 100% of the data to (1,241; 1,356) for 20% sampling, an approximate 80%

Table 3 Maximum read counts at each of the simulated sampling ratio

Sample ratio (%)	$r_{\text{max}}^{\text{naive}}$	r_{noise}^{S1}	r_{max}^{S1}
100	25	29	6,671
80	21	26	5,050
60	19	24	3,768
40	15	21	2,526
20	11	20	1,241

reduction. The dispersed boundary decreases from 29 read count for 100% to 20 for 20%. Thus sequencing depth impacts more significantly compounds with high read counts than the noise threshold, which suggests that higher sequence depth can bring higher confidence to the enriched signal. The maximum counts at each of the simulated ratios are summarized in Table 3. The well-aligned values between r_{max}^{naive} and r_{noise}^{S1} and the relative ratios within r_{max}^{naive} and r_{noise}^{S1}, respectively, support the rationale of read counts decomposition into noise and enriched counts as shown in Fig. 16.

In summary, after transforming read counts to a log scale, a normalization scheme for replicate samples was developed that leads to a deep understanding of enriched signal and random noise in the read counts data. The method has become a component in our DEL selection workflow for the estimation of noise levels. Small errors can accumulate throughout the complex DEL selection, sequencing, and analysis protocol. Each step has a combination of systematic and random errors that eventually lead to the dispersed data pattern observed in replicate plots. A stepwise analysis showed that read counts dispersion across replicates originates from different experimental or data sampling sources with the order of influence being: "different biological samples" > "sequencing conditions of chip load and amount of PhiX" > "sequencing by different service providers for the same samples" > "different demultiplexing for the same sequencing data." Furthermore, based on observed data dispersions of replicate pairs and sequencing depth simulations, a method to determine noise cutoff was proposed. While higher sequencing depth brings in higher confidence to enriched signals, it also adds much more background noise, which is typically eliminated by using a cutoff for the compounds with the fewest read counts in downstream analysis and triage. This implies that maximizing signal to noise ratio by optimizing selection conditions and steps is more critical than increasing the sequencing depth. It should be emphasized that since the uncertainty of enriched signals is read counts-dependent and the randomness level of noise depends on sequencing depth, including replicates in a selection experiment, provides a means for selection QC and noise level estimation. The normalization method and dispersion analysis outlined here is independent of counts cutoffs. Therefore, it can filter out most compounds with low read counts in the decoded results simplifying the analysis without losing information from the selection.

4 DEL Selection Data Triage and Analysis

Although any DEL screening hit can influence a project's direction, teams are most interested in tractable lead matter that can be optimized into a drug that will positively impact human health. Screen reads, a DEL screens primary signal, are only part of puzzle, and other attributes, such as: (1). attractiveness (favorable physicochemical properties and minimal liabilities), (2). evidence of structure–activity relationships, and (3). coverage of a desired biological profile, must be considered when selecting hits to progress. Balancing all these factors requires a flexible analysis tool that can display this complex data in multiple ways as well as a data scientist knowledgeable about biology and medicinal chemistry. This section will describe the tools we have developed and the approach we take in selecting the best hits to progress from DEL screens.

DEL cubic plots: A popular method to analyze DEL primary screening results from 3-cycle libraries is with a 3-D scatterplot commonly referred to as "the cube" or a "DEL cubic plot" [2]. The plot's axes are the building blocks from each library split sorted by 2D structural similarity and markers in the plot represent unique targeted DEL members, assuming all wet chemistry proceeded as planned (unlikely). The markers are sized by the read counts present upon sequencing, so larger markers represent hits with stronger signal. The example in Fig. 19 represents PF-DEL-0001, a triazine library, with three dimensions of diversity. If we select a single marker (blue circle), we can decode the targeted structure (BB-a, red R1 axis; BB-b,

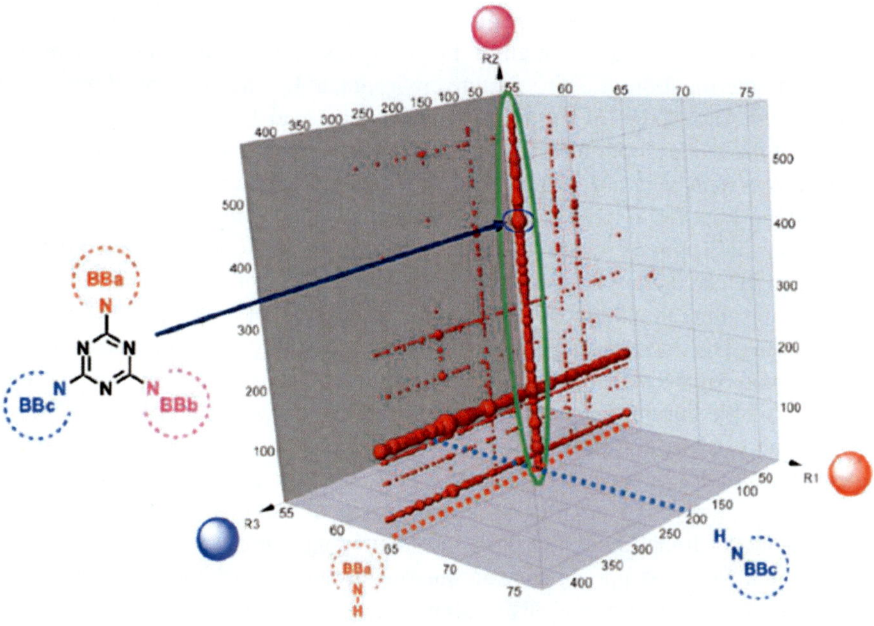

Fig. 19 DEL cubic plot example for PF-DEL-0001

magenta R2 axis; and BB-c, blue R3 axis). Often, lines and planes are observed in this visualization which arise, again assuming only targeted compounds are present, from molecules that have one (corresponding to a plane) or two (corresponding to a line) fixed building blocks. For every analog in the line feature circled in green, R3 is fixed to BB-c and R1 is fixed to BB-a. For our analyses, we characterize the line features for normalized strength, by adding up the total reads of the hits within the line and dividing by the number of monomers present in the corresponding library cycle, and line feature diversity, by calculating the standard deviation of the reads distribution. Line features can be explained in several ways. It is reasonable to assume that closely related molecules will have similar activities for a given target, so a line feature can arise via structure–activity relationships (SAR) related to the variable building block. In most cases, if the building block is making productive target interactions, these line features will be highly diverse. In contrast, if a particular building block is not making productive interactions with the target and is instead aligned outside the binding site, the line feature may look more uniform as shown in Fig. 19, and extend across the entire cube, and these types of features suggest that the variable building block could be truncated to produce hits in better property space (with lower MW) that maintain binding. In addition, line features can arise from failed chemistry during the DEL synthesis [30]. For example, the green circled line feature below could be explained by failed cycle-2 chemistry where R2 may not be present. In that case, the active hit could be the R2-truncated product and the same truncated product could be responsible for all the signal in that feature, and the variability in the signal can be explained by the extent of failed chemistry in each well, which could be influenced by factors like steric bulk of the individual monomers. Often, "null" conditions are included in a DEL synthesis, where specific building blocks are not added at all, to help diagnose issues with failed chemistry. In the end, interrogation of a hit's feature environment by an experienced medicinal chemist with knowledge of the DEL synthesis can lead to hypotheses about the identity of the binder, optimization potential, and even the binding mode of the hit.

DEL physicochemical MPO score: As previously described in Sect. 2.3, we have designed our libraries to have favorable average physicochemical (physchem) properties, but as DEL library synthesis always produces the full combinatorial matrix of possible products, there will always be cases where the "worst" building blocks are combined resulting in unattractive molecules. To ensure that properties are front and center during DEL screen triage, we calculate a set of physchem properties for all hits with reads over a screen-determined threshold as discussed in Sect. 3.2, and this data is available during analysis. To roll up multiple properties into a single score that can be used as a sorting tool or color scale, but at the same time keep the calculation simple, we have derived a DEL physchem multi-parameter optimization (MPO) score derived from four properties deemed important by our medicinal chemists: MW, lipophilicity (cLogP), aromatic ring count, and hydrogen bond donor count. MW and lipophilicity are unarguably the most important factors that control chemical absorption, distribution, metabolism, excretion, and toxicity (ADMET) profiles, and they are correlative with many other properties we chose not to include so as not to impact the contributions from MW and cLogP, but we did

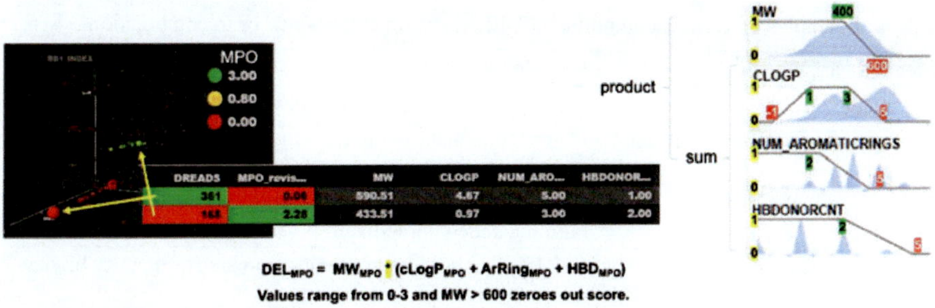

Fig. 20 The DEL physchem MPO score

bring in aromatic ring count due to its negative effect on solubility and hydrogen bond donor count which impacts permeability beyond MW and cLogP alone. Each property is converted to a score between 0 and 1 as described in Fig. 20, then the individual scores are combined by adding up the scores for cLogP, aromatic ring count, and hydrogen bond donor count, then multiplying by the MW score which serves to weigh this property the highest, as a MW score of zero will zero out the entire DEL MPO score. It is important to note that a good DEL MPO score does not mean a hit will be attractive, as many other factors go into that assessment, and the MPO score is rarely used as a filtering tool, but as a visual indicator that a particular feature may contain hits in better property space. In the cubic plot in Fig. 20, hits are sized by read count (DREADS) and colored by the MPO score, and while the green-highlighted line feature contains hits with lower read counts, more favorable physchem properties make the feature more attractive for progression.

4.1 DEL Selection Conditions and Bio-Profiles

One advantage of DEL screening over traditional high-throughput screening is that multiple selection conditions (typically up to 10) can be screened in parallel and differential response of a hit to these varied conditions can provide insights on expected potency, binding sites, likelihood of achieving functional activity, and other biological profiles. Table 4 describes some examples of conditions that can be varied in DEL screens either alone or in combination and knowledge gained by comparing hit signal across conditions.

In planning selection conditions, it is important to predict the biological profiles that might arise in any given screen and rank their relative priorities to the project team, as this information is required for a thorough and fit-for-purpose analysis of the resulting data. Table 5 provides an example of a bio-profile chart for a simplified scenario with three selection conditions involving a target protein alone, competition with a known active site binder, and inclusion of a large molecule protein target binder. The selection conditions are listed in columns while predicted bio-profiles

Table 4 Examples of various conditions in DEL selection either alone or in combination

Selection variable	Knowledge gained
No-target control	Identify support-matrix/bead binders
Competition with small-molecule binder	Identify hits that bind in the same binding site as the small-molecule binder
Competition with cofactors/substrates/protein partners	Identify hits that bind in the same binding site or hits that bind the added components
Different constructs	Can localize binding to specific domains or identify binders that only bind specific constructs
Different protein activity states	Determine if hit binds one preferentially
Different protein loading	Can help prioritize hits based on expected potency

Table 5 An example of a bio-profile chart for a simplified scenario

Selection conditions→ Bio-Profile↓	1: Target	2: w/ active site binder	3: w/ protein partner	Priority
Active site	Yes	No	Yes	1
Protein partner binder	No	No	Yes	3
Allosteric site	Yes	Yes	Yes	2

are in rows, and cell values contain the expectation for signal for a hit from a given bio-profile under the corresponding selection condition.

In a simple screen, like the one depicted in Table 5, a pivot table (one row for each library member) with blank-adjusted reads data as defined by Eq. 6 for each selection condition in columns can be used to plot results from one condition vs. another and select compounds for further analysis. We recommend plotting using a log*10* reads scale and assigning a value of "1" to all null cells so hits that are in one condition but not the other can be plotted and will appear separated from the rest of the data in a line near the appropriate axis. To find "active site" bio-profile hits, one could plot selection condition 2 (y-axis) vs. selection condition 1 (x-axis) and look for hits at high reads in S1 that do not appear in S2 (Fig. 21).

For more complex screens where complex bio-profiles are possible, we have developed a simple method to calculate and cluster bio-profiles automatically using a TIBCO Spotfire heatmap. First, the set of hits of interest is defined and the table is filtered to only show data associated with those hits. Often, we will choose a reads cutoff for the highest read condition for a particular hit, select only the maximum signal example from all line features, and remove frequent hitters as discussed in Sect. 4.3. Next, within each row, the reads for each selection condition are divided by the max reads observed for that row to normalize the data to [0.0, 1.0] scale (i.e., % of max reads for a hit for each selection condition). This normalization removes the absolute signal value and allows hits with similar response to the selection conditions to be clustered together.

After the bio-profiles are calculated, hierarchical clustering of the rows in TIBCO Spotfire groups compounds with similar profiles (Fig. 22, lower right). One can then look for groups of rows that represent a particular profile. In the example below, the

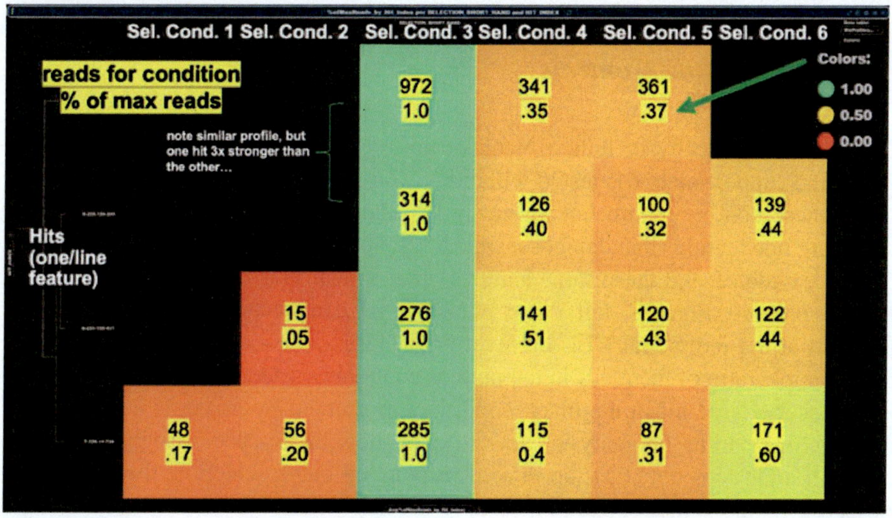

Fig. 21 An example of normalized bio-profile calculation

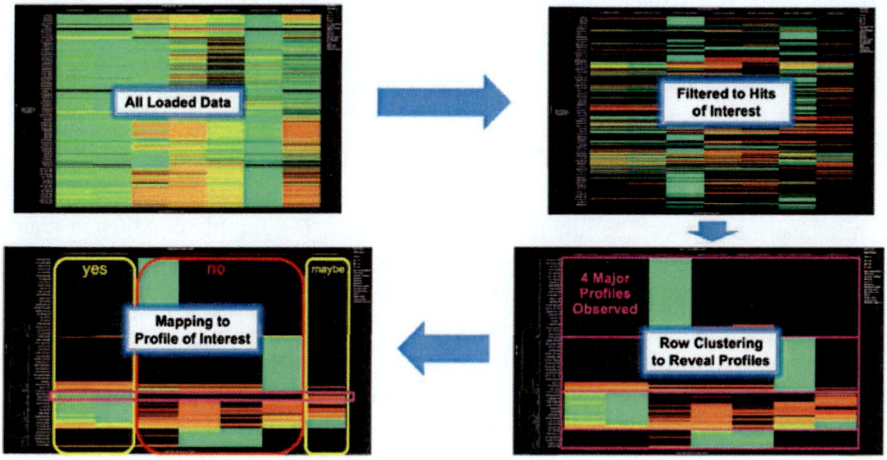

Fig. 22 Bio-profile clustering example

team was interested in a profile where high signal was present in S1 and S2 and no signal was present in S3–6. The rows within the magenta box represent this profile and the heatmap was utilized to select these compounds for further detailed analysis.

4.2 Identifying Bead Binders and Read Count Adjustment Using Blank Controls

DEL screens typically include a blank or no-target control selection condition in which the support matrix is tested without any target protein present. Hits that bind under these conditions are not progressed further. In practice, significant signal, including line features that depict sensible SAR, appear in the blank selection results (Fig. 23, middle), and these same features often appear in selections with the target protein present (Fig. 23, left). The blank signal can be removed using several methods at the individual DEL hit level or at the line/plane feature level. We find that a simple ratio of the reads in the protein-containing selection condition over the reads for the same hit in the blank condition, as defined by Eq. 6, is a convenient method for cleaning up analysis cubes to only show target-specific signal (Fig. 23, right). If a hit does not show up in a corresponding blank selection, then the reads observed in the target-containing sample are left unchanged (the case for most interesting hits). This "blank-adjusted reads" metric can then be used as a cutoff when data is loaded for analysis effectively removing all signal that is equal or stronger in the blank over the target-containing sample (blank-adjusted reads $<=1$) while leaving behind target-specific features.

4.3 Frequent Hitter Identification

While bead binders can be identified and removed using the blank-adjusted reads calculation, another form of problematic signal was identified after screening the same DEL libraries against unrelated targets and seeing the same hits. In these cases, features arose with convincing SAR across target selection conditions (i.e., allosteric bio-profile), the corresponding blank selection conditions had no signal, and these hits looked attractive enough to take forward into our on-DNA confirmation platform [13]. These hits were confirmed as target-specific binders via the BALI-MS

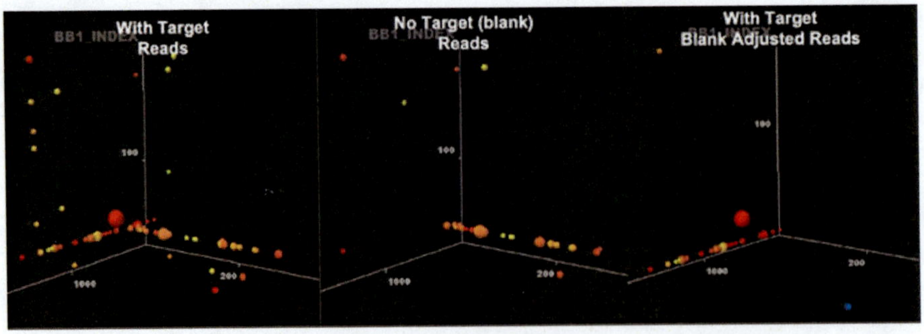

Fig. 23 Cubic plots showing line features. Left: selection condition with target. Middle: selection condition without target. Right: selection condition with target and signal adjusted for the blank

method, but upon off-DNA synthesis, no significant target binding was observed. Upon a closer look at the DEL members associated with these features across all targets that had been screened, it became clear that these hits were present at varying signal levels in most screens pursued. Although it is conceivable that a hit could specifically bind unrelated targets, the level of promiscuity observed, and the fact that the off-DNA synthesized hits did not confirm led us to develop a process for annotating these "frequent hitters," so they are not pursued by project teams in the future. A frequent hitter database was set up wherein hits that appear in any screen at a significant adjusted reads level which is screen dependent are tagged with the screen name. Upon analysis of any new project, a count of the number of times the library member has been a hit in other screens as well as a list of those screens is returned. Cubic and other analysis plots then use an alternate color for frequent hitters to alert the data scientist to drill down and investigate the frequent hitter data further when considering a hit for progression. We do not suggest simply removing the frequent hitters from the analysis for two reasons: 1). related targets may be screened where it is reasonable that the same structures could confirm which would be apparent upon drilling down to the target level; 2). often, frequent hitters appear as line features, and depending on the SAR within that feature, only the strongest hits will be tagged as frequent hitters. If that signal is simply turned off, the remaining line feature would look clean and might be progressed even though closely related compounds are known to be frequent hitters. This frequent hitter database has been very effective for highlighting and removing these problematic hits, but as the simple method is based on a target-specific reads cutoff, borderline cases could slip through the cracks, so we recommend carrying out a database search for all cross-project signal associated with any hit that is progressed to get a full picture of that hit's promiscuity.

4.4 DEL_DV Data Viewer

Analysis of our DEL screens occurs using a custom TIBCO Spotfire-based visualization tool we call DEL_DV (DEL DXP Viewer). The solution allows the user to load DEL primary screening data from any ≥ 1 screening campaign using an appropriate blank-adjusted reads cutoff as estimated by the approach described in Sect. 3.2. The tool presents data at various levels, including (1). unpivoted result table, (2). pivoted to unique library members, (3). pivoted to line features and allows for the bio-profile clustering and selection condition plotting, all linked to the main analysis page (Fig. 24) that allows for thorough interrogation of a single library at a time across all selection conditions investigated and in the context of physchem properties and frequent hitter tagging. Cubic plots appear in the upper right, trellised by selection condition with axes monomers sorted by 2D structural similarity, markers sized by blank-adjusted read count, and colored by stoplight colors for the DEL physchem MPO score and magenta/blue to highlight problematic hits from past screens. Within DEL_DV, line features around each hit in all three directions are

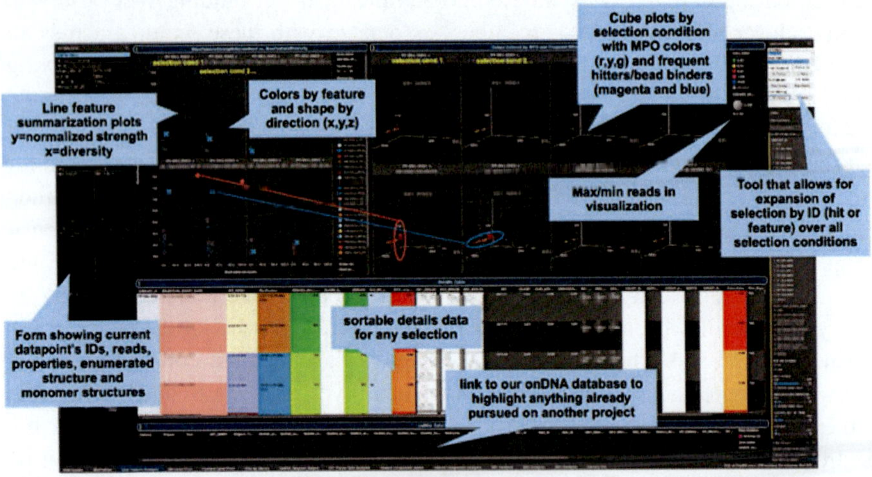

Fig. 24 DEL_DV main visualization and analysis page. Available display components and tools are described

annotated and plots of maximum line feature normalized strength (sum of blank-adjusted reads for the line/number of building blocks present in the library for that split) and diversity (standard deviation of the adjusted reads distribution in the feature) allow the user to track feature strength/diversity across selection conditions, and provide an easy way to select one or multiple line features, which are then highlighted in the cubic plot. Selection of any marker reveals detailed data in a sortable form on the bottom of the screen. A form view that responds to mouse pointer movement of properties, full structure, and building block structures appears on the left of the screen and standard Spotfire filters allow the user to easily adjust the displayed data by any criteria. Finally, a custom tool that allows the user to expand a selection to include all hits with the same product ID or within a feature (across selection conditions) makes tracking bio-profiles facile.

4.5 *Recommended Triage Process for DEL Selection*

A process diagram for triaging a DEL screen is depicted in Fig. 25. During the planning stages of the DEL selection, the data scientist should work with the team to understand the project goals, priorities, and any selection condition variables that will be investigated. Next, the expected bio-profile chart should be generated with clear prioritization, and if the planned selection conditions will not be able to distinguish these profiles, a discussion on changing/augmenting the selection conditions should take place. Once the selection is completed and the primary data is ready for analysis, a high-level analysis to determine the overall data quality should follow. This analysis could include determining the level of blank signal present in

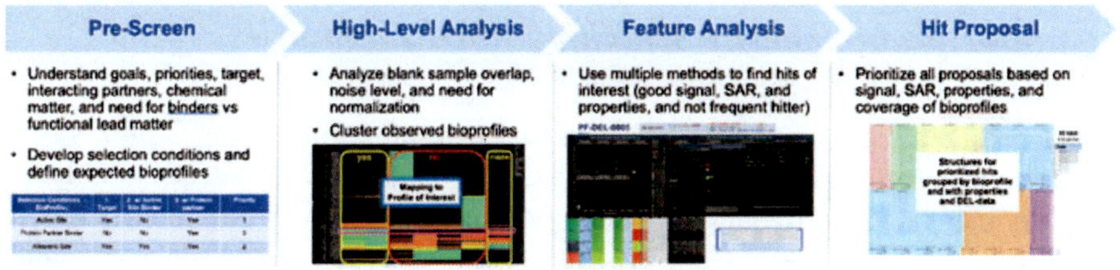

Fig. 25 The DEL selection triage process implemented at Pfizer

the target-containing selection conditions (less is better), evaluating the approximate noise level for the selection, and evaluating the need for normalization across selection conditions. If the data is satisfactory to this point, the blank-adjusted data can be loaded, and bio-profile clustering can provide a high-level view of the biology observed in the selection and be compared to expectations. At this point, an overall hit-property vs. signal analysis is useful to help the analyst understand if hits with adequate signal and suitable properties will be present. Next, multiple entry points to investigate hits of interest, for example, from the bio-profile clusters, from pivot plots, from individual library inspection, and from property filters, should be utilized and a hit's signal, SAR, bio-profile and properties should be thoroughly investigated by a knowledgeable medicinal chemist, always considering the possibility that the observed SAR could result from the originally targeted DEL compounds or from by-products that arise during synthesis. Hit proposals should all be prioritized to provide the project team with the most tractable hits within each bio-profile. The advantage of having a flexible analysis tool and knowledge of the biology, chemistry, and DEL technology is that one can adjust the analysis to a team's needs, unexpected findings, and new information that arises during the triage process. We recommend using all available data for the best outcome.

5 Outlook and Perspectives

While advances in applying cheminformatics and computational approaches to DEL technology are improving applicability in general and success rates in hit discovery, there are still key areas of science being developed. Here we briefly outline perspectives of three topics in the informatics and data analysis field that could further strength the technology to either facilitate hit identification process or improve the hit confirmation rate in DEL technologies. Like the analysis of all DEL results the success of these areas will rely on the active partnership of the experimental scientists running selections along with the computational analysis team, and the Information Technology (IT) support available. The scale of DEL screening results and libraries can easily exceed the capacity of computational systems and databases not designed for the task.

5.1 Structural Similarity Searching in DEL Libraries

An enabled DEL technology platform can quickly result in libraries with billions or even trillions of product compounds. In the data analysis of DEL selections in lead discovery campaigns, a frequently asked question is: are there identical or similar products in any of the screened DELs, and if so what are their product IDs and structures. This question typically arises when (1) hits are identified from a third party's DELs; (2) there are known binders of the target being screened; or (3) hits have been confirmed and additional analogs are being sought. By using a divide-and-conquer approach, we have enabled rapid 2D similarity searching for DELs with billions of products. However, searching larger DEL collections can quickly become impractical using conventional technologies. To address this unmet need from the medicinal design community, we explored a novel hierarchical search approach in which basis products were fully examined and only a small subset of the full array products of large-scale DELs are probed [56]. As illustrated in Fig. 26, by using a fragment-based similarity search against DEL basis products, small subsets of the most structurally similar monomers are identified using a Tversky similarity metric. Then a full enumeration from these small sets of monomers is searched using a Tanimoto similarity of the full set of structural descriptors. With this hierarchical approach, the vast majority of the structural dissimilar compounds can be quickly eliminated vastly speeding the search. This technique balances accuracy and efficiency for searching the unenumerated DEL product space. While the method is under being further investigated, we expected other effective searching algorithms to emerge.

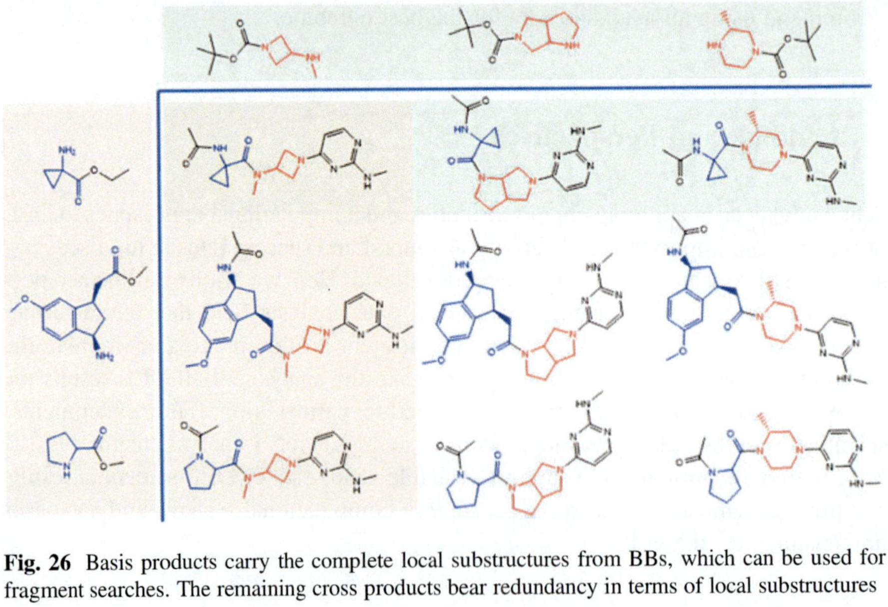

Fig. 26 Basis products carry the complete local substructures from BBs, which can be used for fragment searches. The remaining cross products bear redundancy in terms of local substructures

5.2 Visualizing DEL Chemistry Space

Higher structural diversity in DEL library members is thought to have higher probability of yielding hits with novel structures from DEL selections. The chemical space coverage by DELs has been reported by analyzing reaction schemes and their corresponding BBs [26, 40, 57]. Due to the large scale of DEL libraries, approaches to assess structural diversity within a DEL and overlap between DELs have not yet been published. These analyses can be valuable in guiding new DEL designs to augment already existing DELs.

Visualization of compound libraries in chemical space has been explored with principal components analysis (PCA), a dimensionality reduction approach [58], orthogonal low dimensional BCUTs space [59], and other methods [60]. In the last decade, approaches by applying statistical based nonlinear approaches, for example, self-organizing maps (SOM) and generative topographic mapping (GTM), have been introduced to assess structural diversity of a set of compounds [61, 62]. In addition to their poor visual interpretability, these approaches are not practical for large compound data sets (beyond millions). Therefore, applying these methods to the chemical space of DELs remains challenging. While 2D fingerprints can be used to profile the chemical diversity of compound collections [63], similar applications to DELs become impossible due to the number of pairwise comparisons necessary for all possible products in large DEL libraries.

It is our opinion that the low dimensional BCUT approach should have the potential to address visualization and navigation of compounds in DELs. The method around BCUTs was introduced and developed over two decades ago to deal with large set of library compounds during the combinatorial chemistry boom [59, 64]. The method drew attention within the cheminformatics community with various applications in the pre-DEL era [65–70]. BCUTs are obtained from the lowest and highest eigenvalues of empirical matrixes constructed from bonding adjacent matrix adjusted by bond and atomic properties of a compound structure. The set of metrics are first normalized and re-scaled, selected with maximum χ^2 in the associated distribution to reflect diversity of entire compound set and $r^2 < 0.25$ so that the final set of metrics are not correlated, i.e., orthogonal with each other. In coupling with the representative sub-library defined in Sect. 2.3, applying BCUTs-based applications to DELs might provide a ready method for visualizing and navigating DEL chemistry space.

5.3 Potential ML Applications for DEL Selection Analysis

ML and AI methods have been developed rapidly over the last several years with applications in many business sectors. Healthcare and drug discovery AI/ML applications are building upon the work being done by many of the leading technology companies such as Google, Amazon, and Facebook with the open-source contributions of many of their algorithms. Several effective collaborations have been established between pharmaceutical companies and these large tech companies to further explore drug discovery applications ranging from target discovery to compound design and patient stratification. Additionally, the proliferation of small start-up companies in this space has become enormous as the scale of the opportunity is clear.

DELs provide an ideal use case for ML methods given the scale and the complexity of the problem. The organized nature of DEL data also helps make it an easy target for these algorithms. Building ML models typically involves a significant investment in data engineering to organize and harmonize a dataset such that it can be used to build a model. In contrast the steps necessary for DEL analysis typically leave the data well organized such that the compound structures and labels (read counts) can be fed directly into an ML algorithm. A single ML model can be built using data for multiple DELs that may have been pooled in a single selection. This contrasts with the library-by-library analysis which is now typical. Our initial internal studies have shown the potential for additive model performance as data from DELs were aggregated. However, the inherently noisy nature of the DEL data, as discussed in Sect. 3.2, does introduce challenges to the modeling. While on the surface the enormous DEL data sets are attractive substrate for ML models, the actual training data remains limited. Billions of compounds may be screened but "labels" (active/inactive) may only be known for a small handful of compounds.

McCloskey et al. have recently introduced an ML approach to help identify hits outside the compounds in the DEL used in the target selections [39]. The ML model was trained with disynthon aggregation from DEL selection data set to diminish noise in the selection and sequencing process, and then used for virtual screening of large libraries and knowledge-based filtering to select a set of compounds for experimental validation.

The use of ML techniques to analyze DEL data remains an emerging area of science and there are several trends which should push this work forward quickly. These include rapid advances in ML algorithms, improvements to DEL selection technologies to help create data sets with less noise, increased capacity of NGS sequencers to produce better sampling across DEL selections, and improvements in the synthesis of DEL libraries to help reduce the number of chemical species attached to each DNA tag.

Acknowledgments The authors wish to acknowledge the extended Pfizer DEL team, especially Drs. Mark Flanagan, Timothy Foley, Adam Gilbert, Anthony Harris, Marie-Claire Peakman, Anokha Ratnayake, Sylvie Sakata, Jamison Tuttle, and Jinzhi Zhang for their support and discussions.

Compliance and Ethical Standards

Statement of Informed Consent There are no human subjects in this article and Informed Consent is not applicable.

Funding: The authors received no financial support for the research, authorship, and/or publication of this article.

Declaration of Conflicting Interests: The authors are employed by Pfizer Inc.; their research and authorship of this article were completed within the scope of their employment with Pfizer Inc.

References

1. Brenner S, Lerner RA (1992) Encoded combinatorial chemistry. Proc Natl Acad Sci 89(12): 5381–5383. https://doi.org/10.1073/pnas.89.12.5381
2. Clark MA, Acharya RA, Arico-Muendel CC, Belyanskaya SL, Benjamin DR, Carlson NR, Centrella PA, Chiu CH, Creaser SP, Cuozzo JW, Davie CP, Ding Y, Franklin GJ, Franzen KD, Gefter ML, Hale SP, Hansen NJV, Israel DI, Jiang J, Kavarana MJ, Kelley MS, Kollmann CS, Li F, Lind K, Mataruse S, Medeiros PF, Messer JA, Myers P, O'Keefe H, Oliff MC, Rise CE, Satz AL, Skinner SR, Svendsen JL, Tang L, van Vloten K, Wagner RW, Yao G, Zhao B, Morgan BA (2009) Design, synthesis and selection of DNA-encoded small-molecule libraries. Nat Chem Biol 5:647. https://doi.org/10.1038/nchembio.211
3. Arico-Muendel CC (2016) From haystack to needle: finding value with DNA encoded library technology at GSK. Med Chem Commun 7(10):1898–1909. https://doi.org/10.1039/c6md00341a
4. Favalli N, Bassi G, Scheuermann J, Neri D (2018) DNA-encoded chemical libraries - achievements and remaining challenges. FEBS Lett 592(12):2168–2180. https://doi.org/10.1002/1873-3468.13068
5. Gironda-Martinez A, Donckele EJ, Samain F, Neri D (2021) DNA-encoded chemical libraries: a comprehensive review with successful stories and future challenges. Acs Pharmacol Transl 4(4):1265–1279. https://doi.org/10.1021/acsptsci.1c00118
6. Goodnow RA, Dumelin CE, Keefe AD (2017) DNA-encoded chemistry: enabling the deeper sampling of chemical space. Nat Rev Drug Discov 16(2):131–147. https://doi.org/10.1038/nrd.2016.213
7. Ottl J, Leder L, Schaefer JV, Dumelin CE (2019) Encoded library technologies as integrated Lead finding platforms for drug discovery. Molecules 24(8):1629. https://doi.org/10.3390/molecules24081629
8. Shi Y, Wu YR, Yu JQ, Zhang WN, Zhuang CL (2021) DNA-encoded libraries (DELs): a review of on-DNA chemistries and their output. RSC Adv 11(4):2359–2376. https://doi.org/10.1039/d0ra09889b
9. Song M, Hwang GT (2020) DNA-encoded library screening as core platform technology in drug discovery: its synthetic method development and applications in DEL synthesis. J Med Chem 63(13):6578–6599. https://doi.org/10.1021/acs.jmedchem.9b01782
10. Cai B, Kim D, Akhand S, Sun Y, Cassell RJ, Alpsoy A, Dykhuizen EC, Van Rijn RM, Wendt MK, Krusemark CJ (2019) Selection of DNA-encoded libraries to protein targets within and on living cells. J Am Chem Soc 141(43):17057–17061. https://doi.org/10.1021/jacs.9b08085
11. Chen Q, Li Y, Lin C, Chen L, Luo H, Xia S, Liu C, Cheng X, Liu C, Li J, Dou D (2022) Expanding the DNA-encoded library toolbox: identifying small molecules targeting RNA. Nucleic Acids Res. https://doi.org/10.1093/nar/gkac173
12. Song YA, Li XY (2021) Evolution of the selection methods of DNA-encoded chemical libraries. Acc Chem Res 54(17):3491–3503. https://doi.org/10.1021/acs.accounts.1c00375
13. Ratnayake AS, Flanagan ME, Foley TL, Hultgren SL, Bellenger J, Montgomery JI, Lall MS, Liu B, Ryder T, Kolmel DK, Shavnya A, Feng X, Lefker B, Byrnes LJ, Sahasrabudhe PV, Farley KA, Chen S, Wan J (2021) Toward the assembly and characterization of an encoded library hit confirmation platform: bead-assisted ligand isolation mass spectrometry (BALI-MS). Bioorg Med Chem 41:116205. https://doi.org/10.1016/j.bmc.2021.116205
14. Su WJ, Ge R, Ding DC, Chen WH, Wang WQ, Yan H, Wang WK, Yuan YL, Liu H, Zhang M, Zhang JY, Shu QS, Satz AL, Kuai LT (2021) Triaging of DNA-encoded library selection results by high-throughput Resynthesis of DNA-conjugate and affinity selection mass spectrometry. Bioconjug Chem 32(5):1001–1007. https://doi.org/10.1021/acs.bioconjchem.1c00170
15. Xia B, Franklin GJ, Lu XJ, Bedard KL, Grady LC, Summerfield JD, Shi EX, King BW, Lind KE, Chiu C, Watts E, Bodmer V, Bai XP, Marcaurelle LA (2021) DNA-encoded library hit confirmation: bridging the gap between on-DNA and off-DNA chemistry. ACS Med Chem Lett 12(7):1166–1172. https://doi.org/10.1021/acsmedchemlett.1c00156

16. Belyanskaya SL, Ding Y, Callahan JF, Lazaar AL, Israel DI (2017) Discovering drugs with DNA-encoded library technology: from concept to clinic with an inhibitor of soluble epoxide hydrolase. Chembiochem 18(9):837–842. https://doi.org/10.1002/cbic.201700014
17. Cuozzo JW, Clark MA, Keefe AD, Kohlmann A, Mulvihill M, Ni HH, Renzetti LM, Resnicow DI, Ruebsam F, Sigel EA, Thomson HA, Wang C, Xie ZF, Zhang Y (2020) Novel Autotaxin inhibitor for the treatment of idiopathic pulmonary fibrosis: a clinical candidate discovered using DNA-encoded chemistry. J Med Chem 63(14):7840–7856. https://doi.org/10.1021/acs.jmedchem.0c00688
18. Harris PA, Berger SB, Jeong JU, Nagilla R, Bandyopadhyay D, Campobasso N, Capriotti CA, Cox JA, Dare L, Dong XY, Eidam PM, Finger JN, Hoffman SJ, Kang J, Kasparcova V, King BW, Lehr R, Lan YF, Leister LK, Lich JD, MacDonald TT, Miller NA, Ouellette MT, Pao CS, Rahman A, Reilly MA, Rendina AR, Rivera EJ, Schaeffer MC, Sehon CA, Singhaus RR, Sun HH, Swift BA, Totoritis RD, Vossenkamper A, Ward P, Wisnoski DD, Zhang DH, Marquis RW, Gough PJ, Bertin J (2017) Discovery of a first-in-class receptor interacting protein 1 (RIP1) kinase specific clinical candidate (GSK2982772) for the treatment of inflammatory diseases. J Med Chem 60(4):1247–1261. https://doi.org/10.1021/acs.jmedchem.6b01751
19. Ganem B, Li YT, Henion JD (1991) Detection of noncovalent receptor ligand complexes by mass-spectrometry. J Am Chem Soc 113(16):6294–6296. https://doi.org/10.1021/ja00016a069
20. McLaren DG, Shah V, Wisniewski T, Ghislain L, Liu C, Zhang H, Saldanha SA (2021) High-throughput mass spectrometry for hit identification: current landscape and future perspectives. SLAS Discov 26(2):168–191. https://doi.org/10.1177/2472555220980696
21. Erlanson DA, Fesik SW, Hubbard RE, Jahnke W, Jhoti H (2016) Twenty years on: the impact of fragments on drug discovery. Nat Rev Drug Discov 15(9):605–619. https://doi.org/10.1038/nrd.2016.109
22. Lu WC, Kostic M, Zhang TH, Che JW, Patricelli MP, Jones LH, Chouchani ET, Gray NS (2021) Fragment-based covalent ligand discovery. Rsc Chem Biol 2(2):354–367. https://doi.org/10.1039/d0cb00222d
23. Varela-Rial A, Majewski M, De Fabritiis G (2022) Structure based virtual screening: fast and slow. Wires Comput Mol Sci 12(2). https://doi.org/10.1002/wcms.1544
24. Castan IFSF, Graham JS, Salvini CLA, Stanway-Gordon HA, Waring MJ (2021) On the design of lead-like DNA-encoded chemical libraries. Bioorg Med Chem 43:116273. https://doi.org/10.1016/j.bmc.2021.116273
25. Dickson P, Kodadek T (2019) Chemical composition of DNA-encoded libraries, past present and future. Org Biomol Chem 17(19):4676–4688. https://doi.org/10.1039/c9ob00581a
26. Franzini RM, Randolph C (2016) Chemical space of DNA-encoded libraries. J Med Chem 59(14):6629–6644. https://doi.org/10.1021/acs.jmedchem.5b01874
27. Zhang Y, Clark MA (2021) Design concepts for DNA-encoded library synthesis. Bioorg Med Chem 41. https://doi.org/10.1016/j.bmc.2021.116189
28. Zhu H, Flanagan ME, Stanton RV (2019) Designing DNA encoded libraries of diverse products in a focused property space. J Chem Inf Model 59(11):4645–4653. https://doi.org/10.1021/acs.jcim.9b00729
29. Rama-Garda R, Amigo J, Priego J, Molina-Martin M, Cano L, Dominguez E, Loza MI, Rivera-Sagredo A, de Blas J (2021) Normalization of DNA encoded library affinity selection results driven by high throughput sequencing and HPLC purification. Bioorg Med Chem 40:116178. https://doi.org/10.1016/j.bmc.2021.116178
30. Satz AL (2016) Simulated screens of DNA encoded libraries: the potential influence of chemical synthesis fidelity on interpretation of structure–activity relationships. ACS Comb Sci 18(7):415–424. https://doi.org/10.1021/acscombsci.6b00001
31. Amigo J, Rama-Garda R, Bello X, Sobrino B, de Blas J, Martin-Ortega M, Jessop TC, Carracedo A, Loza MIG, Dominguez E (2018) tagFinder: a novel tag analysis methodology that enables detection of molecules from DNA-encoded chemical libraries. SLAS Discov 23(5):397–404. https://doi.org/10.1177/2472555217753840

32. Kuai L, O'Keeffe T, Arico-Muendel C (2018) Randomness in DNA encoded library selection data can be modeled for more reliable enrichment calculation. SLAS Discov 23(5):405–416. https://doi.org/10.1177/2472555218757718
33. Zhu H, Foley TL, Montgomery JI, Stanton RV (2022) Understanding data noise and uncertainty through analysis of replicate samples in DNA-encoded library selection. J Chem Inf Model 62(9):2239–2247. https://doi.org/10.1021/acs.jcim.1c00986
34. Faver JC, Riehle K, Lancia DR, Milbank JBJ, Kollmann CS, Simmons N, Yu Z, Matzuk MM (2019) Quantitative comparison of enrichment from DNA-encoded chemical library selections. ACS Comb Sci 21(2):75–82. https://doi.org/10.1021/acscombsci.8b00116
35. Kleiner RE, Dumelin CE, Tiu GC, Sakurai K, Liu DR (2010) In vitro selection of a DNA-templated small-molecule library reveals a class of macrocyclic kinase inhibitors. J Am Chem Soc 132(33):11779–11791. https://doi.org/10.1021/ja104903x
36. Buller F, Zhang YX, Scheuermann J, Schafer J, Buhlmann P, Neri D (2009) Discovery of TNF inhibitors from a DNA-encoded chemical library based on Diels-Alder cycloaddition. Chem Biol 16(10):1075–1086. https://doi.org/10.1016/j.chembiol.2009.09.011
37. Satz AL (2015) DNA encoded library selections and insights provided by computational simulations. ACS Chem Biol 10(10):2237–2245. https://doi.org/10.1021/acschembio.5b00378
38. Chen Q, Hall J, Foley TL, Wan J, Li Y, Israel DI (2020b) A method for estimating binding affinity from primary DEL selection data. Biochem Biophys Res Commun 533(2):249–255. https://doi.org/10.1016/j.bbrc.2020.04.029
39. McCloskey K, Sigel EA, Kearnes S, Xue L, Tian X, Moccia D, Gikunju D, Bazzaz S, Chan B, Clark MA, Cuozzo JW, Guie MA, Guilinger JP, Huguet C, Hupp CD, Keefe AD, Mulhern CJ, Zhang Y, Riley P (2020) Machine learning on DNA-encoded libraries: a new paradigm for hit finding. J Med Chem 63(16):8857–8866. https://doi.org/10.1021/acs.jmedchem.0c00452
40. Martin A, Nicolaou CA, Toledo MA (2020) Navigating the DNA encoded libraries chemical space. Commun Chem 3(1). https://doi.org/10.1038/s42004-020-00374-1
41. Helal CJ, Bartolozzi A, Goble SD, Mania NS, Guzman-Perez A, Ohri AK, Shia ZC, Subramanyam C (2018) Increased building block access through collaboration. Drug Discov Today 23(8):1458–1462. https://doi.org/10.1016/j.drudis.2018.03.001
42. OEChem (2018) OEChem toolkit. OpenEye Scientific Software
43. Kolmel DK, Zhu H, Flanagan ME, Sakata SK, Harris AR, Wan J, Morgan BA (2021) Employing Photocatalysis for the design and preparation of DNA-encoded libraries: a case study. Chem Rec 21(4):616–630. https://doi.org/10.1002/tcr.202000148
44. Leo AJ (1993) Calculating log Poct from structures. Chem Rev 93(4):1281–1306. https://doi.org/10.1021/cr00020a001
45. Shi S, Peng Z, Kostrowicki J, Paderes G, Kuki A (2000) Efficient combinatorial filtering for desired molecular properties of reaction products. J Mol Graph Model 18(4–5):478–496. https://doi.org/10.1016/s1093-3263(00)00069-3
46. Zhou JZ, Shi S, Na J, Peng Z, Thacher T (2009) Combinatorial library-based design with basis products. J Comput Aided Mol Des 23(10):725–736. https://doi.org/10.1007/s10822-009-9297-9
47. Carhart RE, Smith DH, Venkataraghavan R (1985) Atom pairs as molecular features in structure-activity studies: definition and applications. J Chem Inf Comput Sci 25(2):64–73. https://doi.org/10.1021/ci00046a002
48. Stahl M, Mauser H (2005) Database clustering with a combination of fingerprint and maximum common substructure methods. J Chem Inf Model 45(3):542–548. https://doi.org/10.1021/ci050011h
49. Martin M (2011) Cutadapt removes adapter sequences from high-throughput sequencing reads. EMBnet J 17(1):3. https://doi.org/10.14806/ej.17.1.200
50. Cock PJ, Antao T, Chang JT, Chapman BA, Cox CJ, Dalke A, Friedberg I, Hamelryck T, Kauff F, Wilczynski B, de Hoon MJ (2009) Biopython: freely available python tools for computational molecular biology and bioinformatics. Bioinformatics 25(11):1422–1423. https://doi.org/10.1093/bioinformatics/btp163

51. Yoo AB, Jette MA, Grondona M (2003) SLURM: simple linux utility for resource management. Lect Notes Comput Sci 2862:44–60
52. Arthur JS, Ley SC (2013) Mitogen-activated protein kinases in innate immunity. Nat Rev Immunol 13(9):679–692. https://doi.org/10.1038/nri3495
53. Chen Q, Cheng X, Zhang L, Li X, Chen P, Liu J, Zhang L, Wei H, Li Z, Dou D (2020a) Exploring the lower limit of individual DNA-encoded library molecules in selection. SLAS Discov 25(5):523–529. https://doi.org/10.1177/2472555219893949
54. Foley TL, Burchett W, Chen Q, Flanagan ME, Kapinos B, Li X, Montgomery JI, Ratnayake AS, Zhu H, Peakman MC (2021) Selecting approaches for hit identification and increasing options by building the efficient discovery of actionable chemical matter from DNA-encoded libraries. SLAS Discov 26(2):263–280. https://doi.org/10.1177/2472555220979589
55. Denton KE, Krusemark CJ (2016) Crosslinking of DNA-linked ligands to target proteins for enrichment from DNA-encoded libraries. Med Chem Commun 7(10):2020–2027. https://doi.org/10.1039/C6MD00288A
56. Zhu H. (2019) A hierarchical structural similarity search approach for large-scale DNA-encoded libraries. 9th international symposium on DNA-encoded chemical libraries, Zurich, Switzerland
57. Pikalyova R, Zabolotna Y, Volochnyuk DM, Horvath D, Marcou G, Varnek A (2022) Exploration of the chemical space of DNA-encoded libraries. Mol Inform 41(6):e2100289. https://doi.org/10.1002/minf.202100289
58. Oprea TI, Gottfries J (2001) Chemography: the art of navigating in chemical space. J Comb Chem 3(2):157–166. https://doi.org/10.1021/cc0000388
59. Pearlman RS, Smith KM (1999) Metric validation and the receptor-relevant subspace concept. J Chem Inf Comput Sci 39(1):28–35. https://doi.org/10.1021/ci980137x
60. Reutlinger M, Schneider G (2012) Nonlinear dimensionality reduction and mapping of compound libraries for drug discovery. J Mol Graph Model 34:108–117. https://doi.org/10.1016/j.jmgm.2011.12.006
61. Lin A, Horvath D, Afonina V, Marcou G, Reymond JL, Varnek A (2018) Mapping of the available chemical space versus the chemical universe of Lead-like compounds. ChemMedChem 13(6):540–554. https://doi.org/10.1002/cmdc.201700561
62. Osolodkin DI, Radchenko EV, Orlov AA, Voronkov AE, Palyulin VA, Zefirov NS (2015) Progress in visual representations of chemical space. Expert Opin Drug Discov 10(9):959–973. https://doi.org/10.1517/17460441.2015.1060216
63. Zhu H, Klug-McLeod J, Bakken GA (2013) A heuristic algorithm for plate selection that maximizes compound diversity. Croat Chem Acta 86(4):435–441. https://doi.org/10.5562/cca2301
64. Pearlman RS, Smith KM (1998) Novel software tools for chemical diversity. Perspect Drug Discov Des 9:339–353
65. Gao H (2001) Application of BCUT metrics and genetic algorithm in binary QSAR analysis. J Chem Inf Comput Sci 41(2):402–407. https://doi.org/10.1021/ci000306p
66. Ma C, Lazo JS, Xie X-Q (2011) Compound acquisition and prioritization algorithm for constructing structurally diverse compound libraries. ACS Comb Sci 13(3):223–231. https://doi.org/10.1021/co100033m
67. Menard PR, Mason JS, Morize I, Bauerschmidt S (1998) Chemistry space metrics in diversity analysis, library design, and compound selection. J Chem Inf Comput Sci 38(6):1204–1213. https://doi.org/10.1021/ci9801062
68. Pirard B, Pickett SD (2000) Classification of kinase inhibitors using BCUT descriptors. J Chem Inf Comput Sci 40(6):1431–1440. https://doi.org/10.1021/ci000386x
69. Shanmugasundaram V, Maggiora GM, Lajiness MS (2005) Hit-directed nearest-neighbor searching. J Med Chem 48(1):240–248. https://doi.org/10.1021/jm0493515
70. Stanton DT (1999) Evaluation and use of BCUT descriptors in QSAR and QSPR studies. J Chem Inf Comput Sci 39(1):11–20. https://doi.org/10.1021/ci980102x

Top Med Chem (2022) 40: 183–216
https://doi.org/10.1007/7355_2022_149

Published online: 11 October 2022

Selection Strategies in DNA-Encoded Libraries

Yuhan Gui and Xiaoyu Li

Contents

Y. Gui
Department of Chemistry and State Key Laboratory of Synthetic Chemistry, The University of Hong Kong, Hong Kong, SAR, China

X. Li (✉)
Department of Chemistry and State Key Laboratory of Synthetic Chemistry, The University of Hong Kong, Hong Kong, SAR, China

Laboratory for Synthetic Chemistry and Chemical Biology of Health @InnoHK, Hong Kong, SAR, China
e-mail: xiaoyuli@hku.hk

Abstract DNA-encoded chemical library (DEL) has emerged as a versatile and innovative technology platform for ligand discovery in chemical biology research and early-stage drug discovery. The rapid development of DEL-compatible reactions has further fueled its applications over the last decade. To date, DELs have been widely adopted by the pharmaceutical industry and are also gaining popularity in academic research. However, a relatively underexplored component of DELs is the selection methodology. DEL selection has been generally considered a massive binding assay, which has been dominated by the so-called bind-wash-elute procedure against immobilized protein targets. Indeed, such a "classical" selection method has seen great success in the selection campaigns against numerous drug targets. Recently, novel DEL selection modalities have emerged, which have not only widened the target scope of DELs to the complex milieu of biological systems but also enabled functional DEL selections beyond identifying physical binders. This chapter furnishes an overview of the current DEL selection methods and concludes with a perspective for future development, aiming to provide a succinct guidance for practitioners who intend to embrace the DEL technology.

Keywords Combinatorial library, DNA-encoded chemical library, Drug discovery, High-throughput screening, Selection method

1 Introduction

Since the 1990s, high-throughput screening (HTS) has become a major technology platform in the pharmaceutical industry. Compared with traditional HTS, the advent of the "high-power screening (HPS)" [1], including the combinatorial biological libraries (phage- [2], ribosome- [3], mRNA- [4], and yeast-display [5] libraries) and DNA-encoded libraries (DELs), is more conducive to expedite the ligand discovery process, mostly because of the dramatically increased screening throughput at a miniaturized assay format [1]. By replacing the spatial encoding in traditional chemical libraries with DNA encoding, DELs can be prepared at a minute scale in a single mixture and the selection can be conducted in a short timeframe. In 2017, Lerner and Brenner proposed an initiative where commercial companies share their DELs to academic researchers, aiming to extend the power of DELs in a wider scope of disciplines and applications [6]. Today, several pharmaceutical companies are also offering pre-made DEL kits to meet the needs of individual researchers [7].

The concept of DEL was originally proposed by Brenner and Lerner in 1992 [8] and rapidly realized experimentally by Brenner and Janda in 1993 [9], as a method to accelerate the traditional one-bead, one-compound (OBOC) combinatorial libraries. Nearly at the same time, an 820,000-member heptapeptide DEL was prepared by Gallop and co-workers using a similar approach [10]. Two years later, Nishigaki reported the iterative enzymatic construction of encoding tags [11]. These pioneering works have laid out the fundamental principles of DELs and also

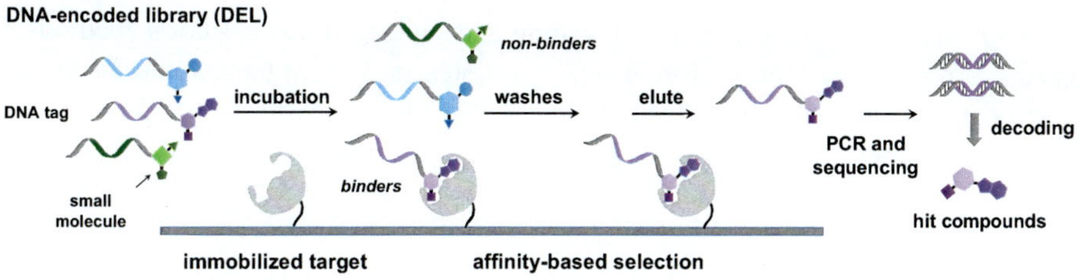

Fig. 1 Schematic illustration of DEL selection against immobilized protein target

preliminarily explored the techniques that later are commonly adopted in DEL selections, such as using a soluble competitor in the selection to identify specific binders and the flow-cytometry-based isolation of hit compounds on individual beads [10]. In 2004, breakthroughs were made in this field by four seminal reports: the encoded self-assembling chemical (ESAC) libraries, the DNA-templated synthesis library (DTS), the DNA-routing library, and the peptide nucleic acid (PNA) encoded library, independently developed by the groups of Neri [12], Liu [13], Harbury [14] and Winssinger [15]. These libraries were synthesized in the solution phase, which circumvented the limit on library size of the bead-based approaches and featured more flexible selection procedures. Indeed, in-solution DELs have become the mainstream till today. In 2009, GlaxoSmithKline (GSK) reported a milestone work of applying DELs on an industrial scale [16]. Since then, momentous technological advances have been made in this field. Notably, with the emergence of the genomic-scale next-generation sequencing (NGS) technologies and the recent efforts to broaden the reaction toolbox of DNA/DEL-compatible chemistry [17–19], multi-billion-compound DELs with diverse chemical structures could be routinely prepared. DELs have also been integrated with other legacy and emerging techniques, such as machine learning/artificial intelligence [20–22], dynamic combinatorial library (DCL) [23–28], fragment-based drug discovery (FBDD) [12, 29–32], one-bead-one-compound (OBOC) libraries [33, 34], and modern instrumentation (flow cytometry [10, 35, 36], microfluidics [37–39], automation, etc.), which have extended DELs to many new applications, particularly through the development of novel selection methods.

In general, the DEL technology can be classified into four major components: (1) chemistry for library synthesis, (2) library encoding, (3) library selection, and (4) hit decoding and data analysis, which have been extensively covered by numerous reviews, books, perspectives, and highlight articles [6, 7, 17–19, 40–57]. Typically, DEL selection is considered as a binding assay, albeit at a massive scale, to identify the physical binders (Fig. 1). After incubating the library with the protein of interest, the binders are separated from the non-binding compounds through a series of washing procedures. The binders are then eluted under strong denaturing conditions, and subsequently, the structures of the hit compounds are decoded by PCR amplification and NGS analysis of the enriched DNA tags in the eluant. In this chapter, we begin with a brief description of how DEL selection has evolved in the past and then highlight the recent developments of new DEL selection methods, especially on the ones that can interrogate complex biological targets and discover functional molecules beyond physical binders. Finally, we express our opinions on the current challenges and future trends in DEL selections.

2 Selection with Immobilized Libraries (Early Works)

Since the concept of DELs originated from traditional OBOC combinatorial libraries, early DEL studies focused on the chemistry that could synthesize the library compound (peptides) and the oligonucleotide tag on the same bead; thus, the library was immobilized and the selections could be straightforwardly conducted with a soluble protein in buffer [8–10]. Gallop and co-workers prepared a heptapeptides library on polystyrene beads and used a fluorescently labeled antibody in the solution as the target [10]. After incubation and repetitive washes, the beads that retained high fluorescence signal were collected and isolated by flow cytometry. The DNA barcodes attached to the isolated beads were amplified and decoded. Later on, in-solution DELs became the mainstream practice, since soluble small-molecule-DNA conjugates can be more feasibly synthesized; for example, the DNA tag is pre-prepared on a DNA synthesizer and then coupled to the small molecule, which avoided the issues in the parallel synthesis of oligonucleotide and small molecules in OBOC-DELs. However, OBOC-DELs have been recently revived with advanced instrumentation, innovative on-bead chemistry, and multi-functional bead design [53, 57]. This research direction has been intensively pursued by the groups of Paegel and Kodadek, whose work has led to many novel applications of DELs and will be discussed in the later part of this chapter.

3 Selection against Immobilized Targets

3.1 General Procedure

Today, most DELs are synthesized in the solution phase; thus, the selections are performed against an immobilized target, usually a purified protein. Such an affinity-based selection procedure is often called *panning*, analogous to the screening technique developed in the biological display library technologies (Fig. 1), e.g., phage display, mRNA display, yeast surface display, and ribosome display libraries, [2, 5, 58–61] where the target immobilization facilitates the separation of the binders from the non-binders. Typically, the target protein of interest is immobilized to the solid matrix via a suitable immobilization protein tag (e.g., GST-tag, His-tag, HA-tag, biotin) or non-specifically to the *N*-hydroxysuccinimide ester (NHS)-activated Sepharose beads, which captures the primary amino group of the lysine side chain on the protein surface. The immobilization is followed by the incubation with the library [62–66]. Alternatively, the library and the target may be incubated in solution and then subjected to immobilization. First, relatively mild washing steps are applied to wash away the non-binders while retaining the binders. Of note, the washing buffer and the protocol (e.g., number of washes) can be empirical, and they are often varied to control the stringency of the selection. Next, strong denaturing condition (e.g., heating, change of pH, or using a high concentration of imidazole or glutathione when the protein is immobilized with His-tag or GST-tag) is applied to elute the binders. The eluent is then PCR-amplified and subjected to sequencing analysis for hit decoding.

A number of the important experimental parameters have to be taken into consideration in the selection. For the copy number of individual library members (library input), studies have shown that large DELs (roughly at a billion-compound scale) require at least 10^3–10^4 copies per compound [67], and 10^5–10^6 copies or more should be used for smaller DELs (roughly at million-compound scale) [68]. In principle, affinity-based selection is a balance between kinetic disassociation (k_{off}) and thermodynamic association (K_a) of the ligand to the target protein [42]. First, it requires an optimized washing condition to better differentiate the binders from the large excess of the non-binders. In theory, high-affinity ligands can be easily enriched and identified over the background and can tolerate a wider range of washing conditions, whereas medium or low-affinity binders, owing to the relatively limited kinetic stability of the binding complex, may be washed away and missed from the selection. Thus, the washing condition should to be carefully controlled, especially for large libraries since low-affinity binders may be obscured by the vast non-specific binding and multiple rounds of selections may be necessary. Other experimental parameters such as incubation time, buffer pH and composition, and the use of different competitors have profound influences on the selection results. In addition, quantitative PCR (qPCR) has been frequently employed for assessing and optimizing the selection performance [68–70] and the UMI (unique molecular identifier) featured in the DNA code is another useful tool to correct the artifacts arising from PCR amplification bias [71]. Recently, the Neri group has provided valuable guidance and detailed protocols to conduct DEL selections [68–70, 72, 73]. Secondly, since DEL selections are performed at a very small scale, high target concentration is required to push the equilibrium toward the formation of the binding complex. However, in practice, researchers have adopted a wide range of target concentrations (0.4–50 μM), presumably because of the specific needs of the selections. Moreover, performing parallel selections at varied target concentrations could also be used [74, 75]. On the one hand, it could be considered as a kind of "replication" of the selection to be used for more reliable hit picking; on the other hand, in theory, lower concentration may identify the ligands of higher binding affinities, while higher concentration tends to identify both the moderate and strong binders [76]. For biological display libraries, the selected library could be eluted, amplified, and subjected to iterated rounds of selections, thereby realizing a true evolution process. For DELs, in principle, multiple rounds of selections are also preferred, as it reduces stochastic noise, and the best of the binders will be distinctively enriched after 2–3 selection cycles [77]. However, in most cases, there is no library amplification since the amplification of small molecules is not straightforward [78]. The lack of the evolution feature is a major difference between DEL selection and the selection of biological display libraries. Moreover, the optimal number of panning steps depends on the size and the nature of the library, as iterated selections will lead to low material recovery and higher amount of library input would be needed. One may save a portion of the selected library from each round of selection and compare the sequencing results.

Despite the multitude of key parameters in DEL selection, there are few systematic reports examining individual variables, especially in one study using the same library and target. Most of the commonly adopted practices, e.g., initial library input, titration of the target concentration, quantitation of library recovery, are scattered among multiple studies across a variety of different libraries and targets, and a direct side-by-side comparison of different selection procedures is still lacking. In fact, although there is consensus on the general workflow of DEL selections, DEL practitioners often develop their own experimental protocols based on their own needs and practical experience. However, it is worth noting that solid-phase DEL selection has been directly compared with crosslinking-based selection in the solution phase [70, 73, 79], and the results showed that covalent crosslinking facilitated the enrichment of low-affinity ligands. Moreover, Neri and co-workers systematically optimized the experimental parameters for live-cell-based DEL selections [73].

DEL selection is a massive binding assay, where millions to billions of compounds are sampled by the target in a single solution. In principle, one would like to select larger libraries to improve the chance of identifying high-affinity binders. However, since DEL synthesis does not remove the side/truncated products and the defectively tagged library compounds, inevitably larger DELs have lower quality and result in significant noise in the selection data [76, 80, 81]. Although many elegant computational and experimental approaches have been developed to address this issue, [75, 82–85] the library size remains to be an important consideration. There is no clear-cut definition on large/small DELs; however, in 2017, a study by Satz and co-workers found that false negative rates outweigh the benefit of increased diversity at library sizes larger than 100 million [86]. With the recent development of computational algorithms to "denoise" selection data, [65, 80, 87–90] ~1 billion might be considered as a rough number separating "large" vs. "small" DELs. Besides data noise, large DELs also require more initial library input to ensure there are enough copies of each compound [67, 68] and deeper sequencing depth to provide sufficient selection coverage [91]. In practice, successful selections of both "large" and "small" DELs have been reported in many studies, and it seems the key is finding the balance of library size, quality, and the resources available for data generation and analysis.

Parallel screens with several target proteins or a single target examined under multiple conditions (different mutants, additives, etc.) are also routinely used in order to identify the specific binders. Neri and co-workers conducted parallel selections against three related proteins (human serum albumin, HSA; rat serum albumin, RSA; and bovine serum albumin, BSA) and provided valuable information concerning the target-specificity with distinctive binding specificities to these targets

[92]. Roche focused on the efficient discovery of a chemical series that showed high selectivity for DDR1 (Discoidin Domain Receptor 1), an important protein in renal fibrosis, over its close isoform DDR2 by using parallel DEL selections [93]. In a very impressive study, 119 bacterial proteins were parallelly selected by the company GSK with a large collection of DELs to evaluate the "*ligandability*" of these targets for further anti-microbial drug discovery [94]. Finally, it is common that a "blank library" with the same DNA sequences but without the small molecules is included in the selection experiments to control for DNA-protein interactions, although few studies reported significant issues [43]. However, when applying DELs to nucleic acid targets, a "blank library" control might be important.

3.2 Selection for Covalent Inhibitors

Recently, covalent drugs have resurged and grown in popularity for drug discovery. The high biochemical efficiency of irreversible binding offers potential pharmacodynamic and pharmacokinetic advantages such as less frequent drug dosing and reduced off-target effects [95]. Covalent DELs usually contain electrophilic building blocks, also termed "warheads" (e.g., α,β-unsaturated carbonyl motif, sulfur(VI)-based reactive center, etc.), that are potentially capable of selectively forming covalent bonds with the nucleophilic side chains of cysteines and lysines on the protein surface. Several research groups have described detailed selection protocols using covalent DELs. The Zhu group profiled the covalent inhibitors of rhinovirus 3C protease by first removing the reversible binders with heat elution and then conducting direct PCR amplification on beads for hit deconvolution [96]. Studies by Neri and co-workers reported the discovery of a covalent *c*-Jun *N*-terminal kinase (JNK-1) inhibitor from several dual-pharmacophore libraries [97]. The Winssinger group disclosed several excellent examples of using PNA-encoded libraries to identify covalent inhibitors for MEK2 (mitogen-activated protein kinase kinase 2), ERBB2 (Erb-B2 receptor tyrosine kinase 2) [98] and bromodomains [99]. In these studies, stringent SDS washes were used to remove the non-covalent binders. Very recently, researchers at Baylor College of Medicine reported CDD-1713, a covalent selective inhibitor of main protease (M^{pro}), a highly important target in the development of anti-COVID-19 drugs [100]. The inhibitor has an aldehyde warhead that forms a covalent bond with the catalytic residue Cys145, which turned out to be imperative for its antiviral activity. We anticipate that the further developments of covalent chemistry in DEL will lead to the expansion of medicinal candidates.

3.3 Selection Data Analysis

After PCR amplification, the DNA amplicons of the isolated binders are subjected to NGS and typically a large amount of data will be generated. Efficient tools for DEL selection data processing and interpretation are highly desirable. Neri and co-workers provided a guide with the detailed protocol and the script for data analysis; they used C++ program to pick the sequences obtained from sequencing, followed by frequency analysis of the occurrence of the individual code combinations [72]. The selection results may be visualized in two- or three-dimensional scatter plot (often called "fingerprints") through the MATLAB software, allowing the visualization of the enrichment fold and the structure-activity relationship (SAR) of the hit compounds. However, to date, the DEL community has not reached a consensus on the general criteria for hit picking and how the post-selection sequence count and the enrichment fold should be evaluated, calculated, and applied in hit ranking, which arguably may be due to the considerable variation in enrichment fold between selection replicates. Given the high importance of identifying the true hits for follow-up validations, a series of powerful statistical and computational approached have been described to enable more reliable hit picking. Eli Lilly published an open-source algorithm called tagFinder to help researchers better deconvolute DEL hits by optimizing sequencing power, enabling fast tag detection, and providing more thorough result characterization [89]. Faver and co-workers modeled the selection outputs by harnessing a binomial distribution and proposed a novel enrichment metric "normalized *Z*-score" as an informatics and analysis pipeline to satisfy key criteria for DEL selection data analysis. This metric enables comparative quantitative enrichment analysis of *n*-synthons from multiple DEL screening and exhibits accuracy and insensitivity respect to library sampling and diversity [87]. Krusemark and co-workers employed the *Z*'-factor statistical analysis to reflect variability associated with the detection of unique DNA barcodes and signal-to-background ratio [90]. Schreiber and co-workers developed a "barcode enrichment" score using a Poisson distribution enrichment model to analyze the results [65]. However, it is worth noting that simple normalization method basing on the total sequencing counts [63, 72] could be used for many selection experiments already and the statistical methods may be more suitable for the "noisier" selections.

A major issue of DEL selection data is that it is very noisy. It is highly beneficial to establish a method for correcting the factors contributing to the noise associated with the raw data. HitGen and Pfizer improved the accuracy of DEL selection data via multi-round selections and serial equilibrium washes, respectively [75, 84]. Kuai and co-workers adopted a count-to-mean count ratio metric, incorporating proper normalization and Poisson confidence interval calculation to avoid the random noise from the low copy counts in the selection data [81]. Roche reported a data aggregation technique to minimize the influence from the truncated products [80]. Totient and co-workers developed and benchmarked a sparse-learning-based "deldenoiser" method for reducing the major noise affecting sequencing results of DELs, taking account into issues surrounding the truncated byproducts, sequencing noise, and the imbalance of DNA barcodes. The authors presented how much noise suppression can be realized via a naive Bayes model and shared the scripts freely with the community as a Python package and command line tool [88].

Collectively, the emergence of these innovative and scalable statistical methods is definitely an exciting trend toward more reliable and accurate data processing;

however, more comprehensive validation, especially with large DELs, is still needed and it seems that there is no conclusion/consensus on which method(s) is the most robust and generally applicable. At this point, performing more selection replicates across a scope of target concentrations combined with adequate controls is still recommended. Arguably, the *Z*-score enrichment metric may be more advantageous for analyzing the selection data of large DELs [42]. In a recent collaboration, X-Chem and Google developed a novel virtual screening platform connecting machine learning with DEL selection output [21]. After training models using affinity-mediated selection data, this approach could automatically consider and incorporate more molecules beyond the ones in the library. It was effective on three different protein targets with an overall hit rate of 30% at 30 μM. Potent hits (IC_{50} <10 nM) for every target were discovered, avoiding the time-consuming process to synthesize the focused library. We expect that machine-learning-based approaches would provide more well-trained and robust predictive pipelines for better hit finding as well as augmenting the chemical diversity of the original DEL.

4 Selection With In-Solution Targets

DEL selection with immobilized protein targets has seen great success in practice; it is virtually suitable for any biological target as long as it can be immobilized and maintain the structural fold. However, some proteins may not be amenable to purification and/or immobilization. Immobilized proteins may also lose their native structures or essential interaction partners (e.g., other protein binders or enzyme co-factors). Some complex targets, such as the whole protein complexes and non-adherent cells, cannot be immobilized. Hence, a number of selection methods for in-solution DEL selections have been developed. On the one hand, they have expanded the target scope of DELs; on the other hand, they have enabled the selection in a more biologically relevant environment. To achieve in-solution selections, distinguishing the binders from the non-binders without physical separation is the key. This section summarizes the recently developed strategies to achieve this goal.

4.1 Target-Binding as the Prerequisite for Tag Amplification

Liu and co-workers described a selective amplification method named Interaction-Dependent PCR (IDPCR), in which complementary ssDNA (single-stranded DNA) tags connect the ligand to the DNA-conjugated target upon ligand binding (Fig. 2a). Because of the enhanced thermal stability, only the DNA duplex with the ligand-target pairs can be formed. The newly formed *pseudo*-hairpin structure contains two primer binding sequences and enables the subsequent selective primer extension and PCR amplification [101]. Later, based on the same principle, the authors developed a

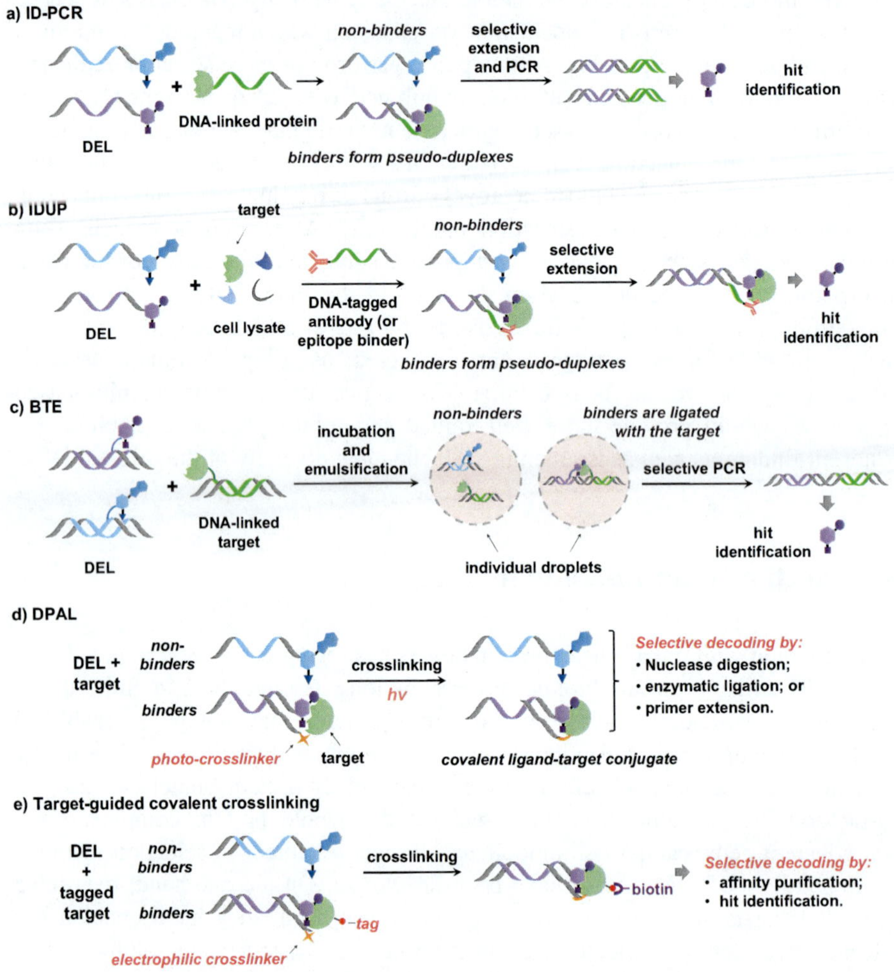

Fig. 2 Strategies have been developed for in-solution DEL selections. (**a**) ID-PCR; (**b**) IDUP; (**c**) BTE; (**d**) DPAL-based methods (nuclease digestion, enzymatic ligation, and primer extension); and (**e**) Target-guided covalent crosslinking

DEL selection method named IDUP (interaction determination using unpurified proteins), which has realized DEL selections in cell lysates by using DNA-linked antibody or His_6/SNAP-tag target protein to form the target-DNA conjugate in situ (Fig. 2b) [102]. Later, IDUP has been applied to interrogate multiple tagged proteins and allowed the simultaneous readout of DNA-barcoded proteins-ligand binding pairs in a multiplexed format [103]. Researchers at Vipergen reported a novel Binder Trap Enrichment (BTE) selection system by using water-in-oil emulsification to trap target-ligand complexes in individual microdroplets and also separate them from the non-binding entities [104]. In the first step of BTE, a DEL is incubated with the DNA-tagged protein and allowed to reach the equilibrium, followed by rapid dilution dominated by kinetic dissociation. Next, the emulsion was formed to trap the binding complexes, and the ligand-target binding pairs were compartmented into separate micelles. Moreover, the DNA tag of the target and the library DNAs were designed so that they can be ligated enzymatically upon ligand binding, which allowed the selective tag amplification for hit identification (Fig. 2c). Recently, Vipergen applied the BTE approach in the selections against p38α MAP kinase [105] and also several targets inside live cells [106].

4.2 Crosslinking-Based Methods

Previously, the Li group has developed an affinity labeling method named DPAL (DNA-Programmed Affinity Labeling), which has been applied in solution phase DEL selections [41, 107–109]. In DPAL, first, a complementary oligonucleotide containing a photo-reactive group is hybridized at the primer-binding site of the library, thereby placing the photo-crosslinker close to the target. When the small molecule binds to the protein, UV irradiation enables the covalent capture of the target protein and forms a stable *pseudo*-duplex structure [110]. Because the large size of the protein molecule, the DNA tag of the binders, but not the non-binders, are protected from the subsequent nuclease digestion by exonuclease I (*exoI*) (Fig. 2d). Thus, the DNA tag of the binders can be enriched and selectively amplified. The authors have shown that the method can be used either in buffer or in cell lysates. However, under/over-digestion by the nuclease appeared to be an issue and it is rather difficult to control. Later, the same group has circumvented this obstacle by covalently ligating the DNA tag with the target protein or using polymerase extension to copy the DNA sequence into the target-DNA conjugate (Fig. 2d) [43, 111]. Recently, the Krusemark group developed a different type of crosslinking-based selection method. First, the target protein is fused with an affinity tag; second, the authors inserted a single-stranded DNA (ssDNA) region between the library compound and the encoding tag, and the ssDNA region was used to hybridize with another DNA strand bearing an electrophilic crosslinker. After ligand binding, the electrophilic crosslinker may covalently capture the target, and the entire binding complex could be selectively isolated via the fused affinity tag (Fig. 2e) [79, 112]. Notably, the purification could be performed under a condition that is denaturing for the protein to disrupt protein-protein interactions while maintaining DNA hybridization. Later, the same group has applied this method in the selection against a membrane protein and an intracellular protein with live cells [113]. Recently, the Neri group presented a critical evaluation of the crosslinking parameters in DEL selection experiments [70, 73]. Consistent with the findings from the Krusemark group [79], it has been proven that covalent crosslinking could facilitate the enrichment of low-affinity ligands in comparison to conventional solid-phase-based selection.

All the crosslinking-based methods require an ssDNA region for hybridization with the DNA strand that presents the crosslinker. However, the majority of DELs are built with double-stranded DNA (dsDNA) tags and are not, at least directly, compatible with these methods. To this end, the Li group recently reported a simple method to convert the dsDNA tags of DELs to ssDNA tags by using *Lambda* exonuclease (λ exo) digestion without the need for library resynthesis and redesign [114]. In addition, another Li group reported a novel approach that can interconvert the dsDNA and ssDNA tags in a DEL by using a reversible inter-strand DNA photo-crosslinker [115, 116].

4.3 Kinetic Separation

Krylov and coworker explored the approach to partition target-ligand binding complex from the unbound DNA library by utilizing nonequilibrium capillary electrophoresis (CE) of equilibrium mixtures (NECEEM) [85, 117–120]. The separation in CE is based on the large disparity of electrophoretic mobilities between the ligand-target complexes and the non-binders; in another word, the binder-target conjugates moved faster than the unbound members toward the negative-electrode end (Fig. 3a). This method is very straightforward and there is no special requirement of the library architecture; however, this method has only been tested with

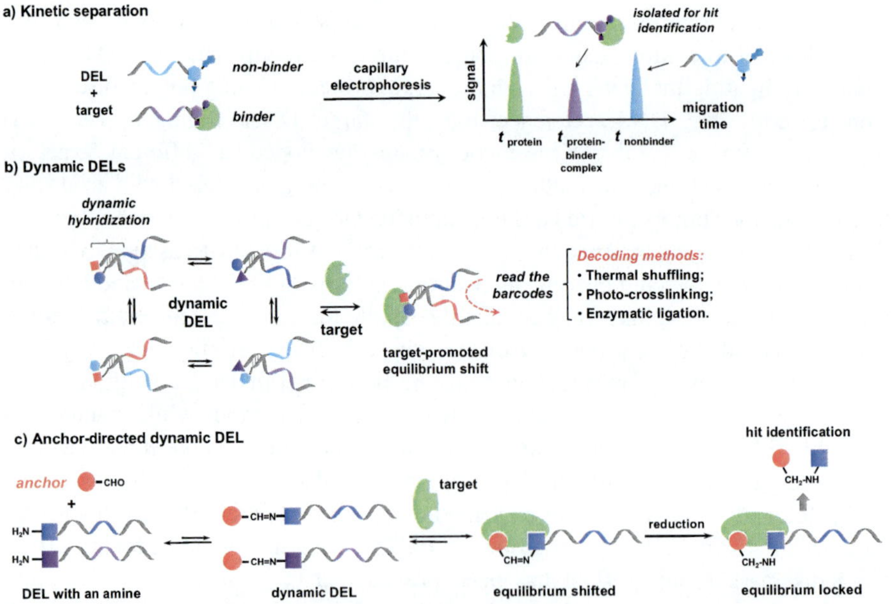

Fig. 3 Approaches have been developed for DEL selections against in-solution targets based on dynamic DNA hybridization or reversible chemical reaction. (**a**) Kinetic separation using CE.; (**b**) Several methods were developed for the synthesis and selection of dynamic DELs; and (**c**) anchor-directed dynamic DEL

several model systems and the effectiveness with large-scale DELs has yet to be proven.

4.4 DNA-Encoded Dynamic Library

Neri and co-workers developed an elegant dual-display DEL method named Encoded Self-Assembling Chemical (ESAC) library, in which high-quality combinatorial libraries of small molecule ("fragment") pairs could be assembled and selected against many biological targets [29, 97, 121]. In 2005, the Hamilton group demonstrated that the protein target could direct the assembly of high-affinity bivalent binders from a dynamic population of DNA duplexes [122]. Recently, Winssinger [28, 123], Zhang [23, 24, 124], Li [25–27, 40, 125] and respective co-workers integrated the two principles and developed a series of dynamic DEL methods (Fig. 3b) [40]. With dynamic DELs, the duplexes contain a short complementary region (6–7 nt), which enables the dynamic exchange/assembly of the DNA strands under physiological conditions. The target protein acts as a template to shift the equilibrium toward the formation of more high-affinity bidentate duplexes or 3-way junction pairs. The equilibrium could be stopped/locked by controlling the temperature, using irreversible photo-crosslinking, or enzymatic ligation. The crosslinked DNA duplexes could be purified for selective PCR-amplification without physical washes, and hit identification is achieved by comparing the library population change with and without target addition [40]. Furthermore, beyond the discovery of individual "fragments", the Li group reported a dynamic DEL method that could convert a "general-purpose" DEL to a biased one by dynamically conjugating a directing group (a known ligand of the target; the "anchor") to the library. In this study, the anchor molecule has an aldehyde group, and it is incubated with a DEL bearing a primary amine, thus forming the dynamic library through reversible imine formation [27]. After the addition of target, this equilibrium is locked by $NaBH_3CN$-mediated imine reduction. Notably, the authors took advantage of the reactivity difference between primary and secondary amines to isolate the target-bound amine species by biotin sulfo-*N*-hydroxy succinimidyl ester for hit identification (Fig. 3c). By switching the anchor molecule, this "*plug-and-play*" feature may convert unbiased libraries to biased ones toward different proteins targets. However, the authors also showed that this method appeared to be limited to the DELs with smaller compounds (e.g., with only two sets of building blocks), presumably because the large size of the library compound may dominate target binding and obscure the anchor's directing effect.

5 Selection with Complex Biological Targets

5.1 Cell Lysates and Protein Complexes

Besides the primary peptide sequence, proteins have many biological characteristics that are important for their cellular functions, such as the native structural folds, ionic charges, non-protein co-factors, protein binding partners, complex formation, post-translational modifications, etc.; however, they may not be faithfully recapitulated in the purified format. Hence, it is desirable to broaden the scope of DEL selections to include the targets in natural biological environment, which is expected to have higher potential to identify biologically relevant ligands. From a drug discovery point view, this may also help identify the candidate compounds with less attrition rate in the later stages of drug development. Previously, the Winssinger group demonstrated the application of PNA-encoded small molecule probes in profiling intact protease activities in crude cell lysates [126–128]. In principle, the IDUP and crosslinking-based methods discussed above (Fig. 2) should be suitable with the selections in cell lysates; so far there are two reported examples, one used the selective exonuclease digestion method [110]; the other one is based on the IDUP approach and notably, multiple targets were simultaneously selected with multiple libraries in cell lysates [103].

Proteins frequently exist in multi-component complexes in cells and their biological activities may rely on the specific composition of the complexes [129]. It is desirable and sometimes even necessary to subject the entire complex, rather than the individual proteins, to the selection. Many protein complexes are stable, and they could be recombinantly expressed, purified, and even immobilized. For instance, coimmunoprecipitation (co-IP) and affinity pulldown are ubiquitously used to study protein complexes [130, 131]. However, so far there is no peer-reviewed report on DEL selection against protein complexes. Recently, the Li group applied the DPAL method to profile histone deacetylases (HDAC) complexes in cell lysates [132]. The authors used a set of 12 DNA-based affinity probes and identified more than 100 potential novel HDAC interactors, including the indirect HDAC-associated proteins that are not tractable to small molecule probes. Although no DEL selection was described in this study, it has shown the potential of using the crosslinking-based method for DEL selection against endogenous protein complexes without purification.

5.2 Membrane Proteins

Membrane proteins on the cell surface perform crucial roles in orchestrating cellular behavior and are attractive pharmacological targets. It is unsurprising that membrane proteins constitute over 60% of the targets of all approved small molecule drugs [133, 134]. However, membrane proteins are extremely difficult to study owing to

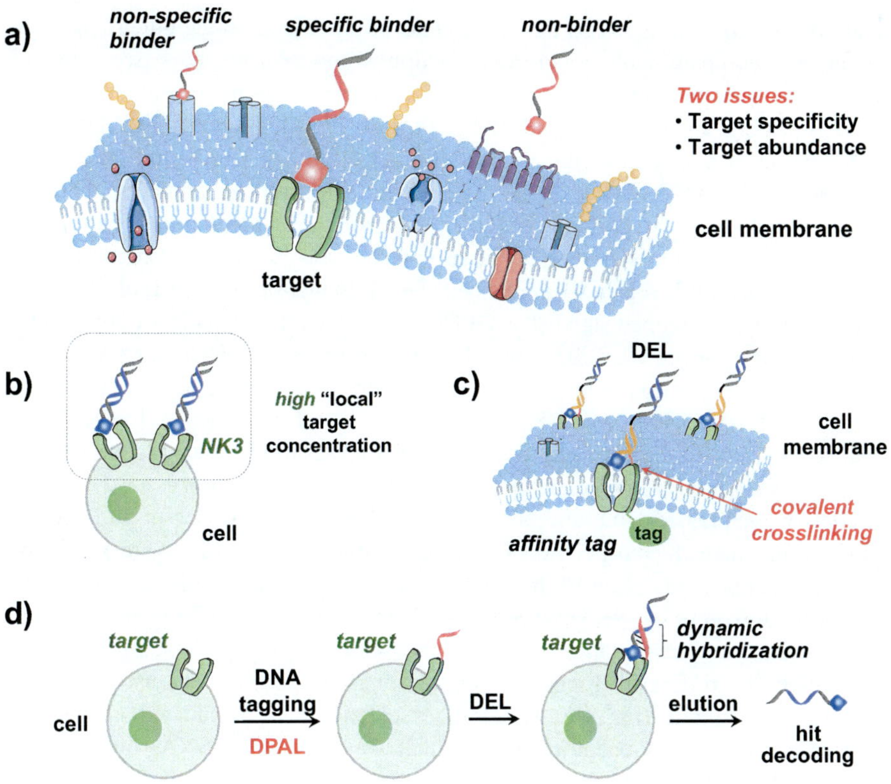

Fig. 4 Approaches have been developed for DEL selections against membrane proteins on live cells

their hydrophobic properties, large molecular size, and their dependence on the native lipid bilayer environment of the cell membrane. In the past, DELs have been used to interrogate the soluble domains of membrane proteins [29, 93, 135–137] as well as the full-length membrane proteins stabilized by nanodisc [138], detergent [139], or point mutations [140]. However, ideally, the selections with membrane proteins should be conducted on live cells in order to maintain their natural structures and properties. To accomplish this goal, two major obstacles need to be overcome: target specificity and target concentration (Fig. 4a). First, cell surface is a complex landscape with numerous non-target proteins and other biomolecules; thus, target binding is inevitably obscured by the vast background noises. Second, a high abundance of the target protein is needed to drive the binding equilibrium, since DEL compounds are in very low concentration in a selection. The typical effective molarity of membrane proteins on a cell is in the low- or sub-nanomolar range [42], which is significantly lower than the target concentration (high μM) typically used DEL selections. Researchers have addressed these issues by overexpressing the membrane protein to a very high level, because on the one hand, overexpression increases the target concentration, and on the other hand, one can always compare with a control selection without target overexpression to identify the specific binders. Early on, Bradley and co-workers pioneered cell-based selections by using PNA-encoded peptide libraries against live D54 and HEK293T cells over-expressing the integrin targets αvβ5, αvβ3, and chemokine receptor CCR6, respectively [141, 142]. The selection experiments with no overexpression or in the presence of a known ligand as the competitor were carried out in the same manner as the negative controls. In 2015, researchers at GSK reported the first DEL selection on live cells [143]. In this study, a DEL of 15 billion compounds was screened against a GPCR target tachykinin receptor neurokinin 3 (NK3) on transduced HEK293 cells. The NK3 protein reached ~500,000 receptors on each cell (10^7 cells/1.0 mL), so that the overall concentration of NK3 in the selected cell suspension was ~8.3 nM, which is still quite low. However, the localized, effective target concentration on each cell should be even higher, considering the NK3 protein molecules are confined in a very small volume (Fig. 4b). Besides overexpression, the authors also used sheared salmon sperm and sodium azide to attenuate the unspecific binding and target internalization, respectively. Importantly, this study showed that, by comparing with the negative control (cells without overexpressing NK3), specific NK3 antagonists could be identified with clear structure-activity relationship from the selection fingerprints. In 2019, the Krusemark group described another example of DEL selection against membrane protein on live cells (Fig. 4c) [113]. The selection was conducted against the overexpressed δ-opioid receptor (DOR), which was fused with a SNAP tag to enable the post-selection purification by using their previously reported crosslinking-based selection method [79]. After the selection, cells were lysed under the protein-denaturing condition to disrupt protein-protein interactions but without denaturing the DNA duplex. The conjugates were then affinity-purified through a BG-biotin probe and used for hit identification. The authors performed a model selection, and two DOR ligands of high affinity were enriched from a 96-member positional scanning library. Although the authors did not specify the exact concentration of DOR on the cells surface, a high effective molarity is expected because of the high cell density (2×10^6 cells/50 μL).

Recently, the Li group developed a DEL selection method that is compatible with live cells but without the need to protein overexpression or other genetic manipulation (Fig. 4d) [109]. The key feature of this method is the specific labeling of the target protein with a DNA tag prior to library selection. The DNA tag acts as a beacon guiding the DEL hybridization to achieve target specificity. Moreover, the hybridization places the library compound in proximity to the target, thereby significantly increasing the local effective target concentration without changing the number of protein molecules on the cell. The authors used the previously described DPAL method to install the DNA tag, and they showed that both small molecules and antibodies could be used to guide the DNA tagging, although small molecules were much more efficient than the antibodies. When antibody is used, a long spacer (18–21 nt) is needed in the capture probe to extend the photo-crosslinker away and avoid labeling the antibody itself. In principle, other types of ligands, such as aptamers, small-engineered proteins (e.g., nanobodies), and peptides, which have been applied to graft DNAs on live cells in other applications, may also be used to guide the labeling to expand the target scope. After tagging, a toehold displacement step is applied to remove the original binding probe and reopen the binding site. Next, the library is incubated with the DNA-tagged cells. Similar to dynamic DELs, the DNA tag is designed to have a short complementary region to the primer bind site of the library. The binders have increased duplex stability and could be retained on the cell surface, while the non-binders only form unstable, dynamic duplexes that are removed during the washing steps. The calculation estimated that a 6–8 nucleotide tag may increase the binding affinity of a weak binder from high μM to low nM, and the length of the DNA tag could be used to tune the selection stringency. A 30.42-million-member DEL was selected against three membrane proteins (carbonic anhydrase 12, folate receptor, and epidermal growth factor receptor) to demonstrate the generality and performance of the method, and ligands of micromolar to nanomolar affinities have been identified. More recently, Neri and co-workers have systematically optimized the experimental parameters implicated in live cell DEL selections and they suggested that multivalency may be an effective approach to drive ligand binding on the cell surface [73].

5.3 Intracellular Selections

Selecting DELs against intracellular targets is highly appealing, since it offers the opportunities to interrogate proteins and other biological targets straightly within the cytosol or other organelles [42]. However, the impermeability and the cellular instability of DNA present two major issues. So far there are two reported intracellular DEL selections. In 2019, the Krusemark group addressed the DEL delivery issue by appending a cyclic cell penetrating peptide (cCPP) to the distal end of the library to facilitate cell entry (Fig. 5a) [113]. Then, the intracellular DEL selection was conducted following their previously developed crosslinking-based protocol [79], where the transient interaction between the target and the ligands was stabilized via covalent crosslinker and the binding complex was isolated via the fused affinity tag on the protein. In 293F cells, the authors overexpressed two proteins with HaloTag, the *Escherichia coli* dihydrofolate reductase (eDHFR) and Chromobox Protein Homolog 7 Chromodomain (CBX7-ChD), respectively, which were used to validate and optimize the method. Furthermore, a 96-member positional scanning library was selected against CBX7-ChD. Although only model systems were tested, the results convincingly demonstrated that target-specific DEL selection can be realized within live cells. More recently in 2021, researchers at Vipergen reported another in-cell DEL selection named cellular binder trap enrichment (cBTE) (Fig. 5b) [106]. In this study, the authors directly delivered the library into *Xenopus* oocyte via microinjection. *Xenopus* oocyte is a gigantic cell suitable for microinjection and also for protein overexpression [106]. The intracellular target was overexpressed and fused to a known "prey" protein tag. A high-affinity binder-

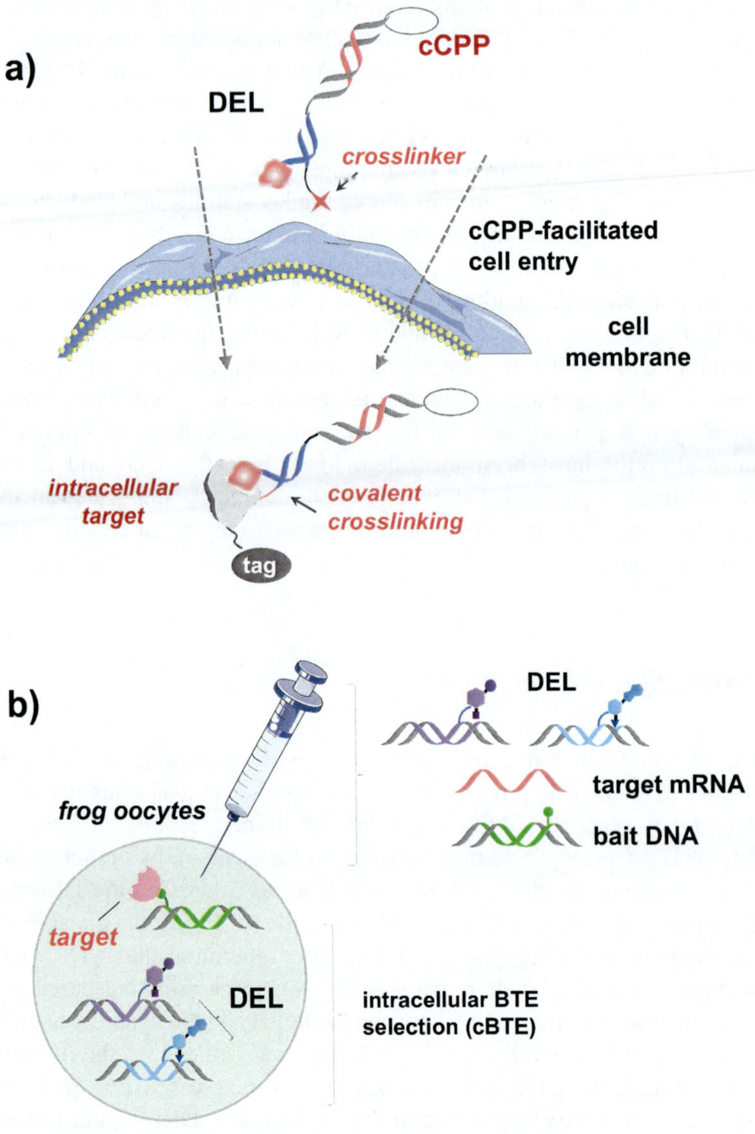

Fig. 5 Two strategies have been developed to deliver DELs into live cells for intracellular selections by using (**a**) cyclic cell-penetrating peptide (cCPP) or (**b**) microinjection

DNA conjugate for the prey protein was used as the "bait". After microinjecting the mRNAs encoding the target-prey fusion for 4 days, the apparent expression level was measured by ELISA. The "bait" and the DEL were subsequently co-injected into the oocyte. Inside the cell, the binders engaged the target and the bait bound to the "prey," which were then subjected to the water-in-oil emulsion, dilution, enzymatic ligation, and hit deconvolution following their previously reported BTE protocol [104, 105]. A 194-million-member DEL was selected against the therapeutically relevant proteins ACSS2, p38α, and DOCK5, and nanomolar binders were identified and validated for p38α. These two studies showed that DEL delivery into live cells could be achieved by using microinjection or cell-penetrating peptides. Other strategies that have been developed for anti-sense oligo delivery might also be explored for intracellular DEL selections.

5.4 Targeting Nucleic Acids

Nucleic acids (RNA and DNA) with complex folded structures tend to contain the pockets that allow for specific binding by small molecule ligands. For instance, 70% of the human genome is transcribed into non-coding RNAs, among which long non-coding RNAs and microRNAs contain highly structured elements and are well-validated therapeutic targets [144]. Researchers in HitGen explored DEL selections with two RNA targets, HIV-1 trans-acting responsive region (TAR) RNA and *Escherichia coli* FMN riboswitch [145]. Apparently, a predominant issue for selecting DELs against RNA is the potential DNA-RNA interactions. To this end, the authors not only optimized the selection strategy by using RNA patch blocking in combination with competitive elution to reduce DNA binding, but also designed an algorithm called *k*-mer to quantify the level of DNA-RNA interactions. This optimized method led to the discovery of two ligands of *E. coli* FMN riboswitch with nanomolar binding affinities. In 2020, a report mentioned DEL selection against Aptamer 21 by using the yoctoReactor DEL platform from Vipergen; however, the details were not described [146]. Recently, in an impressive study, Paegel, Disney, and co-workers have advanced DEL selection against nucleic acids to a new level. The authors explored the multiplexed selections of OBOC-DELs with a library of three-dimensional RNA folds (4,096 RNA targets), thus totaling ~300 million ligand-RNA interactions, simultaneously [147]. The authors first decoded the binders by isolating the beads specifically bound to the RNA targets using a high-throughput two-color FACS sorting. Next, the authors picked 10 representative ligands and identified their targets from the RNA library by using microarray. One DEL-derived ligand was found to bind to pri-miR-27a, an oncogenic miRNA in breast and prostate cancers. Finally, another attractive type of nucleic acid targets is G-quadruplex secondary structures (G4). Formed in nucleic acids by sequences that are rich in guanine, G4 folds into uniquely stacked non-classical secondary structure that plays important functional roles in life cycle of cells [148–150]. Using 33 different DELs, researchers at X-Chem conducted parallel screening against several different types of G4 motifs and identified hit compounds with moderate potency and selectivity for the G4 targets [151].

6 Functional DEL Selections

DELs have been proven to be a powerful technique to identify physical binders. However, it is important to keep in mind that a binder may not always elicit a biological response. In fact, DEL selection is an unbiased binding assay that the entire protein molecule, not only the catalytic pocket or the functionally relevant sites, is sampled by the library compounds. Although a follow-up hit validation step is always performed, it is still highly desirable to be able to straightly identify functionally active compounds from the selection. Even better, if the selection assay could be designed to bias a specific biological function or signal output (e.g., inhibition/activation, antagonism/agonism), it would open many new and exciting opportunities for DEL beyond ligand discovery. Indeed, innovative selection and assay strategies have been developed and also envisioned by many researchers in the DEL field and are summarized in this section.

6.1 Ligand-Guided Selection to a Functionally Relevant Binding Site

In general, DEL selection is not site selective, but site selectivity may be achieved by designing the assay conditions based on the prior knowledge of the target. For instance, conducting the selection in the presence/absence of a known inhibitor may help identify the competitive inhibitors (Fig. 6a). Clark and co-workers selected a DEL against Bruton's tyrosine kinase (BTK) with the addition of ATP or a known inhibitor Dasatinib to identify ATP-competitive inhibitors. Interestingly, the selection also discovered several ligands that bind to novel binding sites [152]. The structure of receptor proteins may be stabilized in an active/inactive form by binding to an agonist/antagonist (Fig. 6b), which could be taken advantage of in DEL selections to identify novel agonists/antagonists. Clark and co-workers demonstrated this strategy in a selection campaign against a GPCR protein protease-activated receptor 2 (PAR2). By leveraging known orthosteric antagonists, novel ligands comprising both agonists and antagonists were identified. The agonists shared a similar structure with known agonists, whereas a novel series of allosteric antagonists were shown to have novel binding modes [93, 140]. Moreover, Lefkowitz and coworker presented the discovery and characterization of the first positive allosteric modulators of the β_2-adrenoceptor (β_2AR) by panning DELs against the purified β_2AR protein occupied by a high-affinity agonist [138].

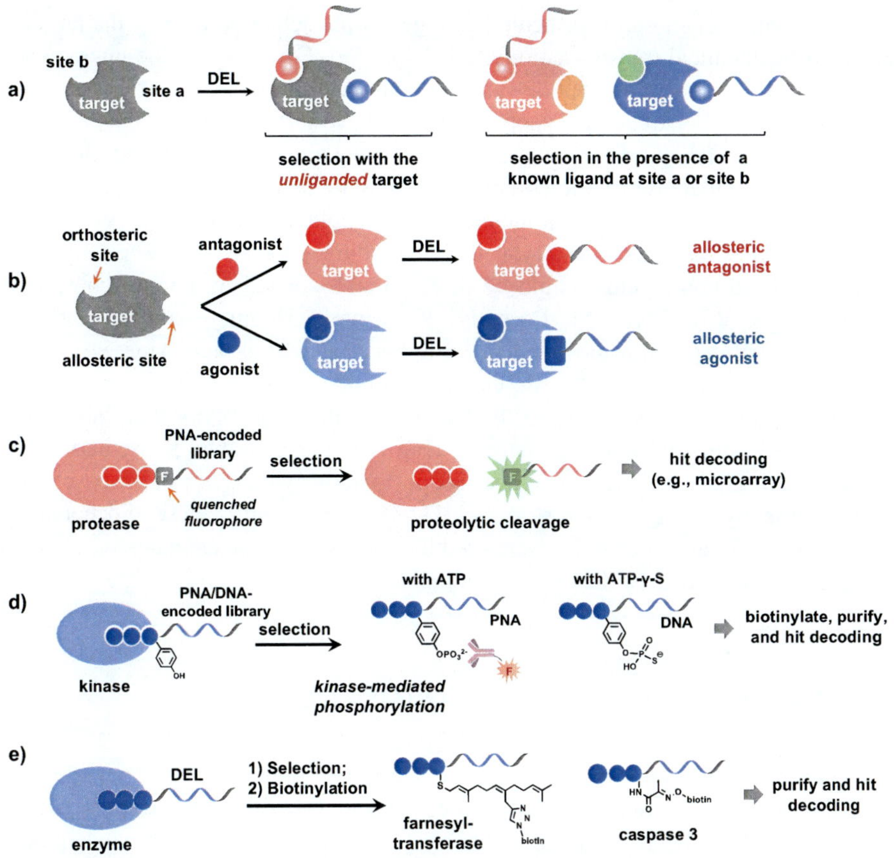

Fig. 6 Many strategies have been developed to realize functional DEL assays. (**a**) Performing DEL selections in the presence and absence of a ligand with known binding site may identify site-selective binders. (**b**) Using a known antagonist or agonist for a receptor protein target could stabilize the conformation of the target in its inactive or active form; thus, the selection may preferably select for allosteric antagonists or agonists. (**c**) A protease target may cleave the peptide in the library and release the fluorescent PNA tag of the substrate peptides, which could be decoded to identify the substrate peptides. (**d**) A kinase target may phosphorylate the tyrosine residue of the ligand, which could be biotinylated and isolated for hit identification. (**e**) For farnesyltransferase and caspase, these enzymes may modify the substrate ligand with biotinylated motifs, which could be further isolated for hit identification

6.2 Encoding the Outcome of the Biochemical Reaction

In a regular selection, the DNA tag of all physical binders are enriched. For a functional DEL assay, ideally it should only enrich the ligands that can elicit the desired biological change/transformation in a form of a specific signal readout. One way to achieve this is using the DNA tag to encode the biochemical outcome. Previously, substrate-based selection has been developed in the field of protein profiling, which relied on the structural changes of the active binders by the target-mediated biochemical transformation [112, 126, 153]. For instance, the interaction between a protease with its peptide substrate may result in a cleavage reaction. By using fluorogenic PNA-encoded peptides, studies have shown that only the protease substrates, of which the bond connecting the latent fluorophore can be cleaved, resulted in large increase in fluorescence and could be identified from the selection (Fig. 6c) [15, 154]. The same principle can be applied to kinases, as the small molecule binders may be phosphorylated upon binding to the enzyme. The active hit (phosphorylated compounds) could be detected and isolated by an anti-phospho-antibody (Fig. 6d) [15, 155]. Recently, Krusemark, Harbury, and co-workers adopted this strategy in a biochemical DEL assay against protein kinase A. The selections were conducted in the presence of a peptide library and ATP-γ-S, so that only the peptide substrates, not just the binders, would be thio-phosphorylated and then purified with biotinylating reagents (Fig. 6e) [156]. Later, Krusemark and co-workers further extended the target scope to farnesyltransferase, caspase and c-Src kinase by using similar strategies [157]. In summary, these biochemical DEL assays were designed to only enrich only the library members whose chemical structures are modified by the enzyme target.

6.3 *One-Bead, One-Compound (OBOC) DELs*

The concept of DEL was originally proposed aiming to introduce DNA encoding to the traditional OBOC combinatorial libraries [8], and the early efforts were focusing on the parallel synthesis of peptides (library compounds) and the oligonucleotides (encoding tags) on the same bead [9, 10]. Although later the solution phase DELs became the mainstream, OBOC-DEL is currently experiencing a renaissance; modern OBOC-DELs are able to spatially confine the individual binding and biological events in microdroplets and thus enable biochemical and even phenotypical DEL assays [38, 39, 57].

The Paegel group has developed a suite of modern instrumentation and elegant methodologies for efficient synthesis and selection of OBOC-DELs, where individual beads could be encapsulated in individual microfluidic droplet; the beads also feature a sophisticated, multi-functional, and photo-cleavable linker connecting the compound with the solid matrix as well as the DNA tag (Fig. 7a) [38, 158, 159]. The library compound can be photochemically cleaved from the bead but still confined within the small volume of the droplet; therefore, it acts as a free ligand, rather than a small-molecule-DNA conjugate as in regular DELs, to interrogate the target and elicit the biological response, such as inhibiting the fluorogenic biochemical reaction of the target enzyme (Fig. 7b) [158, 160]. Essentially, the microdroplet creates a microenvironment that contains all the assay components as well as the DNA tag. The beads are spatially separated so that one can easily attribute the signal readout to the specific compounds. In addition, flow cytometry can be used to efficiently sort the beads for hit decoding [158]. Collectively, these novel features of modern

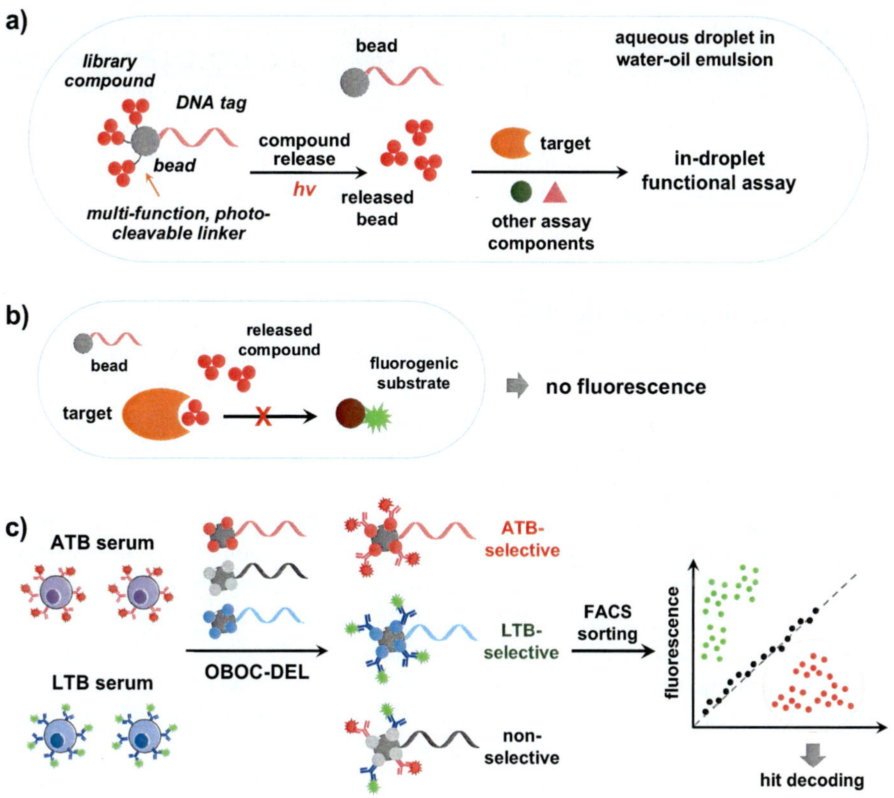

Fig. 7 Functional DEL assays using OBOC-DELs. (**a**) The basic principle of OBOC-DEL-based functional assays. (**b**) An example of using OBOC-DEL in an inhibitor assay; the released compound may inhibit the enzyme in the same microdroplet, thus leading to fluorescence suppression. (**c**) Selection of an OBOC-DEL against human sera to identify the compounds that preferably bind to different disease states

OBOC-DELs, combined with robust on-bead chemistry [35, 161], have paved the way for many novel applications. Paegel and co-workers carried out OBOC-DEL selections against a variety of targets, including Cathepsin D (CatD) [158], discoidin domain receptor 1 (DDR1, 39], and phosphodiesterase autotaxin (ATX) [38], and successfully discovered many potent inhibitors. Moreover, modern OBOC-DEL is poised to increase the target space of DELs to more complex biological systems. Kodadek, Paegel, and co-workers recently described an improved version of this technology by extending the utility of OBOC-DELs in human serum [36]. By using a two-color screening strategy, the ligands that could differentiate the patient sera of two disease states (active tuberculosis and latent tuberculosis; ATB and LTB, respectively) were identified. The author also further identified the potential target, the native antigen Ag85B, that the ligands bound to (Fig. 7c). In addition, Paegel and co-workers have applied OBOC-DELs for antibacterial agent discovery against *Escherichia coli and Bacillus subtili.* Upon inoculated agar plates, a suspension of beads is layered onto the plates using a whole cell bead diffusion assay. The visible beads surrounded with non-opaque growth inhibition zone could be collected for decoding [162]. Overall, it is quite clear that modern OBOC-DEL is beyond simply adding an oligonucleotide to a solid matrix; it is rather an all-around technology platform that can be adapted to become different assay formats. Furthermore, OBOC-DELs have been shown and proposed to have the potential to realize phenotypic DEL assays in recent reviews and patent applications [53, 57, 147, 163].

7 Summary and Outlook

Selection of DNA-encoded libraries has provided promising starting points for the discovery of drug candidates. The maturation of robust DNA-compatible chemical transformations has further provided DELs many efficient tools to explore the chemical space of biological and therapeutic significance [17, 44, 48–50, 164]. Traditional screening against immobilized proteins has already delivered DELs with ample opportunities for new discoveries, which will still be the predominantly adopted and the most productive way of DEL selections in the foreseeable future. However, the expansion of the target scope of DELs to soluble proteins, cell lysates, on the cell surface, inside live cells, nucleic acids, whole bacteria, and human sera, have opened many exciting opportunities for DELs. We envision that methodology development of DEL selection may exploit more complex and dynamic biological systems, such primary cells, organoids, tissue samples, or even live animals [41]. Moreover, phenotypic assay is a powerful approach to identify small molecule modulators that can induce holistic changes in biology and diseases; however, low screening throughout and the high cost have been the major issues. The recent development of OBOC-DELs has certainly made exciting progress at this direction [7, 41, 42, 45, 52, 57]. DEL assays may also be designed to discover modulators for other biochemical transformations other than phosphorylation and peptide cleavage, such as ubiquitination and post-translational modifications, which may even lead to methods for DEL selections against specific signaling pathways. Finally, integration with other techniques, such as affinity selection-mass spectrometry (ASMS) [165], DNA origami [166], PROteolysis Targeting Chimera (PROTAC) [7], cancer theranostics [167], and machine learning [20–22], will further empower DELs to new paradigms.

Compliance with Ethical Standards
This chapter is a review and no humans or animals were involved in the preparation of the chapter. The authors declare no conflict of interest.

Funding: This work was supported by grants from the Shenzhen Bay Laboratory, Shenzhen, China (SZBL2020090501008), the Research Grants Council of Hong Kong SAR, China (AoE/P-705/16, 17301118, 17111319, 17303220, 17300321, and C7005-20G), NSFC of China (21877093 and 91953119). We acknowledge the support from "Laboratory for Synthetic Chemistry and Chemical Biology" under the Health@InnoHK Program and State Key Laboratory of Synthetic Chemistry by Innovation and Technology Commission, Hong Kong SAR, China. Correspondence should be addressed to Y. G or X. L.

References

1. Sunkari YK, Siripuram VK, Nguyen TL, Flajolet M (2022) High-power screening (hps) empowered by DNA-encoded libraries. Trends Pharmacol Sci 43:4–15
2. McCafferty J, Griffiths AD, Winter G, Chiswell DJ (1990) Phage antibodies: filamentous phage displaying antibody variable domains. Nature 348:552–554
3. Hanes J, Plückthun A (1997) In vitro selection and evolution of functional proteins by using ribosome display. Proc Natl Acad Sci U S A 94:4937–4942
4. Wilson DS, Keefe AD, Szostak JW (2001) The use of mrna display to select high-affinity protein-binding peptides. Proc Natl Acad Sci U S A 98:3750–3755
5. Boder ET, Wittrup KD (1997) Yeast surface display for screening combinatorial polypeptide libraries. Nat Biotechnol 15:553–557
6. Lerner RA, Brenner S (2017) DNA-encoded compound libraries as open source: a powerful pathway to new drugs. Angew Chem Int Ed 56:1164–1165
7. Huang Y, Li Y, Li X (2022) Strategies for developing DNA-encoded libraries beyond binding assays. Nat Chem 14:129–140
8. Brenner S, Lerner RA (1992) Encoded combinatoria. Proc Natl Acad Sci U S A 89:5381–5383
9. Nielsen J, Brenner S, Janda KD (1993) Synthetic methods for the implementation of encoded combinatorial chemistry. J Am Chem Soc 115:9812–9813
10. Needels MC, Jones DG, Tate EH, Heinkel GL, Kochersperger LM, Dower WJ, Barrett RW, Gallop MA (1993) Generation and screening of an oligonucleotide-encoded synthetic peptide library. Proc Natl Acad Sci U S A 90:10700–10704
11. Kinoshita Y, Nishigaki K (1995) Enzymatic synthesis of code regions for encoded combinatorial chemistry (ecc). Nucleic Acids Symp Ser:201–202
12. Melkko S, Scheuermann J, Dumelin CE, Neri D (2004) Encoded self-assembling chemical libraries. Nat Biotechnol 22:568–574
13. Gartner ZJ, Tse BN, Grubina R, Doyon JB, Snyder TM, Liu DR (2004) DNA-templated organic synthesis and selection of a library of macrocycles. Science 305:1601–1605
14. Halpin DR, Harbury PB (2004) DNA display i. Sequence-encoded routing of DNA populations. PLoS Biol 2(E173)
15. Winssinger N, Damoiseaux R, Tully DC, Geierstanger BH, Burdick K, Harris JL (2004) Pna-encoded protease substrate microarrays. Chem Biol 11:1351–1360
16. Clark MA, Acharya RA, Arico-Muendel CC, Belyanskaya SL, Benjamin DR, Carlson NR, Centrella PA, Chiu CH, Creaser SP, Cuozzo JW, Davie CP, Ding Y, Franklin GJ, Franzen KD, Gefter ML, Hale SP, Hansen NJ, Israel DI, Jiang J, Kavarana MJ, Kelley MS, Kollmann CS, Li F, Lind K, Mataruse S, Medeiros PF, Messer JA, Myers P, O'Keefe H, Oliff MC, Rise CE, Satz AL, Skinner SR, Svendsen JL, Tang L, van Vloten K, Wagner RW, Yao G, Zhao B, Morgan BA (2009) Design, synthesis and selection of DNA-encoded small-molecule libraries. Nat Chem Biol 5:647–654
17. Fitzgerald PR, Paegel BM (2021) DNA-encoded chemistry: drug discovery from a few good reactions. Chem Rev 121:7155–7177
18. Dickson P, Kodadek T (2019) Chemical composition of DNA-encoded libraries, past present and future. Org Biomol Chem 17:4676–4688
19. Goodnow Jr RA, Dumelin CE, Keefe AD (2017) DNA-encoded chemistry: enabling the deeper sampling of chemical space. Nat Rev Drug Discov 16:131–147
20. Guasch L, Reutlinger M, Stoffler D, Wichert M (2021) Augmenting chemical space with DNA-encoded library technology and machine learning. Chimia (Aarau) 75:105–107
21. McCloskey K, Sigel EA, Kearnes S, Xue L, Tian X, Moccia D, Gikunju D, Bazzaz S, Chan B, Clark MA, Cuozzo JW, Guie MA, Guilinger JP, Huguet C, Hupp CD, Keefe AD, Mulhern CJ, Zhang Y, Riley P (2020) Machine learning on DNA-encoded libraries: a new paradigm for hit finding. J Med Chem 63:8857–8866
22. Martin A, Nicolaou CA, Toledo MA (2020) Navigating the DNA encoded libraries chemical space. Commun Chem 3:1–9

23. Reddavide FV, Lin W, Lehnert S, Zhang Y (2015) DNA-encoded dynamic combinatorial chemical libraries. Angew Chem Int Ed 54:7924–7928
24. Reddavide FV, Cui M, Lin W, Fu N, Heiden S, Andrade H, Thompson M, Zhang Y (2019) Second generation DNA-encoded dynamic combinatorial chemical libraries. Chem Commun 55:3753–3756
25. Li G, Zheng W, Chen Z, Zhou Y, Liu Y, Yang J, Huang Y, Li X (2015) Design, preparation, and selection of DNA-encoded dynamic libraries. Chem Sci 6:7097–7104
26. Zhou Y, Li C, Peng J, Xie L, Meng L, Li Q, Zhang J, Li XD, Li X, Huang X, Li X (2018) DNA-encoded dynamic chemical library and its applications in ligand discovery. J Am Chem Soc 140:15859–15867
27. Deng Y, Peng J, Xiong F, Song Y, Zhou Y, Zhang J, Lam FS, Xie C, Shen W, Huang Y, Meng L, Li X (2020) Selection of DNA-encoded dynamic chemical libraries for direct inhibitor discovery. Angew Chem Int Ed 59:14965–14972
28. Machida T, Novoa A, Gillon E, Zheng S, Claudinon J, Eierhoff T, Imberty A, Romer W, Winssinger N (2017) Dynamic cooperative glycan assembly blocks the binding of bacterial lectins to epithelial cells. Angew Chem Int Ed 56:6762–6766
29. Wichert M, Krall N, Decurtins W, Franzini RM, Pretto F, Schneider P, Neri D, Scheuermann J (2015) Dual-display of small molecules enables the discovery of ligand pairs and facilitates affinity maturation. Nat Chem 7:241–249
30. Daguer JP, Zambaldo C, Ciobanu M, Morieux P, Barluenga S, Winssinger N (2015) DNA display of fragment pairs as a tool for the discovery of novel biologically active small molecules. Chem Sci 6:739–744
31. Daguer JP, Ciobanu M, Alvarez S, Barluenga S, Winssinger N (2011) DNA-templated combinatorial assembly of small molecule fragments amenable to selection/amplification cycles. Chem Sci 2:625–632
32. Barluenga S, Zambaldo C, Ioannidou HA, Ciobanu M, Morieux P, Daguer JP, Winssinger N (2016) Novel ptp1b inhibitors identified by DNA display of fragment pairs. Bioorg Med Chem Lett 26:1080–1085
33. Bobers J, Skopic MK, Dinter R, Sakthithasan P, Neukirch L, Gramse C, Weberskirch R, Brunschweiger A, Kockmann N (2020) Design of an automated reagent-dispensing system for reaction screening and validation with DNA-tagged substrates. ACS Comb Sci 22:101–108
34. Castanon J, Roman JP, Jessop TC, de Blas J, Haro R (2018) Design and development of a technology platform for DNA-encoded library production and affinity selection. SLAS Discov 23:387–396
35. Shin MH, Lee KJ, Lim HS (2019) DNA-encoded combinatorial library of macrocyclic peptoids. Bioconjug Chem 30:2931–2938
36. Mendes KR, Malone ML, Ndungu JM, Suponitsky-Kroyter I, Cavett VJ, McEnaney PJ, MacConnell AB, Doran TM, Ronacher K, Stanley K, Utset O, Walzl G, Paegel BM, Kodadek T (2017) High-throughput identification of DNA-encoded igg ligands that distinguish active and latent mycobacterium tuberculosis infections. ACS Chem Biol 12:234–243
37. MacConnell AB, McEnaney PJ, Cavett VJ, Paegel BM (2015) DNA-encoded solid-phase synthesis: encoding language design and complex oligomer library synthesis. ACS Comb Sci 17:518–534
38. Cochrane WG, Malone ML, Dang VQ, Cavett V, Satz AL, Paegel BM (2019) Activity-based DNA-encoded library screening. ACS Comb Sci 21:425–435
39. Hackler AL, FitzGerald FG, Dang VQ, Satz AL, Paegel BM (2020) Off-DNA DNA-encoded library affinity screening. ACS Comb Sci 22:25–34
40. Shi B, Zhou Y, Li X (2022) Recent advances in DNA-encoded dynamic libraries. RSC Chem Biol 3:407–419
41. Song Y, Li X (2021) Evolution of the selection methods of DNA-encoded chemical libraries. Acc Chem Res 54:3491–3503
42. Huang Y, Li X (2021) Recent advances on the selection methods of DNA-encoded libraries. Chembiochem 22:2384–2397

43. Shi B, Zhou Y, Huang Y, Zhang J, Li X (2017) Recent advances on the encoding and selection methods of DNA-encoded chemical library. Bioorg Med Chem Lett 27:361–369
44. Satz AL, Brunschweiger A, Flanagan ME, Gloger A, Hansen NJV, Kuai L, Kunig VBK, Lu X, Madsen D, Marcaurelle LA, Mulrooney C, O'Donovan G, Sakata S, Scheuermann J (2022) DNA-encoded chemical libraries. Nat Rev Methods Primers 2:1–17
45. Gironda-Martinez A, Donckele EJ, Samain F, Neri D (2021) DNA-encoded chemical libraries: a comprehensive review with succesful stories and future challenges. ACS Pharmacol Transl Sci 4:1265–1279
46. Goodnow RA, Davie CP (2017) DNA-encoded library technology: a brief guide to its evolution and impact on drug discovery, platform technologies in drug discovery and validation, pp 1–15
47. Madsen D, Azevedo C, Micco I, Petersen LK, Hansen NJV (2020) An overview of DNA-encoded libraries: a versatile tool for drug discovery. Prog Med Chem 59:181–249
48. Plais L, Scheuermann J (2022) Macrocyclic DNA-encoded chemical libraries: a historical perspective. RSC Chem Biol 3:7–17
49. Shi Y, Wu Y-r, Yu J-q, Zhang W-n, Zhuang C-l (2021) DNA-encoded libraries (dels): a review of on-DNA chemistries and their output. RSC Adv 11:2359–2376
50. Reiher CA, Schuman DP, Simmons N, Wolkenberg SE (2021) Trends in hit-to-lead optimization following DNA-encoded library screens. ACS Med Chem Lett 12:343–350
51. Song M, Hwang GT (2020) DNA-encoded library screening as core platform technology in drug discovery: its synthetic method development and applications in del synthesis. J Med Chem 63:6578–6599
52. Conole D, Hunter JH, Waring MJ (2021) The maturation of DNA encoded libraries: opportunities for new users. Future Med Chem 13:173–191
53. Satz AL, Kuai L, Peng X (2021) Selections and screenings of DNA-encoded chemical libraries against enzyme and cellular targets. Bioorg Med Chem Lett 39:127851
54. Flood DT, Kingston C, Vantourout JC, Dawson PE, Baran PS (2020) DNA encoded libraries: a visitor's guide. Isr J Chem 60:1–14
55. Kunig VBK, Potowski M, Klika Skopic M, Brunschweiger A (2021) Scanning protein surfaces with DNA-encoded libraries. ChemMedChem 16:1048–1062
56. Götte K, Chines S, Brunschweiger A (2020) Reaction development for DNA-encoded library technology: from evolution to revolution? Tetrahedron Lett 61:151889
57. Kodadek T, Paciaroni NG, Balzarini M, Dickson P (2019) Beyond protein binding: recent advances in screening DNA-encoded libraries. Chem Commun 55:13330–13341
58. Kang AS, Barbas CF, Janda KD, Benkovic SJ, Lerner RA (1991) Linkage of recognition and replication functions by assembling combinatorial antibody fab libraries along phage surfaces. Proc Natl Acad Sci 88:4363–4366
59. Clackson T, Hoogenboom HR, Griffiths AD, Winter G (1991) Making antibody fragments using phage display libraries. Nature 352:624–628
60. Wilson DS, Keefe AD, Szostak JW (2001) The use of mrna display to select high-affinity protein-binding peptides. Proc Natl Acad Sci 98:3750–3755
61. Hanes J, Plückthun A (1997) In vitro selection and evolution of functional proteins by using ribosome display. Proc Natl Acad Sci 94:4937–4942
62. Wang S, Denton KE, Hobbs KF, Weaver T, McFarlane JMB, Connelly KE, Gignac MC, Milosevich N, Hof F, Paci I, Musselman CA, Dykhuizen EC, Krusemark CJ (2020) Optimization of ligands using focused DNA-encoded libraries to develop a selective, cell-permeable cbx8 chromodomain inhibitor. ACS Chem Biol 15:112–131
63. Kleiner RE, Dumelin CE, Tiu GC, Sakurai K, Liu DR (2010) In vitro selection of a DNA-templated small-molecule library reveals a class of macrocyclic kinase inhibitors. J Am Chem Soc 132:11779–11791
64. Maianti JP, McFedries A, Foda ZH, Kleiner RE, Du XQ, Leissring MA, Tang WJ, Charron MJ, Seeliger MA, Saghatelian A, Liu DR (2014) Anti-diabetic activity of insulin-degrading enzyme inhibitors mediated by multiple hormones. Nature 511:94–98

65. Gerry CJ, Wawer MJ, Clemons PA, Schreiber SL (2019) DNA barcoding a complete matrix of stereoisomeric small molecules. J Am Chem Soc 141:10225–10235
66. Usanov DL, Chan AI, Maianti JP, Liu DR (2018) Second-generation DNA-templated macrocycle libraries for the discovery of bioactive small molecules. Nat Chem 10:704–714
67. Chen Q, Cheng X, Zhang L, Li X, Chen P, Liu J, Zhang L, Wei H, Li Z, Dou D (2020) Exploring the lower limit of individual DNA-encoded library molecules in selection. SLAS Discov 25:523–529
68. Sannino A, Gabriele E, Bigatti M, Mulatto S, Piazzi J, Scheuermann J, Neri D, Donckele EJ, Samain F (2019) Quantitative assessment of affinity selection performance by using DNA-encoded chemical libraries. Chembiochem 20:955–962
69. Li Y, Zimmermann G, Scheuermann J, Neri D (2017) Quantitative pcr is a valuable tool to monitor the performance of DNA-encoded chemical library selections. Chembiochem 18: 848–852
70. Sannino A, Gironda-Martinez A, Gorre EMD, Prati L, Piazzi J, Scheuermann J, Neri D, Donckele EJ, Samain F (2020) Critical evaluation of photo-cross-linking parameters for the implementation of efficient DNA-encoded chemical library selections. ACS Comb Sci 22: 204–212
71. Kivioja T, Vaharautio A, Karlsson K, Bonke M, Enge M, Linnarsson S, Taipale J (2011) Counting absolute numbers of molecules using unique molecular identifiers. Nat Methods 9: 72–74
72. Decurtins W, Wichert M, Franzini RM, Buller F, Stravs MA, Zhang Y, Neri D, Scheuermann J (2016) Automated screening for small organic ligands using DNA-encoded chemical libraries. Nat Protoc 11:764–780
73. Oehler S, Catalano M, Scapozza I, Bigatti M, Bassi G, Favalli N, Mortensen MR, Samain F, Scheuermann J, Neri D (2021) Affinity selections of DNA-encoded chemical libraries on carbonic anhydrase ix-expressing tumor cells reveal a dependence on ligand valence. Chem A Eur J 27:8985–8993
74. Zhou Y, Shen W, Peng J, Deng Y, Li X (2021) Identification of isoform/domain-selective fragments from the selection of DNA-encoded dynamic library. Bioorg Med Chem 45:116328
75. Hall J, Foley TL, Chen Q, Israel DI, Xu Y, Ford KK, Xie P, Fan J, Wan J (2020) A simple method for determining compound affinity and chemical yield from DNA-encoded library selections. Biochem Biophys Res Commun 527:250–256
76. Satz AL (2015) DNA encoded library selections and insights provided by computational simulations. ACS Chem Biol 10:2237–2245
77. Disch JS, Duffy JM, Lee ECY, Gikunju D, Chan B, Levin B, Monteiro MI, Talcott SA, Lau AC, Zhou F, Kozhushnyan A, Westlund NE, Mullins PB, Yu Y, von Rechenberg M, Zhang J, Arnautova YA, Liu Y, Zhang Y, McRiner AJ, Keefe AD, Kohlmann A, Clark MA, Cuozzo JW, Huguet C, Arora S (2021) Bispecific estrogen receptor alpha degraders incorporating novel binders identified using DNA-encoded chemical library screening. J Med Chem 64: 5049–5066
78. Weisinger RM, Wrenn SJ, Harbury PB (2012) Highly parallel translation of DNA sequences into small molecules. PLoS One 7:e28056
79. Denton KE, Krusemark CJ (2016) Crosslinking of DNA-linked ligands to target proteins for enrichment from DNA-encoded libraries. MedChemComm 7:2020–2027
80. Satz AL (2016) Simulated screens of DNA encoded libraries: the potential influence of chemical synthesis fidelity on interpretation of structure-activity relationships. ACS Comb Sci 18:415–424
81. Kuai L, O'Keeffe T, Arico-Muendel C (2018) Randomness in DNA encoded library selection data can be modeled for more reliable enrichment calculation. SLAS Discov. https://doi.org/10.1177/24725552187577182472555218757718
82. Xia B, Franklin GJ, Lu X, Bedard KL, Grady LC, Summerfield JD, Shi EX, King BW, Lind KE, Chiu C, Watts E, Bodmer V, Bai X, Marcaurelle LA (2021) DNA-encoded library hit

confirmation: bridging the gap between on-DNA and off-DNA chemistry. ACS Med Chem Lett 12:1166–1172
83. Su W, Ge R, Ding D, Chen W, Wang W, Yan H, Wang W, Yuan Y, Liu H, Zhang M, Zhang J, Shu Q, Satz AL, Kuai L (2021) Triaging of DNA-encoded library selection results by high-throughput resynthesis of DNA-conjugate and affinity selection mass spectrometry. Bioconjug Chem 32:1001–1007
84. Chen Q, Hall J, Foley TL, Wan J, Li Y, Israel DI (2020) A method for estimating binding affinity from primary del selection data. Biochem Biophys Res Commun 533:249–255
85. Le ATH, Krylova SM, Beloborodov SS, Wang TY, Hili R, Johnson PE, Li F, Veedu RN, Belyanskaya S, Krylov SN (2021) How to develop and prove high-efficiency selection of ligands from oligonucleotide libraries: a universal framework for aptamers and DNA-encoded small-molecule ligands. Anal Chem 93:5343–5354
86. Satz AL, Hochstrasser R, Petersen AC (2017) Analysis of current DNA encoded library screening data indicates higher false negative rates for numerically larger libraries. ACS Comb Sci 19:234–238
87. Faver JC, Riehle K, Lancia Jr DR, Milbank JBJ, Kollmann CS, Simmons N, Yu Z, Matzuk MM (2019) Quantitative comparison of enrichment from DNA-encoded chemical library selections. ACS Comb Sci 21:75–82
88. Komar P, Kalinic M (2020) Denoising DNA encoded library screens with sparse learning. ACS Comb Sci 22:410–421
89. Amigo J, Rama-Garda R, Bello X, Sobrino B, de Blas J, Martin-Ortega M, Jessop TC, Carracedo A, Loza MIG, Dominguez E (2018) Tagfinder: a novel tag analysis methodology that enables detection of molecules from DNA-encoded chemical libraries. SLAS Discov 23: 397–404
90. Denton KE, Wang S, Gignac MC, Milosevich N, Hof F, Dykhuizen EC, Krusemark CJ (2018) Robustness of in vitro selection assays of DNA-encoded peptidomimetic ligands to cbx7 and cbx8. SLAS Discov 23:417–428
91. McCarthy KA, Franklin GJ, Lancia Jr DR, Olbrot M, Pardo E, O'Connell JC, Kollmann CS (2020) The impact of variable selection coverage on detection of ligands from a DNA-encoded library screen. SLAS Discov 25:515–522
92. Franzini RM, Nauer A, Scheuermann J, Neri D (2015) Interrogating target-specificity by parallel screening of a DNA-encoded chemical library against closely related proteins. Chem Commun 51:8014–8016
93. Richter H, Satz AL, Bedoucha M, Buettelmann B, Petersen AC, Harmeier A, Hermosilla R, Hochstrasser R, Burger D, Gsell B, Gasser R, Huber S, Hug MN, Kocer B, Kuhn B, Ritter M, Rudolph MG, Weibel F, Molina-David J, Kim JJ, Santos JV, Stihle M, Georges GJ, Bonfil RD, Fridman R, Uhles S, Moll S, Faul C, Fornoni A, Prunotto M (2019) DNA-encoded library-derived ddr1 inhibitor prevents fibrosis and renal function loss in a genetic mouse model of alport syndrome. ACS Chem Biol 14:37–49
94. Machutta CA, Kollmann CS, Lind KE, Bai X, Chan PF, Huang J, Ballell L, Belyanskaya S, Besra GS, Barros-Aguirre D, Bates RH, Centrella PA, Chang SS, Chai J, Choudhry AE, Coffin A, Davie CP, Deng H, Deng J, Ding Y, Dodson JW, Fosbenner DT, Gao EN, Graham TL, Graybill TL, Ingraham K, Johnson WP, King BW, Kwiatkowski CR, Lelievre J, Li Y, Liu X, Lu Q, Lehr R, Mendoza-Losana A, Martin J, McCloskey L, McCormick P, O'Keefe HP, O'Keeffe T, Pao C, Phelps CB, Qi H, Rafferty K, Scavello GS, Steiginga MS, Sundersingh FS, Sweitzer SM, Szewczuk LM, Taylor A, Toh MF, Wang J, Wang M, Wilkins DJ, Xia B, Yao G, Zhang J, Zhou J, Donahue CP, Messer JA, Holmes D, Arico-Muendel CC, Pope AJ, Gross JW, Evindar G (2017) Prioritizing multiple therapeutic targets in parallel using automated DNA-encoded library screening. Nat Commun 8:1–11
95. Dalton SE, Campos S (2020) Covalent small molecules as enabling platforms for drug discovery. Chembiochem 21:1080–1100

96. Zhu Z, Grady LC, Ding Y, Lind KE, Davie CP, Phelps CB, Evindar G (2019) Development of a selection method for discovering irreversible (covalent) binders from a DNA-encoded library. SLAS Discov 24:169–174
97. Zimmermann G, Rieder U, Bajic D, Vanetti S, Chaikuad A, Knapp S, Scheuermann J, Mattarella M, Neri D (2017) A specific and covalent jnk-1 ligand selected from an encoded self-assembling chemical library. Chemistry 23:8152–8155
98. Zambaldo C, Daguer JP, Saarbach J, Barluenga S, Winssinger N (2016) Screening for covalent inhibitors using DNA-display of small molecule libraries functionalized with cysteine reactive moieties. Med Chem Commun 7:1340–1351
99. Daguer JP, Zambaldo C, Abegg D, Barluenga S, Tallant C, Muller S, Adibekian A, Winssinger N (2015) Identification of covalent bromodomain binders through DNA display of small molecules. Angew Chem Int Ed 54:6057–6061
100. Chamakuri S, Lu S, Ucisik MN, Bohren KM, Chen YC, Du HC, Faver JC, Jimmidi R, Li F, Li JY, Nyshadham P, Palmer SS, Pollet J, Qin X, Ronca SE, Sankaran B, Sharma KL, Tan Z, Versteeg L, Yu Z, Matzuk MM, Palzkill T, Young DW (2021) DNA-encoded chemistry technology yields expedient access to sars-cov-2 m(pro) inhibitors. Proc Natl Acad Sci U S A 118:e2111172118
101. McGregor LM, Gorin DJ, Dumelin CE, Liu DR (2010) Interaction-dependent pcr: identification of ligand-target pairs from libraries of ligands and libraries of targets in a single solution-phase experiment. J Am Chem Soc 132:15522–15524
102. McGregor LM, Jain T, Liu DR (2014) Identification of ligand-target pairs from combined libraries of small molecules and unpurified protein targets in cell lysates. J Am Chem Soc 136: 3264–3270
103. Chan AI, McGregor LM, Jain T, Liu DR (2017) Discovery of a covalent kinase inhibitor from a DNA-encoded small-molecule library x protein library selection. J Am Chem Soc 139: 10192–10195
104. Blakskjaer P, Heitner T, Hansen NJ (2015) Fidelity by design: Yoctoreactor and binder trap enrichment for small-molecule DNA-encoded libraries and drug discovery. Curr Opin Chem Biol 26:62–71
105. Petersen LK, Blakskjær P, Chaikuad A, Christensen AB, Dietvorst J, Holmkvist J, Knapp S, Kořínek M, Larsen LK, Pedersen AE, Röhm S, Sløk FA, Hansen NJV (2016) Novel p38α map kinase inhibitors identified from yoctoreactor DNA-encoded small molecule library. Medchemcomm 7:1332–1339
106. Petersen LK, Christensen AB, Andersen J, Folkesson CG, Kristensen O, Andersen C, Alzu A, Slok FA, Blakskjaer P, Madsen D, Azevedo C, Micco I, Hansen NJV (2021) Screening of DNA-encoded small molecule libraries inside a living cell. J Am Chem Soc 143:2751–2756
107. Li G, Liu Y, Yu X, Li X (2014) Multivalent photoaffinity probe for labeling small molecule binding proteins. Bioconjug Chem 25:1172–1180
108. Li G, Liu Y, Liu Y, Chen L, Wu S, Liu Y, Li X (2013) Photoaffinity labeling of small-molecule-binding proteins by DNA-templated chemistry. Angew Chem Int Ed 52:9544–9549
109. Huang Y, Meng L, Nie Q, Zhou Y, Chen L, Yang S, Fung YME, Li X, Huang C, Cao Y, Li Y, Li X (2021) Selection of DNA-encoded chemical libraries against endogenous membrane proteins on live cells. Nat Chem 13:77–88
110. Zhao P, Chen Z, Li Y, Sun D, Gao Y, Huang Y, Li X (2014) Selection of DNA-encoded small molecule libraries against unmodified and non-immobilized protein targets. Angew Chem Int Ed 53:10056–10059
111. Shi B, Deng Y, Li X (2019) Polymerase-extension-based selection method for DNA-encoded chemical libraries against nonimmobilized protein targets. ACS Comb Sci 21:345–349
112. Cai B, Krusemark CJ (2022) Multiplexed small-molecule-ligand binding assays by affinity labeling and DNA sequence analysis. Angew Chem Int Ed Engl 61:e202113515
113. Cai B, Kim D, Akhand S, Sun Y, Cassell RJ, Alpsoy A, Dykhuizen EC, Van Rijn RM, Wendt MK, Krusemark CJ (2019) Selection of DNA-encoded libraries to protein targets within and on living cells. J Am Chem Soc 141:17057–17061

114. Gui Y, Wong CS, Zhao G, Xie C, Hou R, Li Y, Li G, Li X (2022) Converting double-stranded DNA-encoded libraries (dels) to single-stranded libraries for more versatile selections. ACS Omega 7:11491–11500
115. Zhao G, Zhong S, Zhang G, Li Y, Li Y (2022) Reversible covalent headpiece enables interconversion between double- and single-stranded DNA-encoded chemical libraries. Angew Chem Int Ed 61:e202115157
116. Zhao G, Fan X, Li Y, Zhang G, Li Y (2022) Switchable DNA-encoded chemical library: interconversion between double- and single-stranded DNA formats. Chembiochem. https://doi.org/10.1002/cbic.202200025e202200025
117. Drabovich AP, Berezovski MV, Musheev MU, Krylov SN (2009) Selection of smart small-molecule ligands: the proof of principle. Anal Chem 81:490–494
118. Bao J, Krylova SM, Cherney LT, Hale RL, Belyanskaya SL, Chiu CH, Arico-Muendel CC, Krylov SN (2015) Prediction of protein-DNA complex mobility in gel-free capillary electrophoresis. Anal Chem 87:2474–2479
119. Bao J, Krylova SM, Cherney LT, Hale RL, Belyanskaya SL, Chiu CH, Shaginian A, Arico-Muendel CC, Krylov SN (2016) Predicting electrophoretic mobility of protein-ligand complexes for ligands from DNA-encoded libraries of small molecules. Anal Chem 88:5498–5506
120. Kochmann S, Le ATH, Hili R, Krylov SN (2018) Predicting efficiency of neceem-based partitioning of protein binders from nonbinders in DNA-encoded libraries. Electrophoresis 39: 2991–2996
121. Bassi G, Favalli N, Vuk M, Catalano M, Martinelli A, Trenner A, Porro A, Yang S, Tham CL, Moroglu M, Yue WW, Conway SJ, Vogt PK, Sartori AA, Scheuermann J, Neri D (2020) A single-stranded DNA-encoded chemical library based on a stereoisomeric scaffold enables ligand discovery by modular assembly of building blocks. Adv Sci (Weinh) 7:2001970
122. Sprinz KI, Tagore DM, Hamilton AD (2005) Self-assembly of bivalent protein-binding agents based on oligonucleotide-linked organic fragments. Bioorg Med Chem Lett 15:3908–3911
123. Farrera-Soler L, Daguer JP, Raunft P, Barluenga S, Imberty A, Winssinger N (2020) Pna-based dynamic combinatorial libraries (pdcl) and screening of lectins. Bioorg Med Chem 28:115458
124. Reddavide FV, Thompson M, Mannocci L, Zhang Y (2019) DNA-encoded fragment libraries: dynamic assembly, single-molecule detection, and high-throughput hit validation. Aldrichimica Acta 52:63–74
125. Zhou Y, Peng J, Shen W, Li X (2020) Psoralen as an interstrand DNA crosslinker in the selection of DNA-encoded dynamic chemical library. Biochem Biophys Res Commun 533: 215–222
126. Winssinger N, Harris JL (2005) Microarray-based functional protein profiling using peptide nucleic acid-encoded libraries. Expert Rev Proteomics 2:937–947
127. Harris J, Mason DE, Li J, Burdick KW, Backes BJ, Chen T, Shipway A, Van Heeke G, Gough L, Ghaemmaghami A, Shakib F, Debaene F, Winssinger N (2004) Activity profile of dust mite allergen extract using substrate libraries and functional proteomic microarrays. Chem Biol 11:1361–1372
128. Winssinger N, Ficarro S, Schultz PG, Harris JL (2002) Profiling protein function with small molecule microarrays. Proc Natl Acad Sci U S A 99:11139–11144
129. Marsh JA, Teichmann SA (2015) Structure, dynamics, assembly, and evolution of protein complexes. Annu Rev Biochem 84:551–575
130. Berggard T, Linse S, James P (2007) Methods for the detection and analysis of protein-protein interactions. Proteomics 7:2833–2842
131. Pardo M, Choudhary JS (2012) Assignment of protein interactions from affinity purification/mass spectrometry data. J Proteome Res 11:1462–1474
132. Zhang J, Peng J, Huang Y, Meng L, Li Q, Xiong F, Li X (2020) Identification of histone deacetylase (hdac)-associated proteins with DNA-programmed affinity labeling. Angew Chem Int Ed 59:17525–17532

133. Yin H, Flynn AD (2016) Drugging membrane protein interactions. Annu Rev Biomed Eng 18: 51–76
134. Gulezian E, Crivello C, Bednenko J, Zafra C, Zhang Y, Colussi P, Hussain S (2021) Membrane protein production and formulation for drug discovery. Trends Pharmacol Sci 42:657–674
135. Kollmann CS, Bai X, Tsai CH, Yang H, Lind KE, Skinner SR, Zhu Z, Israel DI, Cuozzo JW, Morgan BA, Yuki K, Xie C, Springer TA, Shimaoka M, Evindar G (2014) Application of encoded library technology (elt) to a protein-protein interaction target: discovery of a potent class of integrin lymphocyte function-associated antigen 1 (lfa-1) antagonists. Bioorg Med Chem 22:2353–2365
136. Xie J, Wang S, Ma P, Ma F, Li J, Wang W, Lu F, Xiong H, Gu Y, Zhang S, Xu H, Yang G, Lerner RA (2020) Selection of small molecules that bind to and activate the insulin receptor from a DNA-encoded library of natural products. iScience 23:101197
137. Favalli N, Bassi G, Pellegrino C, Millul J, De Luca R, Cazzamalli S, Yang S, Trenner A, Mozaffari NL, Myburgh R, Moroglu M, Conway SJ, Sartori AA, Manz MG, Lerner RA, Vogt PK, Scheuermann J, Neri D (2021) Stereo- and regiodefined DNA-encoded chemical libraries enable efficient tumour-targeting applications. Nat Chem 13:540–548
138. Ahn S, Pani B, Kahsai AW, Olsen EK, Husemoen G, Vestergaard M, Jin L, Zhao S, Wingler LM, Rambarat PK, Simhal RK, Xu TT, Sun LD, Shim PJ, Staus DP, Huang LY, Franch T, Chen X, Lefkowitz RJ (2018) Small-molecule positive allosteric modulators of the beta2-adrenoceptor isolated from DNA-encoded libraries. Mol Pharmacol 94:850–861
139. Ahn S, Kahsai AW, Pani B, Wang QT, Zhao S, Wall AL, Strachan RT, Staus DP, Wingler LM, Sun LD, Sinnaeve J, Choi MJ, Cho T, Xu TT, Hansen GM, Burnett MB, Lamerdin JE, Bassoni DL, Gavino BJ, Husemoen G, Olsen EK, Franch T, Costanzi S, Chen X, Lefkowitz RJ (2017) Allosteric "beta-blocker" isolated from a DNA-encoded small molecule library. Proc Natl Acad Sci U S A 114:1708–1713
140. Brown DG, Brown GA, Centrella P, Certel K, Cooke RM, Cuozzo JW, Dekker N, Dumelin CE, Ferguson A, Fiez-Vandal C, Geschwindner S, Guie MA, Habeshian S, Keefe AD, Schlenker O, Sigel EA, Snijder A, Soutter HT, Sundstrom L, Troast DM, Wiggin G, Zhang J, Zhang Y, Clark MA (2018) Agonists and antagonists of protease-activated receptor 2 discovered within a DNA-encoded chemical library using mutational stabilization of the target. SLAS Discov 23:429–436
141. Svensen N, Diaz-Mochon JJ, Bradley M (2011) Decoding a pna encoded peptide library by pcr: the discovery of new cell surface receptor ligands. Chem Biol 18:1284–1289
142. Svensen N, Diaz-Mochon JJ, Bradley M (2011) Encoded peptide libraries and the discovery of new cell binding ligands. Chem Commun 47:7638–7640
143. Wu Z, Graybill TL, Zeng X, Platchek M, Zhang J, Bodmer VQ, Wisnoski DD, Deng J, Coppo FT, Yao G, Tamburino A, Scavello G, Franklin GJ, Mataruse S, Bedard KL, Ding Y, Chai J, Summerfield J, Centrella PA, Messer JA, Pope AJ, Israel DI (2015) Cell-based selection expands the utility of DNA-encoded small-molecule library technology to cell surface drug targets: identification of novel antagonists of the nk3 tachykinin receptor. ACS Comb Sci 17: 722–731
144. Warner KD, Hajdin CE, Weeks KM (2018) Principles for targeting rna with drug-like small molecules. Nat Rev Drug Discov 17:547–558
145. Chen Q, Li Y, Lin C, Chen L, Luo H, Xia S, Liu C, Cheng X, Liu C, Li J, Dou D (2022) Expanding the DNA-encoded library toolbox: identifying small molecules targeting rna. Nucleic Acids Res. https://doi.org/10.1093/nar/gkac173
146. Mukherjee H, Blain JC, Vandivier LE, Chin DN, Friedman JE, Liu F, Maillet A, Fang C, Kaplan JB, Li J, Chenoweth DM, Christensen AB, Petersen LK, Hansen NJV, Barrera L, Kubica N, Kumaravel G, Petter JC (2020) Pearl-seq: a photoaffinity platform for the analysis of small molecule-rna interactions. ACS Chem Biol 15:2374–2381
147. Benhamou RI, Suresh BM, Tong Y, Cochrane WG, Cavett V, Vezina-Dawod S, Abegg D, Childs-Disney JL, Adibekian A, Paegel BM, Disney MD (2022) DNA-encoded library versus

rna-encoded library selection enables design of an oncogenic noncoding rna inhibitor. Proc Natl Acad Sci U S A 119:e2114971119
148. Sun ZY, Wang XN, Cheng SQ, Su XX, Ou TM (2019) Developing novel g-quadruplex ligands: from interaction with nucleic acids to interfering with nucleic acid(−)protein interaction. Molecules 24:396
149. Hansel-Hertsch R, Di Antonio M, Balasubramanian S (2017) DNA g-quadruplexes in the human genome: detection, functions and therapeutic potential. Nat Rev Mol Cell Biol 18:279–284
150. Bochman ML, Paeschke K, Zakian VA (2012) DNA secondary structures: stability and function of g-quadruplex structures. Nat Rev Genet 13:770–780
151. Litovchick A, Tian X, Monteiro MI, Kennedy KM, Guie MA, Centrella P, Zhang Y, Clark MA, Keefe AD (2019) Novel nucleic acid binding small molecules discovered using DNA-encoded chemistry. Molecules 24:2026
152. Cuozzo JW, Centrella PA, Gikunju D, Habeshian S, Hupp CD, Keefe AD, Sigel EA, Soutter HH, Thomson HA, Zhang Y, Clark MA (2017) Discovery of a potent btk inhibitor with a novel binding mode by using parallel selections with a DNA-encoded chemical library. Chembiochem 18:864–871
153. Fang H, Peng B, Ong SY, Wu Q, Li L, Yao SQ (2021) Recent advances in activity-based probes (abps) and affinity-based probes (afbps) for profiling of enzymes. Chem Sci 12:8288–8310
154. Diaz-Mochon JJ, Bialy L, Bradley M (2006) Dual colour, microarray-based, analysis of 10,000 protease substrates. Chem Commun:3984–3986
155. Pouchain D, Díaz-Mochón JJ, Bialy L, Bradley M (2007) A 10,000 member pna-encoded peptide library for profiling tyrosine kinases. ACS Chem Biol 2:810–818
156. Krusemark CJ, Tilmans NP, Brown PO, Harbury PB (2016) Directed chemical evolution with an outsized genetic code. PLoS One 11:e0154765
157. Jetson RR, Krusemark CJ (2016) Sensing enzymatic activity by exposure and selection of DNA-encoded probes. Angew Chem Int Ed 55:9562–9566
158. MacConnell AB, Price AK, Paegel BM (2017) An integrated microfluidic processor for DNA-encoded combinatorial library functional screening. ACS Comb Sci 19:181–192
159. Price AK, MacConnell AB, Paegel BM (2014) Microfluidic bead suspension hopper. Anal Chem 86:5039–5044
160. Price AK, Paegel BM (2016) Discovery in droplets. Anal Chem 88:339–353
161. Roy A, Koesema E, Kodadek TJ (2021) High-throughput quality control assay for the solid-phase synthesis of DNA-encoded libraries of macrocycles. Angew Chem Int Ed 60:11983–11990
162. Cochrane WG, Fitzgerald PR, Paegel BM (2021) Antibacterial discovery via phenotypic DNA-encoded library screening. ACS Chem Biol 16:2752–2756
163. Koesema E, Roy A, Paciaroni NG, Coito C, Tokmina-Roszyk M, Kodadek T (2022) Synthesis and screening of a DNA-encoded library of non-peptidic macrocycles. Angew Chem Int Ed 61:e202116999
164. Patel S, Badir SO, Molander GA (2021) Developments in photoredox-mediated alkylation for DNA-encoded libraries. Trends Chem 3:161–175
165. Prudent R, Annis DA, Dandliker PJ, Ortholand JY, Roche D (2021) Exploring new targets and chemical space with affinity selection-mass spectrometry. Nat Rev Chem 5:62–71
166. Kielar C, Reddavide FV, Tubbenhauer S, Cui M, Xu X, Grundmeier G, Zhang Y, Keller A (2018) Pharmacophore nanoarrays on DNA origami substrates as a single-molecule assay for fragment-based drug discovery. Angew Chem Int Ed 57:14873–14877
167. Kolodny G, Li X, Balk S (2018) Addressing cancer chemotherapeutic toxicity, resistance, and heterogeneity: novel theranostic use of DNA-encoded small molecule libraries. Bioessays 40:e1800057

Top Med Chem (2022) 40: 217–258
https://doi.org/10.1007/7355_2022_150

Published online: 16 September 2022

From DEL Selections to Validated Hits to Clinical Leads

Lisa Marcaurelle, Westley Tear, and Gang Yao

Contents

Abstract DNA-encoded libraries (DELs) have become an increasingly utilized screening technology for identifying chemical matter for drug discovery campaigns. While DELs enable billions of compounds to be screened simultaneously, the combinatorial nature of DELs can often yield hits that fall outside of desired property space; meaning that DEL hit follow-up has typically not followed trends from hits identified from traditional screening techniques that employ highly curated compound collections. This chapter focuses on: (1) important factors to consider when designing and analyzing a DEL screen, (2) follow-up strategies to use once on-DNA hits are identified, (3) analysis of hit-to-lead trends from DEL screens, and (4) case studies of hits that have been developed into clinical candidates.

Keywords Copies, DNA-encoded libraries, Enrichment, Hit-to-lead, Medicinal chemistry, Screening techniques, Sequencing

L. Marcaurelle (✉), W. Tear, and G. Yao
Encoded Library Technologies, NCE Molecular Discovery, GSK, 200 Cambridge Park Drive, Cambridge, MA, USA
e-mail: lisa.a.marcaurelle@gsk.com; westley.x.tear@gsk.com; gang.x.yao@gsk.com

Abbreviations

ADAMTS-4	Aggrecanase-1
ADME	Absorption, distribution, metabolism, and excretion
ASMS	Affinity-selection mass spectrometry
ATAD2	ATPase family AAA domain containing protein 2
ATX	Autotaxin
AUDA	12-[[(Tricyclo-[3.3.1.13,7]dec-1-ylamino)carbonyl]amino] dodecanoic acid
BALI-MS	Bead-assisted ligand isolation mass spectrometry
BB	Building block
BCATm	Mitochondrial branched chain aminotransferase
BD	Bromodomain (domain)
BET	Bromodomain and extraterminal domain
BRD	Bromodomain (protein)
COPD	Chronic obstructive pulmonary disease
DDR	Discoidin domain receptor
DEL	DNA-encoded library
DHETs	Dihydroxyeicosatrienoic acids
DMP	Dimethyl pyridone
EETs	Epoxyeicosatrienoic acids
ERα	Estrogen receptor α
FaSSIF	Fasted state simulating intestinal fluid
FL	Full length
FRET	Fluorescence resonance energy transfer
FTE	Full time employee
GSK3β	Glycogen synthase kinase-3β
HAO1	Hydroxyacid oxidase 1
HTS	High-throughput screen
IDO1	Indoleamine 2,3-dioxygenase-1
KAc	Acetylated lysine
LE	Ligand efficiency
LLE	Lipophilic ligand efficiency
LPA	Lysophosphatidic acid
LPC	Lysophosphatidylcholine
Mcl-1	Myeloid cell leukemia 1
MPO	Multi-parameter optimization
MW	Molecular weight
NTC	No target control
OXA-48	Oxacillinase-48 carbapenemase
p38α	p38α mitogen-activated-protein kinase
PAD4	Protein arginine deiminase 4
PCR	Polymerase chain reaction
PFI	Property forecast index

PI3Kα	Phosphoinositide 3-kinase α
PK	Pharmacokinetic
PPBO	Physiochemical property-based optimization
QD	*Quaque die* (once per day)
RIP1	Receptor interacting protein 1
Ro5	Rule of five
RotB	Rotatable bonds
SAR	Structure–activity relationship
SBDD	Structure-based drug design
sEH	Soluble epoxide hydrolase
SIRT	Sirtuin
TAB1	TAK1 binding protein 1
TAK1	Transforming growth factor β-activated kinase 1
TR-FRET	Time-resolved fluorescence resonance energy transfer
Wip1	Wild-type p53 induced phosphatase

1 DEL Selections

There is a large amount of information available from DEL selections that can be highly impactful on a program's ability to both identify hits and progress them to clinical candidates. This chapter will focus on how DEL screening data can be used to inform decisions during this process, as well as highlight key trends and learnings from successful DEL campaigns. While there is no one correct approach to designing and following up on a DEL screen, historical examples can help to accelerate the drug discovery process.

Following the completion of a DEL screen, DNA sequences are amplified via PCR, and then sequenced using next generation sequencing technology [1–4]. These sequences are used to identify the structure of the enumerated molecule using building block (BB), library, and pool tags [1, 5–7]. The degenerate tags of the DNA sequence allow for the number of unique copies of each fully enumerated structure to be quantified [2, 5, 8, 9]. Before or after the structure corresponding to the DNA sequences has been enumerated, the copy numbers for each tag can be compared to the expected noise threshold for the selection [2, 4, 5, 9–11]. This allows for the statistical determination of which compounds are enhanced in the selection condition as opposed to noise from the selection output [2, 4, 5, 10–12]. Negative binomial, binomial, and Poisson distributions have been used in the interpretation of DEL screening data [2, 5, 13, 14].

There are several factors that can be incorporated into the design of a DEL selection to help identify molecules that fit the desired profile for a program (Fig. 1). To help further identify on-DNA compounds that are high-affinity binders, the selection output is often compared to a No target control (NTC) selection [5, 9, 15–18]. This allows for the removal of compounds from the data set that are binding

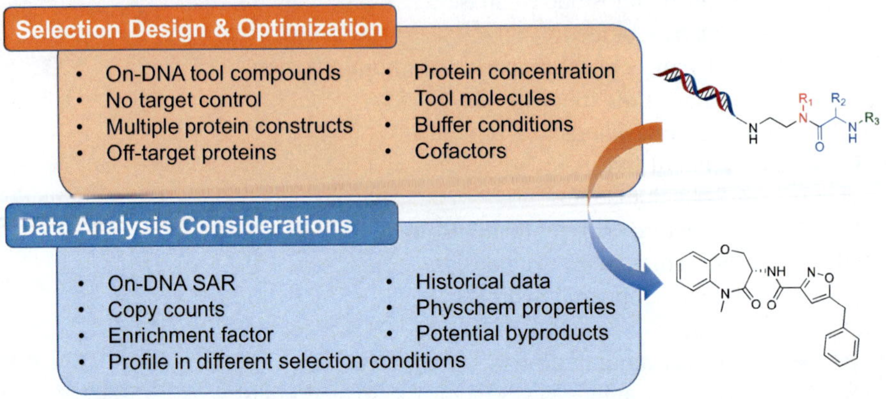

Fig. 1 Factors to consider when designing, optimizing, and analyzing a DEL screen

to other elements of the selection, such as the affinity tagging system or beads, allowing the analysis to focus on compounds which bind specifically to the target protein [1]. To help further distinguish binders, multiple selections can be performed varying many different experimental factors including protein constructs, off-target proteins, protein concentration, tool molecules, and buffer conditions. The use of these different selection conditions often varies, depending on the availability of reagents and the desired target product profile.

A useful method for identifying potential binding sites for compounds is the use of multiple constructs of the same protein [19, 20]. This confers several advantages. For one, different constructs of the same protein can fold dissimilarly from one another under selection conditions; by screening against multiple constructs the likelihood of screening against a physiologically relevant form of the protein is increased. Screening against different protein constructs helps to inform on likely binding sites for a DEL molecule. For example, it is not uncommon to screen against both the full length (FL) protein and known truncated constructs (often catalytic domain or crystallography constructs). Comparison of the selection outputs against these constructs allows for the identification of compounds that are likely binding in the active site (i.e., binds to the truncated construct as well as FL) or allosteric binders (i.e., binds only to the FL construct). In the case of the former, identifying binders that show binding against multiple constructs of a protein provides increased confidence in the validity of the binders. Another potential application is the screening of known mutants. For example, while targeting phosphoinositide 3-kinase α (PI3Kα) Yang et al. sought to identify wild-type binders that also showed binding against the H1047R mutant [21]. Depending on the aims of the program, comparison across multiple constructs can help greatly with prioritizing series identified for potential follow-up.

Tool molecules can also be used to help determine the likely binding site of a compound [15, 19, 20, 22–24]. These tool compounds can be endogenous substrates, or synthetically derived molecules. In selections tool molecules are normally preincubated with the target protein to saturate the binding site [15, 18, 19]. This was done in a selection for soluble epoxide hydrolase (sEH) using a known inhibitor,

12-[[(tricyclo-[3.3.1.13,7]dec-1-ylamino)carbonyl]amino]dodecanoic acid (AUDA), as well as in a screen of PI3Kα inhibitors using an ATP competitive inhibitor (ZSTK474) [16, 18, 21]. With tool molecules highly occupying their binding site, the probability of DEL compounds binding in the same site is low. Comparison of protein plus tool selection to one containing protein *sans* tool gives clues to the binding nature of the DEL molecules. On-DNA compounds that show up in both selections likely bind to a different pocket of the protein, whereas compounds that are binders against the protein alone, but not present in the protein plus tool selection data are likely binding in the same pocket as the tool compound. Again, this information can be useful for prioritizing which series to follow up on from a DEL screen depending on the desired molecule profile [19].

Tool molecules have also been conjugated to DNA to help with selection condition optimization and validation [18]. On-DNA tool compounds facilitate optimization of factors such as buffer conditions, inclusion of cofactors, and appropriate concentration of off-DNA tool molecules to use. Changes to buffer conditions and inclusion of cofactors used in a selection can cause changes to the protein shape, and available binding pockets, changing the tractability of a protein [19, 20, 23, 25–27]. Multiple groups have screened different cofactors and buffers in parallel [15, 19, 20, 23, 25, 26, 28]. When analyzing the output of selections with multiple conditions, series can often be grouped together based on their presence or absence in separate conditions, allowing for appropriate prioritization of compounds.

Perhaps one of the most useful ways to perform DEL screens is to include a counter screen against off-target proteins that are identified as potential selectivity concerns [15, 19, 29]. For example, this strategy was successfully applied by Richter et al. to identify a discoidin domain receptor 1 (DDR1) series selective over DDR2 [6]. Comparison of the selection results from the target protein compared to the off-target protein can help distinguish series that are more likely to be selective for the target versus off-target. When comparing different proteins in this manner, it is advisable to have as similar of constructs as possible. For instance, if screening against the catalytic core of one protein, and the full length of another, binders may be identified that appear to selective for the FL construct, but are in fact binding at an allosteric site not present in the former protein construct.

Another important item to consider when comparing selection data while looking for selectivity is that DELs are an affinity-based screening method. While a compound may have much lower affinity for one protein compared to the other, it could still be seen to bind to both target and off-target protein in the selection data. Analysis of both copy numbers and selection trends within a cluster (discussed in Sect. 2) are important for identifying series that have higher affinity for the target, but still exhibit some off-target binding.

One way to help determine the relative affinities of compounds for a construct during the selection process is to vary the concentration of protein present [6, 15, 19, 20, 23, 24]. There is generally more protein present in a DEL selection relative to the number individual DEL molecules [15, 23, 27]. By varying the amount of protein in the selection, a sense of higher or lower affinity compounds can be determined. As the protein concentration is decreased only higher affinity binders should still be present in the selection data. This useful screening information has been used on programs for targets such as DDR1 [6] and sEH [16, 18].

2 Data Analysis and Hit Triage

When analyzing the selection output of a DEL library, there are several different factors that should be taken into consideration (Fig. 1). The use of copy counts can be incredibly impactful to the analysis of a DEL screen. The number of copies of any given on-DNA compound corresponds to the number of times a unique DNA sequence for the compound was sequenced post selection. Copy counts do not directly correlate to the binding affinity of the corresponding compound off-DNA, particularly when screened at a single protein concentration [10, 12]. This is due to myriad factors including varying yields in on-DNA chemistry, differences in lipophilicity within a library, and varying selection conditions [30, 31]. Copy counts do however give higher confidence that the observed binding event is real, and not an artifact of the selection process. For example, in the analysis of a PI3Kα screen, Yang et al. removed low copy number binders from the analysis to focus on high copy count binders [21]. In general, differences in copy counts are best thought of on a logarithmic scale to avoid the conflation of differences in signal and experimental noise. For instance, the difference between a molecule with five copies and a molecule with eight copies is negligible, however the difference between a molecule with five copies and one with 50 should give higher confidence to the compound with more copies. With all else being equal, when selecting compounds for follow-up within a library, molecules and series with higher copy numbers are generally preferable to those with lower copy numbers. Multiple programs have used copy number when visualizing or prioritizing compounds follow-up including: DDR1 [6] and mitochondrial branched chain aminotransferase (BCATm) [32].

When running DEL selections, it is not uncommon to pool multiple libraries together to enable screening libraries simultaneously [5, 9, 28, 32]. Historically libraries have been pooled with the same amount for each library, regardless of library size. This was done in part to simplify the pooling process, as well as ensure that larger libraries were not consumed at a much higher rate in screening than smaller libraries. Comparing selection data across multiple libraries makes the direct comparison of copy counts more difficult [2, 5, 33]. This is due to the different sizes of DELs, arising from different numbers of cycles and BBs used in a particular library. These variations lead to vastly different potential numbers of compounds in each library (Table 1). For instance, one theoretical DEL library may be compromised of one million compounds (library 1), whereas another DEL library may have 50 million individual compounds (library 2). If both libraries are screened at equal concentrations, there are theoretically 50 more of each compound from library 1 compared to each library 2 molecule. This means that if there is no enrichment in the selection, for every copy observed for library 2, there should be 50 observed for library 1. This can cause many larger, more diverse libraries, to

Table 1 Comparison of two theoretical libraries of different sizes

Library	# of cpds in library	# of each individual molecule at equal library concentration	Theoretical enrichment for 10 copies of a molecule
1	1,000,000	50	0.2× enriched
2	50,000,000	1	10× enriched

appear to have smaller numbers of copies for individual library molecules when compared to smaller libraries [2, 33]. To compare direct copy numbers across these two libraries would be to unfairly weight toward library 1, and potentially miss out on binders from library 2.

To help bridge the gap between different starting numbers of molecules between libraries, enrichment factors are often used [2, 5]. In its simplest form, the enrichment factor balances the number of expected copies across libraries by assigning a multiplier based on the library size [2]. In the above example, to compare between library 1 and 2, if 10 copies of a molecule were seen from library 1, and 10 copies were seen from library 2, the library 2 compound would be 50-fold more enriched than the library 1 compound, because it was at a lower concentration in the library pool. In this case, comparison of copy numbers would indicate both compounds are equally enriched, whereas in actuality a much larger percentage of the library 2 compound bound the target during selection. There are multiple ways to calculate enrichment factor, which we shall not cover in this chapter, but care should be taken when choosing which method to use for cross-library comparison. It is important to note that enrichment factor is not needed for cross-library comparison if the libraries are equal in size, or if the library pooling has been balanced to have the same number of copies of molecules across libraries, as opposed to pooling to have equal concentrations of each library.

Copy number and enrichment factor alone are not sufficient for the analysis of DEL selection data. In particular, individual DEL molecules that show high copy counts/enrichment but have no closely associated molecules present in the selection data are often false positives [20]. Given the split-and-pool method of synthesis used for most DEL library production, it is highly unlikely for any one DEL molecule to not have structurally similar compounds present in the library [1, 10, 20]. Therefore, individual DEL compounds that show high enrichment without similar compounds present should be considered highly suspect [20]. These singly enriched compounds can arise from myriad factors, including DNA-driven binding [34], poor protein quality, or a sequencing error.

When looking at DEL selection data, evidence of on-DNA structure–activity relationships (SAR) is often the best predictor of compounds that will confirm off-DNA [20]. The most prevalent way to find and distinguish on-DNA SAR is to look at scatterplots of individual libraries in 2 or 3-D (cube plot), depending on the number of cycles of chemistry used in the library synthesis [15, 22]. This sort of analysis has been undertaken for targets such as receptor interacting protein 1 (RIP1 [35]), sEH [16, 18], autotaxin (ATX) [17], BCATm [9, 32], bromodomain 4 (BRD4 [36]), ATPase family AAA-domain containing protein 2 (ATAD2) (using

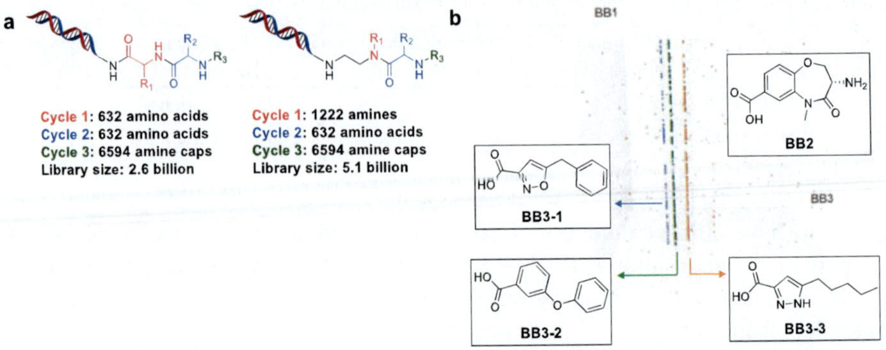

Fig. 2 (**a**) Three-cycle amino acid library, (**b**) Cube view of the selection output from the library against RIP1, highlighting the BB2, BB3 combination that gave on-DNA SAR [35]

di-synthon enrichment as the z-axis) [37], PI3Kα [21], sirtuin 3 (SIRT3 [28]), and enoyl-acyl-carrier protein reductase InhA [25]. These cube plots often will graph the building blocks used for a round of chemistry on each access (i.e., x = BB1, y = BB2, z = BB3), where the point on the plot corresponds to the fully enumerated compound. Viewing library selection data this way makes identifying selection patterns much easier. If BBs from two separate cycles of chemistry were more enriched in a selection, a line will generally appear where the two BBs are conserved, with the third BB showing variation [25, 35, 36]. This is demonstrated in Fig. 2 from a selection against RIP1 by Harris et al., wherein one BB2 shows SAR with three separate BB3s, while BB1 is varied in the selection output [35]. Examining these lines can indicate what BBs are driving the selection result (i.e., the conserved BBs) as well as what sort of substitution is tolerated for the third BB position. Sizing by copy count/enrichment factor allows for trends to be observed for the variable BB [36]. This can be used to help prioritize compounds for follow-up that likely have stronger interactions with the target protein. When selecting compounds for follow-up, generally those that have higher copy counts demonstrate on-DNA SAR are best to test first.

Another useful way to triage DEL selection data is to examine the chemical clusters that emerge from a selection output. The different types of clustering that are available to use are beyond the scope of this chapter, and as such we will broadly define chemical clustering as a way in which to group similar chemical structures together. By clustering similar compounds together, analysis can be performed to determine which clusters are more enriched than others, as well as determining which selection profiles within the cluster look like across different selection conditions. This sort of analysis can be helpful in determining trends in affinity, property space, and potential selectivity. For example, in a DDR1 campaign, Richter et al. examined different clusters to determine compounds for follow-up instead of employing cube plots [6]. Clustering is particularly powerful to help investigate series that are made up of separate DEL libraries, though it should be noted that clusters often come from within the same library due to their shared BBs and reaction scheme. While many clusters will contain compounds from only one given DEL library, analysis of clusters that contain multiple libraries can help identify series that may be missed when simply analyzing libraries one at a time. As always, a balance of copy counts and on-DNA SAR should be considered when analyzing clusters of compounds.

Analysis of any DEL selection data should also include an examination of historical selection data (if available) to determine whether compounds are truly target-specific binders or are promiscuous under DEL screening conditions. Given the number of components present in a DEL screen (beads, protein tags, additives, etc.) there will invariably be compounds that are false positives. Some compounds may also be favored to bind when proteins are not folded correctly, and not be true binders against the active form of the protein. Examination of available historical screening data can help to greatly cut down on these false positives. Again, in this case, consideration on a logarithmic scale is encouraged to separate signal from noise. Given that compounds off-DNA may not confirm on-DNA data, a compound with historical on-DNA signal against a few targets may not be indicative that it is a promiscuous binder. However, if a compound has been present in screens against many targets, it likely should be deprioritized.

Another matter to consider when interpreting the output of many separate libraries in a selection is the molecular shape of the libraries. In general, DEL libraries can be scaffold based, branched, linear, or cyclic. These libraries may give different selection results depending on the nature of the binding pocket of the target of interest [27, 38]. Pooling of DEL libraries consisting of all these molecular shapes increases the probability of finding binders against any given target [35].

Further consideration when analyzing DEL selections should be given to the property space that series occupy. Given the combinatorial nature of DEL synthesis, invariably some compounds end up outside of the classically defined property space for small molecule inhibitors [23, 27]. Depending on the selection output, these large compounds can significantly obscure binders that are in a more desirable property space. Filters on properties such as molecular weight (MW), clogP, and rotatable bonds (RotB) can be applied to remove enumerated compounds that fall outside of the desired space, as well as minimum copy count threshold to remove low affinity binders. Recently, there have been efforts to apply multi-parameter optimization (MPO) scores to enumerated compounds [23]. These MPO scores allow for the removal of hard cut-offs from property filters, meaning that compounds that fall slightly outside of a desired range for one property are not excluded from the data set. Applying property filters allows analysts to more easily identify compounds that would be better starting points for drug discovery campaigns [23]. It is important, however, to not be overly rigorous when applying filters, as they can remove valuable on-DNA SAR information from the analysis. Follow-up on DEL compounds often involves truncation of compounds to identify the minimum pharmacophore (discussed in Sect. 5.1), thus initial compound design should be more focused on identifying compounds that are representative of the on-DNA data [17, 20]. For instance, in a selection against RIP1, Harris et al. determined that the first cycle of chemistry was unnecessary, due to the variability at that cycle of chemistry in the selection data (Fig. 2) [35]. This has also been found in other programs, often with cycle 1 truncation proving to be productive when it is not selected for, likely because it is located closer to the DNA attachment point, and thus more likely to be solvent exposed without significant interactions with the protein [6, 16, 21, 24, 25, 28, 31]. As a general practice, applying mild filters to remove egregiously poor compounds can be helpful for initial analysis, with strict filters being applied once the on-DNA SAR is understood to identify the best starting points which balance copy counts with property space.

When performing analysis of DEL selections, it is important to keep in mind that the compounds on-DNA for a particular sequence may not solely correspond to the enumerated structure [20, 23]. In an ideal scenario, all BBs for a compound would give 100% conversion to product for each step, resulting in each DNA sequence corresponding solely to the enumerated compound. In actuality, the diversity of DEL libraries is driven by the array of BBs that can be used in each step, and while this gives DELs a higher chance of finding binders to a given target, it also means that the chemistry involved at each step is not quantitative, even if library conditions are optimized to be as broadly applicable as possible [10, 23, 27, 33]. Invariably, this means that each DNA sequence encapsulates both the enumerated product and a mixture of by-products from each step involved in the synthesis that cannot be purified away [16]. The percentage of these by-products varies compound to compound and cannot be accurately quantified in library settings, though the structures likely can be predicted from pre-validation chemistry work [16]. Particularly in later cycles of DEL library synthesis, pools contain thousands of compounds and thus determination of relative amounts of any given compound is not quantifiable.

Potential by-products should be kept in mind when looking at the selection results from a library, as they can influence interpretation of the selection data in several ways. Library conditions can cause functional group transformations that are not enumerated in the final product; examining by-products observed during library validation can give some insight into the probability of these unintended reactions occurring. In addition to functional group transformations, some BBs may react more than once in a given step of chemistry or may react in a subsequent step to form a product that has not been enumerated. These compounds can result in on-DNA compounds that are much larger than the enumerated product. When on-DNA SAR indicates selection for a group of BBs that could react more times than intended, that series of compounds should be regarded with increased scrutiny. Conversely, some BBs may not react as well with the on-DNA system, which could result in smaller molecules, or a different product being formed if important functionality is not installed.

One potentially useful way of sleuthing out these non-reactive compounds is to examine for the presence of a "null" in the selection data. Nulls are often included in library designs, containing some or all the reagents used in a library step, but not including any BB to react with. The presence of a null in selection data for a cycle of chemistry can be indicative that the BBs that are selected for did not react or that substitution at that position may minimally impact the binding of the series. Where nulls are identified for series with on-DNA SAR, they can prove to be productive positions to truncate the molecule and improve the physicochemical profile (discussed in Sect. 5.1).

The output of DEL selection data can be very large and at times overwhelming. Useful tools such as statistical filtering, NTC, comparison across multiple conditions and constructs, comparison of enrichment factors, property filters, clustering, historical data, and examination of on-DNA SAR can help to home in on series that have a high chance of confirming off-DNA. When used in combination these tools can help to simplify DEL data sets into useful packets of information.

3 Hit Confirmation

The most common practice for DEL hit confirmation involves the synthesis of compounds "off-DNA" for testing in biochemical and/or biophysical assays. The number of compounds selected for synthesis varies depending on target tractability and the success of the DEL selections. Selected compounds may range from a single series to dozens; with an important consideration being available chemistry resources for compound synthesis. Compounds are synthesized by traditional organic synthesis methods and chemistry plans often include the synthesis of "truncates" and/or anticipated side products, to account for non-enumerated structures producing the binding event on-DNA. While it is advised to measure off-DNA binding/activity prior to embarking on hit optimization, some risk may be taken to synthesize analogs during hit confirmation to probe SAR. For example, parallel synthesis methods may be feasible depending on the DEL chemistry to enable rapid analog generation and decrease overall cycle time.

An underappreciated challenge of DEL hit follow-up is the complexity associated with on-DNA chemistry where reactions are run in the presence of a large excess of reagents and water. On-DNA library synthesis most often yields a mixture of compounds including intermediates and by-products, and rarely (if ever) a pure desired final product. The DEL selection process is extremely sensitive, and even minor impurities in a final library mixture can be detected as binders. By-product hits could indeed contribute to perceived false positive rates observed with off-DNA hit resynthesis. To overcome this challenge and improve hit confirmation rates, methods have been developed in which the library "recipe" is employed for on-DNA hit confirmation. GSK devised an on-DNA hit resynthesis strategy enabled by photocleavable and acid-labile linkers, where upon cleavage from the DNA headpiece, affinity-selection mass spectrometry (ASMS) was employed to identify true binders from the reaction mixture [39]. The exact library protocol and reaction sequence was employed to mimic the library production conditions. Alternately, Pfizer demonstrated that binding affinity can be directly measured on-DNA without cleavage by bead-assisted ligand isolation mass spectrometry (BALI-MS) [23, 40]. RIP2 and BRD4 were presented as case studies by GSK and Pfizer, respectively, and in both examples side products were shown to be the true binders. More recently Wuxi demonstrated that on-DNA hit resynthesis could be conducted in a plate-based format to increase compound throughput. In this example ASMS was also employed as the hit confirmation method. With just one full time employee (FTE), >100 DNA-conjugated compounds could be produced and tested in ~1 month [41]. These on-DNA hit confirmation strategies provide confidence in the commitment of resources to off-DNA synthesis for further hit characterization and optimization. One limitation of on-DNA hit confirmation strategies is that it may extend project timelines compared to direct off-DNA synthesis. ASMS is only semiquantitative and no precise K_D can be derived. That said, employing small-scale, high-throughput chemistry methods for on-DNA synthesis offers the advantage of interrogating multiple chemotypes compared to direct off-DNA hit synthesis when chemistry resources (or BB quantities) are limited.

4 Property Analysis of DEL-Derived Hits and Leads

In recent years, a growing number of DEL-derived hit molecules have been reported for various therapeutic targets [42, 43]. A comprehensive 2016 review from the analysis of 155 published DEL hits showed that there is a direct correlation between the MW of DEL hits with the number of cycles of DEL chemistry [38]. The MW of the DEL hits tends to increase with the cycles of library chemistry, where the majority of hits from 4-cycle libraries (and above) have a MW > 500. In contrast to MW, DEL cycle numbers do not seem to have a direct effect on cLogP values. Analysis of ligand efficiencies (LEs) [44] revealed that reported hits tend to have the highest LEs for compounds derived from 3-cycle DELs. LEs dropped for 4- and 5-cycle libraries as a result of the increasing number of non-productive atoms in the DEL molecules. Although the number of encoded structures can be increased exponentially by increasing the cycles of DEL chemistry, the physicochemical properties of DEL molecules suffer when employing more than three cycles of chemistry.

The rule of five (Ro5 [45]) drug-like space is mostly populated with hits from 2- and 3-cycle libraries which offer a good compromise between structural diversity and physicochemical properties. Analysis of published DEL hits suggests that in order to produce drug-like compounds (MW < 500 Da; cLogP <5; LE ≥ 3, [45–48]) directly from DELs, it is important to construct libraries with compact and polar scaffolds and incorporate BBs which increase polar interactions and minimize hydrophobic driven binding for DEL molecules [49]. Hit molecules from earlier published DEL screenings tend to have higher MW and are located outside the preferred chemical space. DEL hits from more recent reports have increasingly showed favorable characteristics which reflects a conscious effort to design and construct DELs with better physicochemical properties. Many highly potent molecules with single-digit nanomolar or picomolar IC_{50} values have been identified using DNA-encoded technology while demonstrating physicochemical properties similar to those derived from traditional discovery platforms [49–51].

High-quality leads and several clinical candidates [17, 18, 35] derived from DEL technology have been reported in the past decade and several excellent reviews on the optimization of DEL-derived hits have been published [42, 43, 51]. Here we will focus on the analysis of DEL hits that have been optimized successfully via medicinal chemistry in order to evaluate their drug-likeness and potential for progressing into clinical candidates. Common hit-to-lead optimization strategies applied for DEL-derived hits are also discussed.

Examples of the DEL-derived hits and their corresponding leads or clinical candidates developed in recent years are listed in Table 2. The majority of these hits were derived from 2- and 3-cycle DELs. There are a total of 7 hits derived from 2-cycle DELs, 13 from 3-cycle DELs, and only two from 4-cycle DELs based on this list. We analyzed the physicochemical properties: MW, clogD, property forecast index (PFI) [64], RotB, LE, and lipophilic ligand efficiency (LLE) of these hits and their resulting leads to identify trends in the hit-to-lead optimization process [65]. First, we evaluated the MW and clogD of these ELT hits (Fig. 3). The MW and clogD of the DEL hits span a wide range from 322 to 673 for MW and 1.8 to 9.9 for clogD. Of the total 22 hits reported, 9 fall into the drug-like space (MW $\leq$ 500; cLogD $\leq$ 5; RotB $\leq$ 10), 6 failed the MW filter (MW $\leq$ 500) alone, 4 failed the clogD filter alone (clogD $\leq$ 5), and none failed the RotB filter alone (RotB $\leq$ 10). The remaining 3 failed more than one of the three filters. Clearly, high MW and clogD are major challenges for DEL-derived hits that need to be addressed in the subsequent lead optimization process to progress further in the drug discovery process.

A box plot of MW of DEL-derived compounds vs. the number of DEL chemistry cycles is shown in Fig. 4a. There is a clear trend of higher MW for hits derived from DELs with higher chemistry cycles. The median MW are 394, 497, and 592 for compounds derived from 2-, 3-, and 4-cycle DELs, respectively. In contrast, there is not a clear correlation between the clogD value or correspondingly the PFI value of compounds with the number of DEL chemistry cycles (Fig. 4b, c). The median clogD are 2.6, 4.5, and 3.4 for compounds derived from 2-, 3-, and 4-cycle DELs, respectively. This analysis indicates that the high clogD and PFI values for some DEL hits are probably a consequence of the lipophilic BBs used in the library synthesis instead of the higher number of DEL chemistry cycles. The binding of these DEL hits with high clogD values is driven in large part by hydrophobic interactions. Figure 4d shows a box plot of the number of rotatable bonds of compounds vs. the number of DEL chemistry cycles. There is a clear trend for a greater number of rotatable bonds with increased cycles of chemistry. While the median number of rotatable bonds for molecules from 2- and 3-cycle libraries fall within the desired space (RotB $\leq$ 10) [46], the median number of rotatable bonds for compounds from 4 cycle DELs is much higher at 11.5, which lies outside the desired space. The LE and LLE vs. DEL cycle number plots for the compounds shows there is a small drop of LE with increasing number of DEL cycles (Fig. 4e), and no significant change in the LLE with increasing numbers of DEL cycles (Fig. 4f). Overall, the LE drop is very small with a median LE of 0.29 observed for 2-cycle DELs, 0.27 for 3-cycle DELs, and 0.25 for 4-cycle DELs. Detailed analysis indicates molecules from DELs with higher number of chemistry cycles tend to have higher

Table 2 Published examples of hits and lead pairs identified using DEL technology

Target	DEL hit molecule	Properties	Lead molecule	Properties
ADAMTS-4 [52]	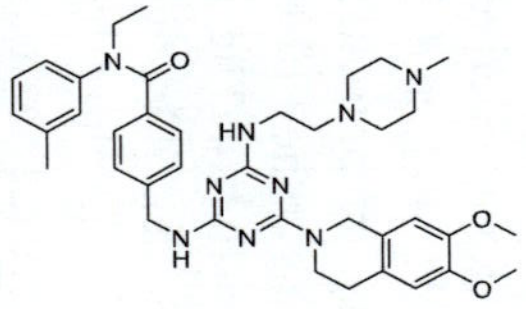	MW = 680 clogD/PFI = 4.7/ 8.7 RotB = 14 IC_{50} = 10 nM LE/LLE = 0.22/ 3.4	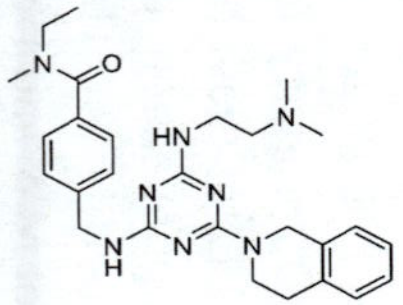	MW = 489 clogD/PFI = 3.8/ 6.8 RotB = 11 IC_{50} = 43 nM LE/LLE = 0.29/ 3.8
ATX [17]	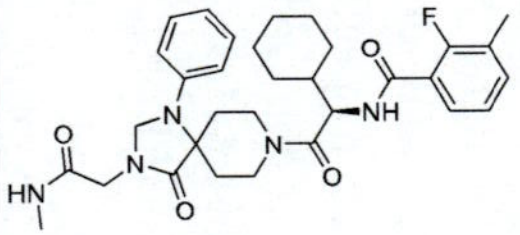	MW = 578 clogD/PFI = 4.9/ 6.9 RotB = 8 IC_{50} = 86 nM LE/LLE = 0.24/ 3.3	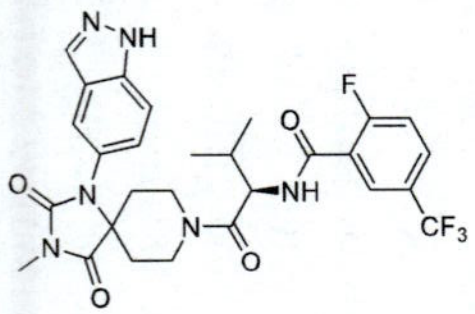	MW = 589 clogD/PFI = 4.6/ 7.6 RotB = 6 IC_{50} = 55 nM LE/LLE = 0.24/ 4.6
BCATm [32]	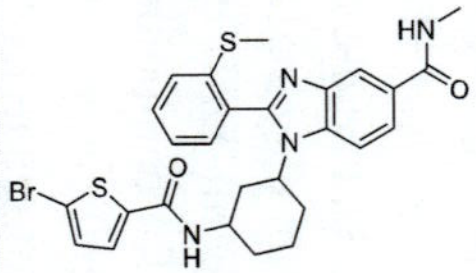	MW = 584 clogD/PFI = 4.6/ 8.6 RotB = 6 IC_{50} = 251 nM LE/LLE = 0.26/ 0.71	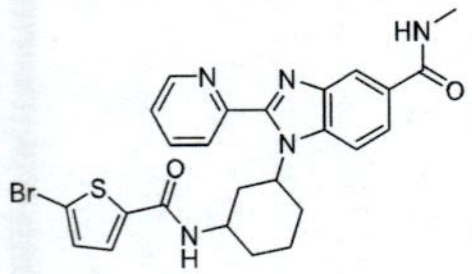	MW = 539 clogD/PFI = 3.8/ 7.8 RotB = 5 IC_{50} = 50 nM LE/LLE = 0.30/ 3.2
BRD4 [36]	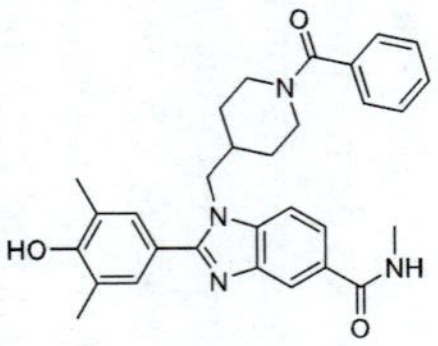	MW = 497 clogD/PFI = 3.1/ 7.1 RotB = 5 IC_{50} = 199 nM LE/LLE = 0.25/ 2.3	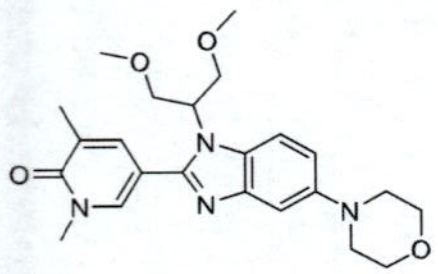	MW = 427 clogD/PFI = 2.6/ 5.6 RotB = 7 IC_{50} = 12.6 nM LE/LLE = 0.37/ 6.4

BRD4 BD2 [53]	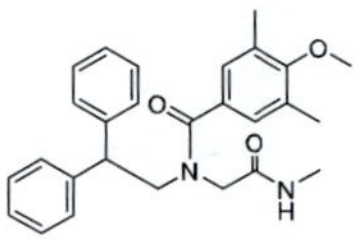	MW = 431 clogD/PFI = 6.4/9.4 RotB = 9 IC_{50} = 251 nM LE/LLE = 0.29/1.8	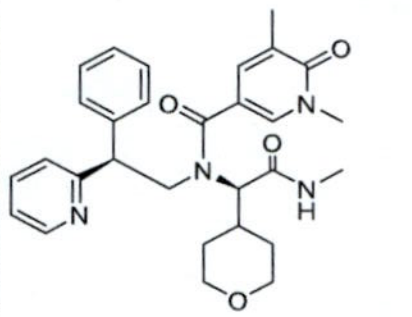	MW = 503 clogD/PFI = 2.7/5.7 RotB = 9 IC_{50} = 5 nM LE/LLE = 0.31/7.5
ERα [54]	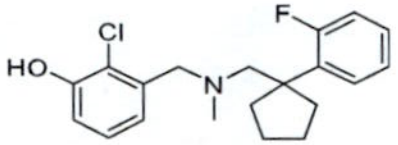	MW = 348 clogD/PFI = 9.9/12 RotB = 5 IC_{50} = 15 nM LE/LLE = 0.46/2.0	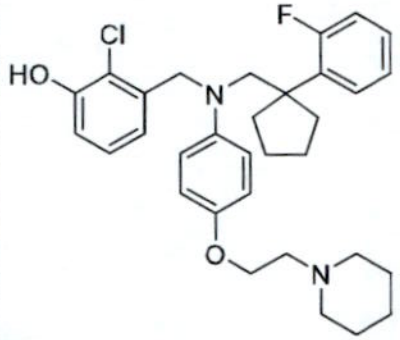	MW = 537 clogD/PFI = 8.2/11 RotB = 9 IC_{50} = 8 nM LE/LLE = 0.30/–1.1
Glycogen synthase kinase-3β (GSK3β) [55]	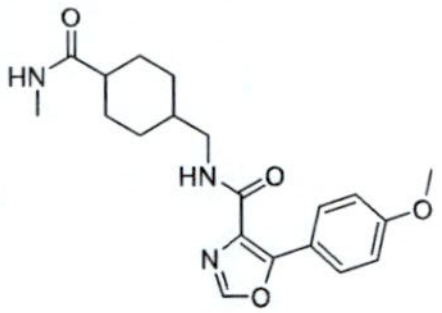	MW = 371 clogD/PFI = 2.8/4.8 RotB = 5 IC_{50} = 316 nM LE/LLE = 0.34/5.0	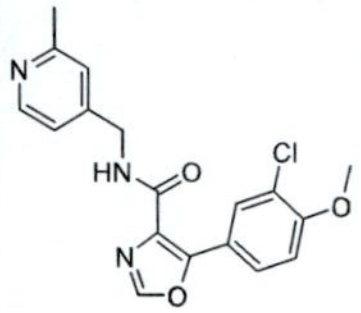	MW = 358 clogD/PFI = 4.1/7.1 RotB = 5 IC_{50} = 31 nM LE/LLE = 0.42/4.7
Hydroxyacid oxidase 1 (HAO1) [56]	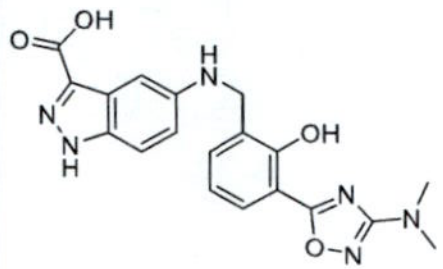	MW = 394 clogD/PFI = 2.0/6.0 RotB = 4 IC_{50} = 37 nM LE/LLE = 0.36/4.6	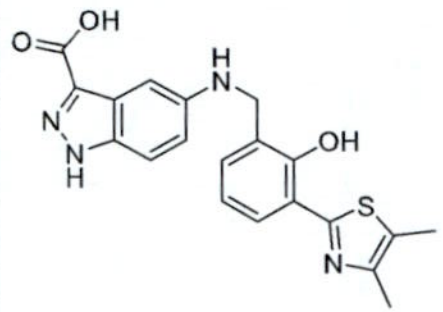	MW = 394 clogD/PFI = 2.6/6.6 RotB = 4 IC_{50} = 25 nM LE/LLE = 0.38/3.8
Indoleamine 2,3-dioxygenase-1 (IDO1) [57]		MW = 462 clogD/PFI = 7.8/11 RotB = 3		MW = 513 clogD/PFI = 6.1/9.1 RotB = 4

(continued)

Table 2 (continued)

Target	DEL hit molecule	Properties	Lead molecule	Properties
	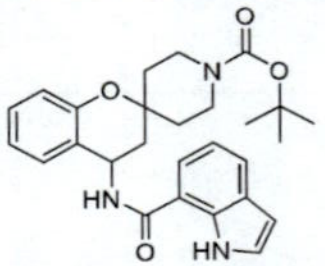	IC_{50} = 251 nM LE/LLE = 0.27/ 2.2	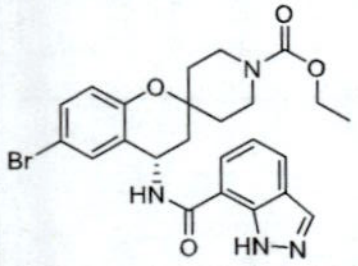	IC_{50} = 5.4 nM LE/LLE = 0.35/ 3.6
InhA [7]	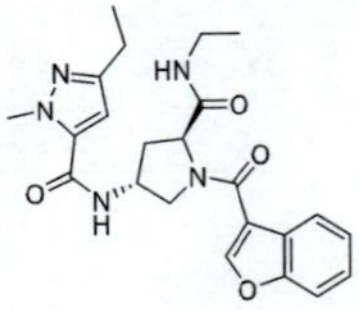	MW = 438 clogD/PFI = 3.3/ 6.3 RotB = 7 IC_{50} = 34 nM LE/LLE = 0.33/ 5.8	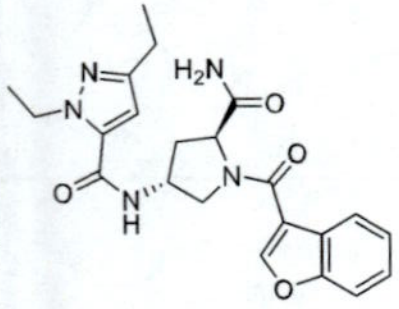	MW = 424 clogD/PFI = 3.0/ 6.0 RotB = 7 IC_{50} = 12 nM LE/LLE = 0.36/ 5.6
Mcl-1 [58]	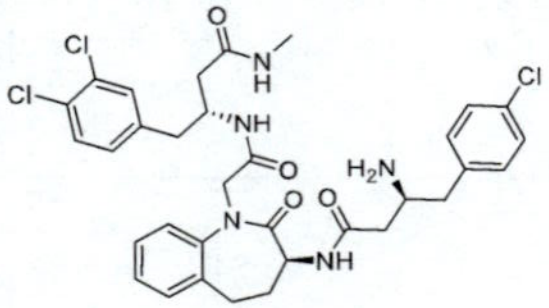	MW = 673 clogD/PFI = 4.5/ 7.5 RotB = 12 IC_{50} = 2000 nM LE/LLE = 0.18/ 0.25	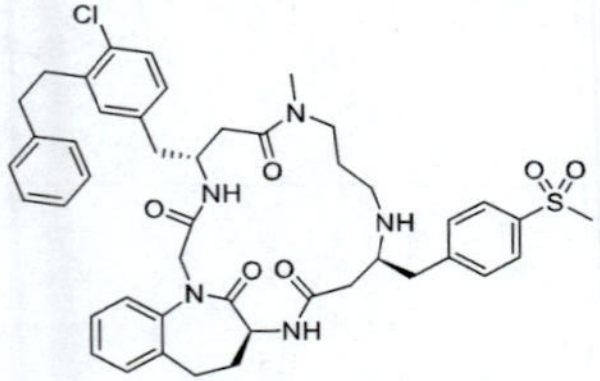	MW = 827 clogD/PFI = 3.0/ 7.0 RotB = 8 IC_{50} = 3 nM LE/LLE = 0.21/ 2.0
Protein arginine deiminase 4(PAD4) [26]	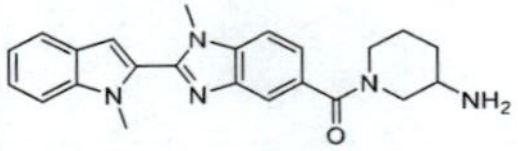	MW = 388 clogD/PFI = 2.1/ 6.1 RotB = 3 IC_{50} = 3,200 nM LE/LLE = 0.27/ 2.2	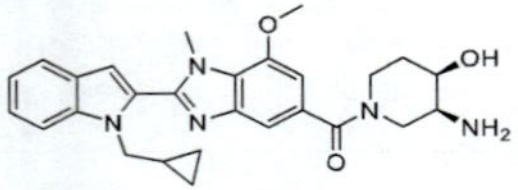	MW = 474 clogD/PFI = 3.4/ 7.4 RotB = 6 IC_{50} = 50 nM LE/LLE = 0.29/ 4.0

PI3Ka [21]	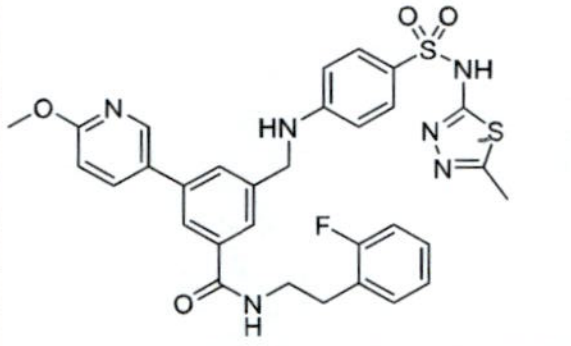	MW = 633 clogD/PFI = 4.1/ 9.1 RotB = 10 IC_{50} = 6.5 nM LE/LLE = 0.26/ 3.3	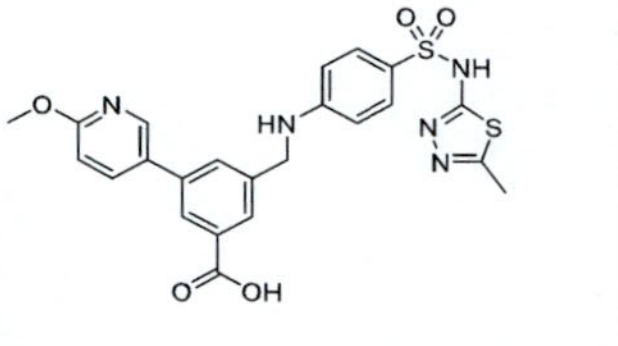	MW = 512 clogD/PFI = 1.0/ 5.0 RotB = 6 IC_{50} = 10 nM LE/LLE = 0.32/ 4.4
RIP1 [35, 59]	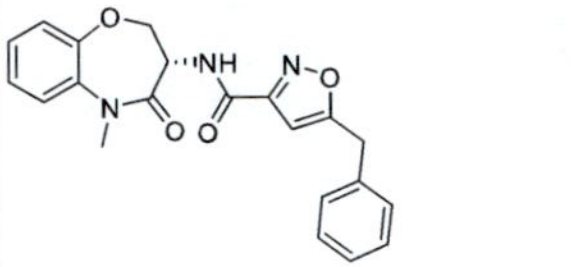	MW = 377 clogD/PFI = 6.0/ 9.0 RotB = 4 IC_{50} = 1.6 nM LE/LLE = 0.44/ 6.2	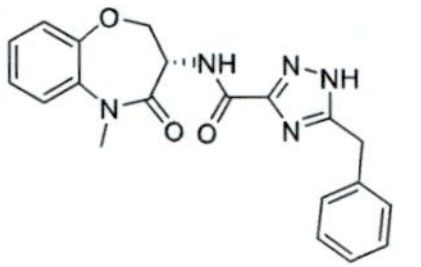	MW = 377 clogD/PFI = 3.9/ 6.9 RotB = 4 IC_{50} = 1 nM LE/LLE = 0.45/ 6.4
sEH [16, 18]	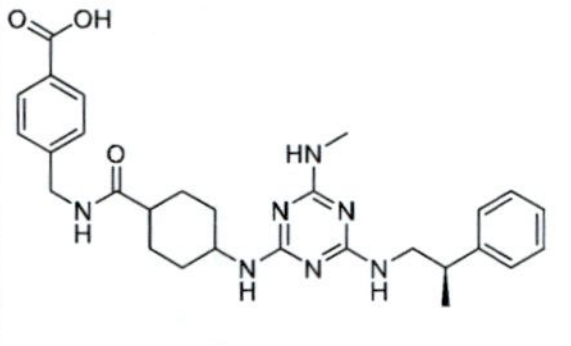	MW = 518 clogD/PFI = 2.3/ 5.3 RotB = 10 IC50 = 24 nM LE/LLE = 0.28/ 2.2	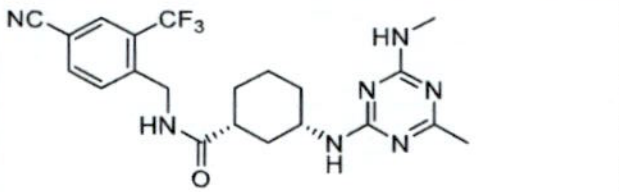	MW = 448 clogD/PFI = 3.6/ 5.6 RotB = 6 IC50 = 0.027 nM LE/LLE = 0.46/ 6.8
SIRT1 [28]	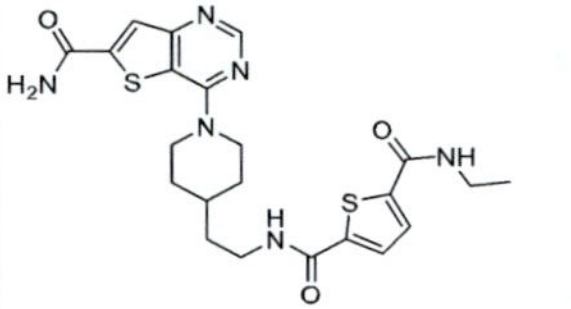	MW = 487 clogD/PFI = 2.3/ 5.3 RotB = 8 IC_{50} = 3.6 nM LE/LLE = 0.36/ 6.5	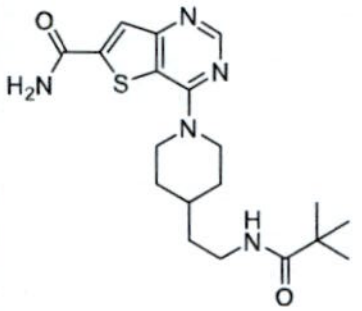	MW = 390 clogD/PFI = 2.7/ 4.7 RotB = 5 IC_{50} = 15 nM LE/LLE = 0.41/ 6.1

(continued)

Table 2 (continued)

Target	DEL hit molecule	Properties	Lead molecule	Properties
TAK1 [24]	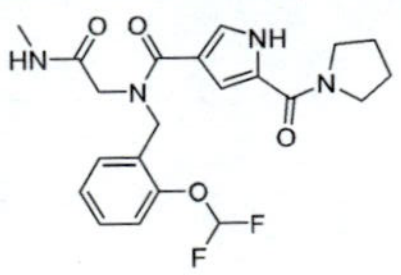	MW = 434 clogD/PFI = 2.6/4.6 RotB = 10 IC_{50} = 1,300 nM LE/LLE = 0.27/5.1	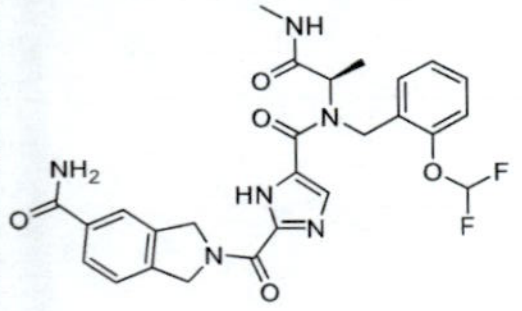	MW = 540 clogD/PFI = 2.2/5.2 RotB = 11 IC_{50} = 2 nM LE/LLE = 0.31/8.3
Wild-type p53 induced phosphatase (Wip1) [60]	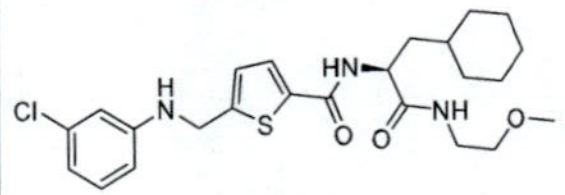	MW = 478 clogD/PFI = 6.6/8.6 RotB = 11 IC_{50} = 13 nM LE/LLE = 0.35/2.7	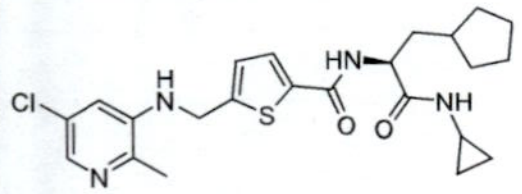	MW = 461 clogD/PFI = 5.2/7.2 RotB = 9 IC_{50} = 6 nM LE/LLE = 0.37/3.6
Z α_1-antitrypsin [61]	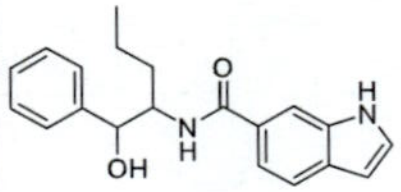	MW = 322 clogD/PFI = 4.6/7.6 RotB = 6 IC_{50} = 316 nM LE/LLE = 0.38/3.1	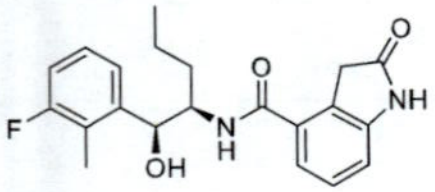	MW = 370 clogD/PFI = 3.9/5.9 RotB = 6 IC_{50} = 5 nM LE/LLE = 0.43/5.6
DDR1 [6]	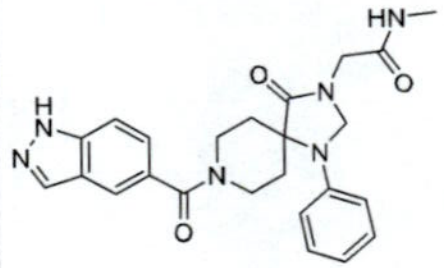	MW = 447 clogD/PFI = 1.8/4.8 RotB = 5 IC_{50} = 1,400 nM LE/LLE = 0.25/4.6	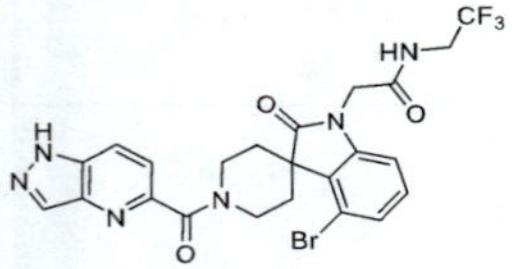	MW = 565 clogD/PFI = 3.1/6.1 RotB = 5 IC_{50} = 29 nM LE/LLE = 0.29/5.9

OXA-48 [31]	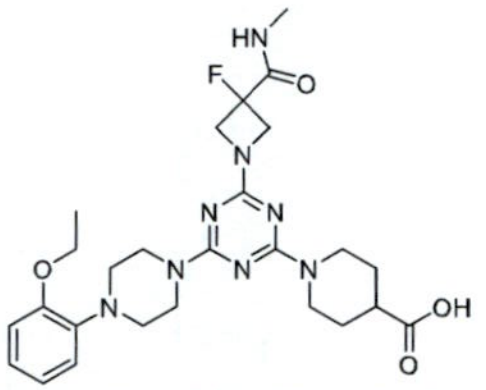	MW = 543 clogD/PFI = 2.8/ 4.8 RotB = 7 K_i = 900 nM LE/LLE = 0.22/ 3.0	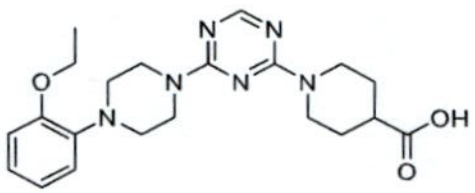	MW = 413 clogD/PFI = 2.8/ 4.8 RotB = 5 K_i = 530 nM LE/LLE = 0.29/ 4.6
p38α mitogen-activated-protein kinase (p38α) [62, 63]	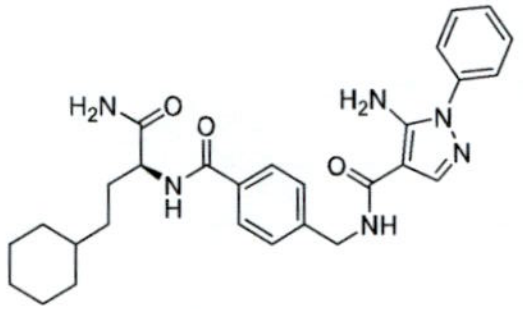	MW = 503 clogD/PFI = 4.5/ 7.5 RotB = 9 IC_{50} = 39 nM LE/LLE = 0.28/ 3.5	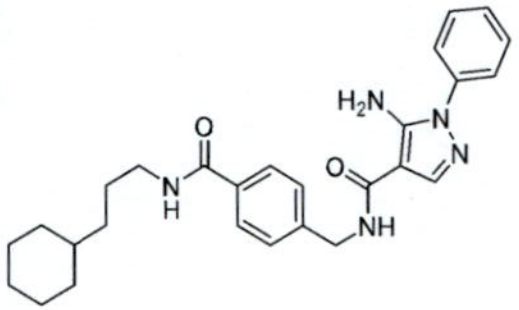	MW = 460 clogD/PFI = 6.2/ 9.2 RotB = 8 IC_{50} = 5 nM LE/LLE = 0.34/ 3.1

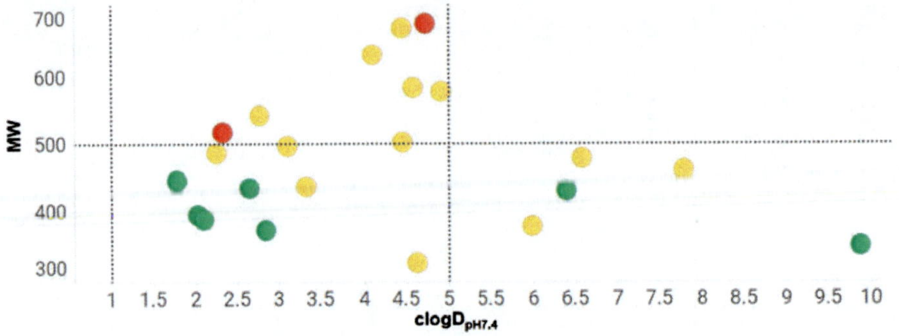

Fig. 3 Physicochemical properties of DEL hits. MW (y-axis, horizontal line at 500 Da) plotted versus ChromlogD (x-axis, vertical line at values of 1 and 5). Number of DEL chemistry cycles are coded by color (green: two cycles; yellow: three cycles; red: four cycles)

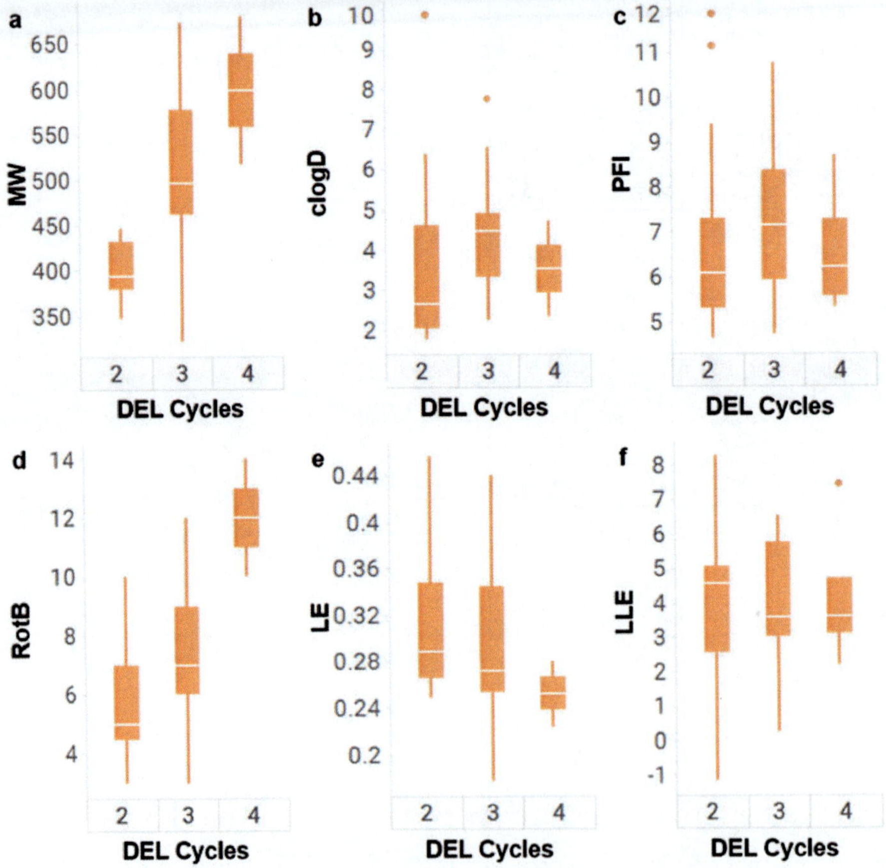

Fig. 4 Box plot distribution of DEL-derived compounds versus number of DEL chemistry cycles, based on: (**a**) MW, (**b**) cLogD, (**c**) PFI, (**d**) RotB, (**e**) LE, (**f**) LLE

potency, which compensates for the increased MW, resulting in minimum impact on the LE and LLE from the number of DEL chemistry cycles.

Hits identified from different sources (e.g., high-throughput screen (HTS), fragment-based screening, and DEL) tend to follow different physicochemical trajectories during hit optimization [66, 67]. Unlike HTS compounds which were generated with no inherent limitations on synthetic chemistry and no linker-based effects during the screen [68], DEL molecules have historically been made with a few DEL-compatible reactions (e.g., amidation, sp^2-sp^2 cross-couplings, nucleophilic substitution, and reductive amination) [69] via split-and-pool "DNA-recorded" synthesis. This resulted in decreased chemotype diversity but deeper SAR within a particular chemical space. As such, there is a great effort to define the "minimum pharmacophore" from a DEL hit during follow-up, to reduce the MW and improve the LE.

In addition to high MW, high lipophilicity is another common challenge for DEL-derived hits. One common hit-to-lead strategy is to introduce polar functionality to the DEL hits to reduce clogD and improve PFI. This is often assisted by structure-based drug design (SBDD). LLE has been frequently applied as an important metric in the lead optimization process which is often guided by physiochemical property-based optimization (PPBO) [65]. A unique feature of DEL hits is that they contain a linker attachment site that is covalently connected to the encoding DNA tags. The large, highly polar, and charged DNA tags are thermodynamically favored to be exposed to the solvent. While the DNA tag may limit the freedom of DEL molecules to sample different binding poses, it may also provide a clue on how the DEL molecules interact with the target protein. This tag information has been utilized in docking studies of DEL hits to facilitate the hit-to-lead process [54].

5 Strategies in DEL Hit-to-Lead Optimization

To examine common hit-to-lead strategies for DEL-derived hits, the molecular properties of the DEL hits and their resulting leads were compared. In this analysis, target estrogen receptor α (ERα) was excluded as it employed the nontraditional PROTACs modality approach, which is not consistent with Lipinski's Ro5 [70]. Fig. 5a, b showed that both increases and decreases of MW and clogD are observed in the lead optimization process. Unsurprisingly, PFI (Fig. 5c) showed a close correlation with clogD, with 19 of the 22 series showing a related increase (6 series) or decrease (12 series) in both clogD and PFI, with one series (OXA-48) showing no change in either value between hit and lead. In contrast to MW, clogD, and PFI, there is a clear upward trend for LE, and in particular LLE values (Fig. 5e, f), which is consistent with other recent analyses of hit-to-lead optimization [71–73]. This is mostly the result of hit-to-lead efforts which simultaneously improved the potencies and reduced the clogD values of the lead molecules toward acceptable ranges, in order to achieve desired potency, selectivity, and ADME properties to improve the probability of success in the clinic.

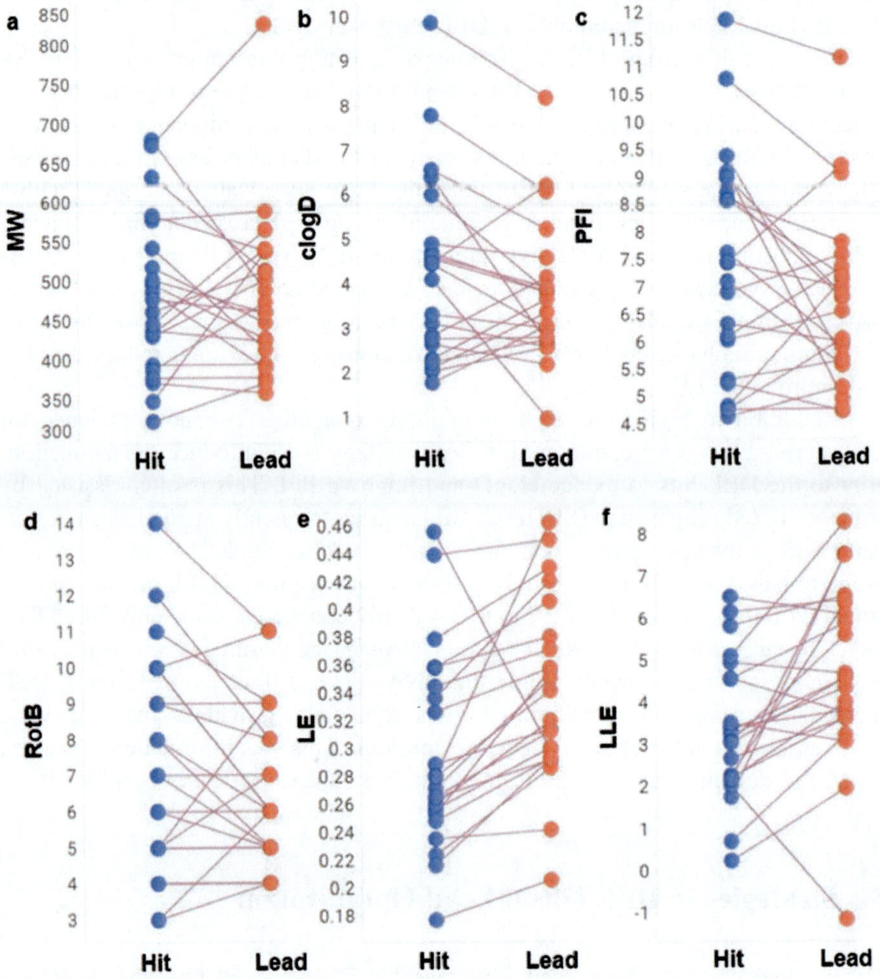

Fig. 5 Comparison of matched DEL hits (blue) and their resulting lead compounds (red), based on: **(a)** MW, **(b)** cLogD, **(c)** PFI, **(d)** RotB, **(e)** LE, **(f)** LLE

Due to the combinatorial nature of DEL technology, DEL hits are often more complex and potent than traditional HTS hits [48]. During the hit-to-lead optimization process, several common techniques are frequently used in order to optimize the potency and modulate physicochemical properties. Some of the most frequently used techniques are: (1) truncations of inefficient features to reduce MW, (2) potency and physiochemical property optimization guided by SBDD, and (3) preorganization of the compound into the bound conformation to improve potency and cell permeability through conformational restrictions.

5.1 Truncation of Inefficient Features

Since DEL hits are derived from libraries with large numbers of structurally related members, rich on-DNA SAR information is usually available from the affinity-selection data. This on-DNA SAR information can become very useful in designing off-DNA compounds for hit confirmation and optimization. The enriched combinations of BBs within hit molecules generally indicate importance for interaction with a target protein, while the variable BBs may not be essential for binding. The variable BB portion of the molecules can often be modified and even removed completely without scarifying the hit molecule's potency resulting in hits with improved LE and/or LLE. In the case of ADAMTS-4 (aggrecanase-1), an inhibitor with a biochemical potency of 10 nM was identified from a 4-cycle triazine DEL (Fig. 6) [52]. On-DNA SAR indicated the trisynthon combination of 4-(aminomethyl)benzoic acid (red), 2,6-disubstituted tetrahydroisoquinoline (green), and N-ethyl-3-methylaniline (orange) were highly enriched while no preferred substituent was observed at the cycle 1 substituent (blue). Truncation and modification of the DEL hit molecule (**1**) while maintaining the selected core structure resulted in lead molecule (**2**) with significantly reduced MW (from 680 to 489), clogD (from 4.7 to 3.8), and improved LE (from 0.22 to 0.29).

The discovery of oxacillinase-48 carbapenemase (OXA-48) inhibitors is another example where truncation was used to reduce MW and improve LE and LLE (Fig. 7). Screening of a 3-cycle triazine DEL identified a trisynthon combination of 3-fluoro-N-methylazetidine-3-carboxamide (orange), 1-(2-ethoxyphenyl)piperazine (blue), and piperidine-4-carboxylic acid (green), which led to the hit compound **3** after truncation of the DNA tag [31]. Compound **3** exhibits a sub-micromolar potency ($K_i = 0.9 \pm 0.1$ μM) against OXA-48. Further truncation of cycle 1 BB (orange) resulted in the lead molecule CDD-97 (**4**) with an improved potency ($K_i = 0.53 \pm 0.08$ μM). This result indicates that the cycle 1 BB fluoro-N-methylazetidine-3-carboxamide group did not contribute to binding. X-ray

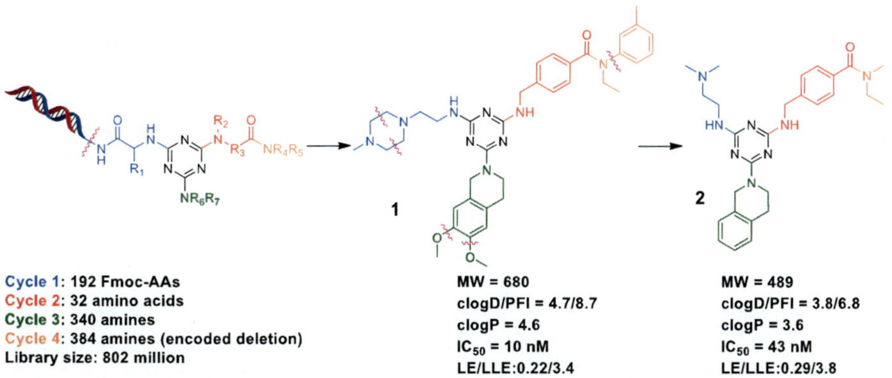

Fig. 6 Hit-to-lead optimization for ADAMTS-4. Pink lines indicate key points of truncation during hit-to-lead optimization [52]

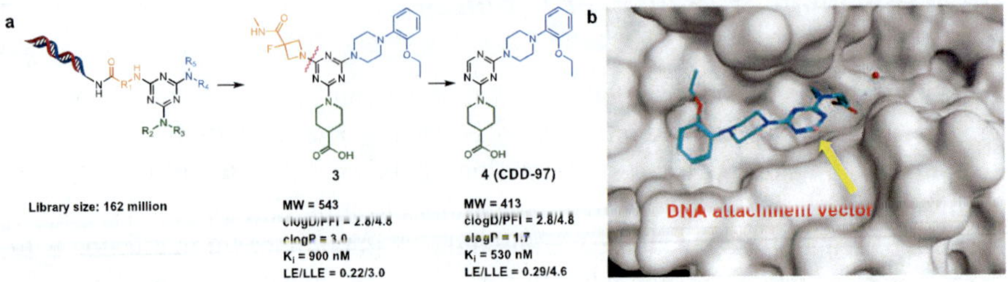

Fig. 7 (**a**) Hit-to-lead optimization for OXA-48, with pink lines indicating where truncation was undertaken in follow-up, (**b**) X-ray crystal structure of CDD-97 bound to OXA-48 [31]

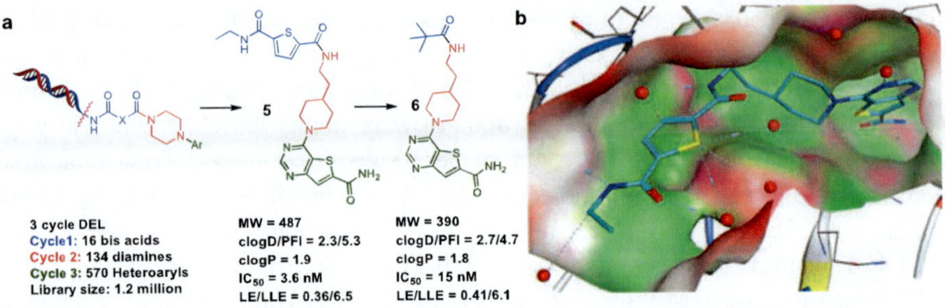

Fig. 8 (**a**) Hit-to-lead optimization for sirtuins, (**b**) Crystal structure of SIRT3 (118–399) bound with **5**, highlighting the solvent exposure of BB1 and the binding of the thienopyrimidine deeper in the pocket. Surface colors: red, solvent exposed; green, hydrophobic; magenta, polar [28]

crystallography showed that **4** binds non-covalently in the active site of OXA-48. The DNA tag linking cycle 1 BB is solvent exposed. In this case, the cycle 1 BB only served as an extension of the DNA linker and did not interact with OXA-48 and thus could be truncated without losing activity.

The sirtuins, SIRT1, SIRT2, and SIRT3, are NAD+ dependent deacetylases, a family of signaling proteins involved in metabolic regulation. They are potential targets for metabolic, inflammatory, oncologic, and neurodegenerative disorders. Screening of a 3-cycle heterocycle enriched DEL against a Flag-SBP-tagged SIRT3 construct identified pan-inhibitor **5** of SIRT1/2/3 with nanomolar potency (IC_{50} = 3.6, 2.7 and 4.0 nM for SIRT1, SIRT2, and SIRT3, respectively) (Fig. 8) [28]. This hit series was defined by the cycle 3 BB 4-chlorothieno[3,2-d]pyrimidine-6-carboxamide (green). The off-DNA synthesis was performed with an ethyl amide at the DNA attachment point. Subsequent SAR studies by replacing the variable BB1 group 5-(ethylcarbamoyl)thiophene-2-carboxylic acid (blue) with a smaller pivalic acid group resulted in a smaller lead molecule **6** (MW: 390) with improved LE (0.41 for lead molecule **6** vs. 0.36 for hit molecule **5**). The co-crystal structure of the DEL hit **5** bound to SIRT 3 was solved and the conserved carboxamide was shown to form four hydrogen bonds with the protein surface, similar to the analogous nicotinamide portion of carba-NAD. This observation is consistent with the selection data which highlights the importance of the conserved BB3 4-chlorothieno

[3,2-d]pyrimidine-6-carboxamide group for binding to SIRT3. Again, the DNA linking BB1 was found to be exposed to solvent.

5.2 *Potency and Physiochemical Property Optimization Guided by SBDD*

Introduction of polar groups to improve potency and reduce clogD is another common strategy for optimizing DEL hits. Often this was achieved with the assistance of SBDD. Wellaway et al. described the successful discovery of a candidate-quality dimethylpyridone benzimidazole bromodomain and extraterminal domain (BET) inhibitor from the hybridization of a dimethylphenol benzimidazole series, identified using DEL technology, with an N-methyl pyridone series identified through fragment screening [36]. In this case, DEL hits were identified from a 3-cycle benzimidazole library (Fig. 9). This hit series was defined by the invariable BB2 4-hydroxy-3,5-dimethylbenzaldehyde colored in red. Some variation on BB1 (colored in blue) and BB3 (colored in green) were tolerated. The off-DNA compound **7** derived from the most enriched BB1 (piperidin-4-ylmethanamine) and BB2 (4-hydroxy-3,5-dimethylbenzaldehyde) disynthon was prepared and confirmed to be a BRD4 inhibitor with an IC_{50} value of 199 nM in the biochemical assay. The variable BB3 group was further modified to give a smaller acetamide **8** (MW = 435) without losing significant activity. The crystal structure of acetamide **8** bound to BRD4 bromodomain 1 (BD1) was obtained and revealed the dimethylphenol moiety positioned in the KAc binding site (Fig. 10a). Specifically, the hydroxyl accepted a

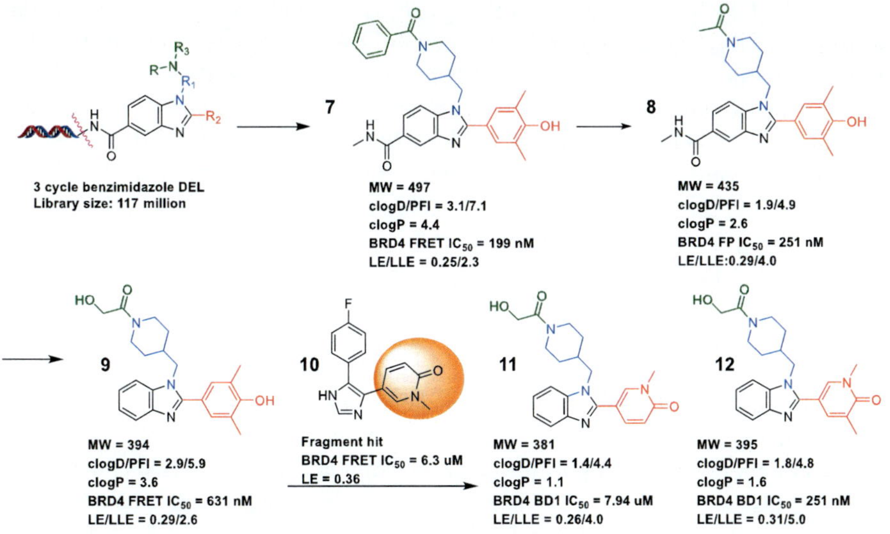

Fig. 9 Off-DNA hit confirmation and identification of lead compound **12** via hybridization of DEL hit **9** and fragment screening hit **10** [36]

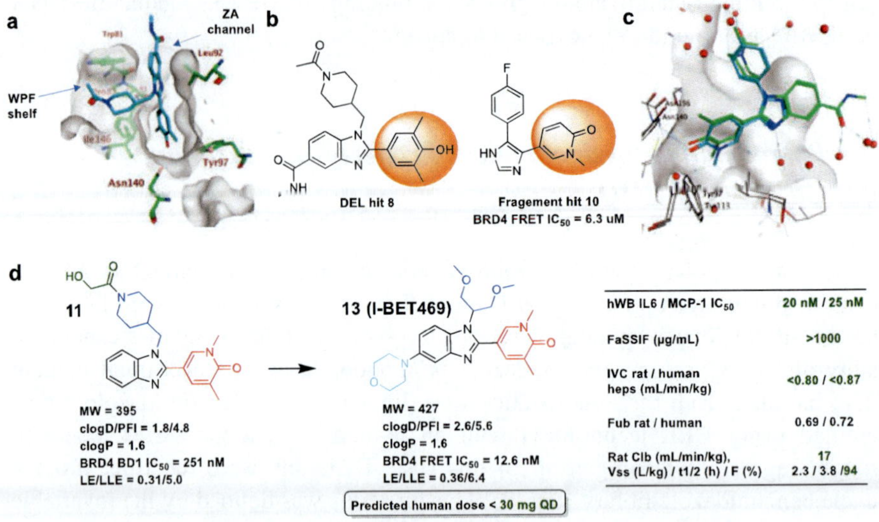

Fig. 10 (**a**) X-ray crystal structure of acetamide **8** in BRD4, (**b**) Structure of the DEL hit **8** and fragment hit **10**, (**c**) Superposition of X-ray crystal structures of N-methyl pyridone imidazole **10** (cyan) in BRD2 BD1 and 2,6-dimethylphenol **8** (green) in BRD4 BD1, (**d**) Discovery of the candidate-quality dimethylpyridone **13** (**I-BET469**) with favorable oral pharmacokinetic properties [36]

hydrogen bond from Asn140 and donated a hydrogen bond via water to Tyr97, while an ortho-methyl substituent occupied a lipophilic pocket adjacent to Phe83.

A substructure search for related N-methyl pyridone analogs in the GSK compound collection identified the imidazole **10** (BRD4 fluorescence resonance energy transfer (FRET) IC_{50} = 6.3 uM, LE = 0.36) (Fig. 10b). A 2.3 Å X-ray crystal structure of this compound was obtained in BRD2 BD1 and confirmed that N-methylpyridone also binds in the acetlyated lysine (KAc) binding site and shares the same interactions as the dimethylphenol group. The superimposed crystal structures of phenol **8** in BRD4 BD1 and pyridone **10** in BRD2 BD1 (Fig. 10c) suggested the pyridone group may be a more efficient KAc mimetic than the dimethylphenol group. The pyridones **11** and **12** were prepared and the 1,3-dimethylpyridone **12** showed higher BRD4 activity compared to the dimethylphenol compound **9** and the lipophilicity (clogP) was lowered by ~100-fold resulting in a higher LLE (ΔLLE = 2.4) (Fig. 9). Further optimization of pyridone **12** was accomplished by replacing the acyl piperidine moiety (colored in blue and green) with a 1,3-dimethoxypropan-2-yl group which improved the selectivity against the related bromodomain BAZ2A (Fig. 10d). Appendage of a morpholine group at the 5-position improved both the biochemical and cellular activity of the BET inhibitors. It is worth noting that the 5-position of the benzimidazole was the DNA attachment vector and is exposed to the solvent as indicated in the X-ray crystal structure. In the end a candidate-quality molecule **13** (**I-BET469**) was discovered which possesses favorable oral pharmacokinetic properties, displays activity in vivo, and is projected to have a low human efficacious dose of <30 mg once per day (QD).

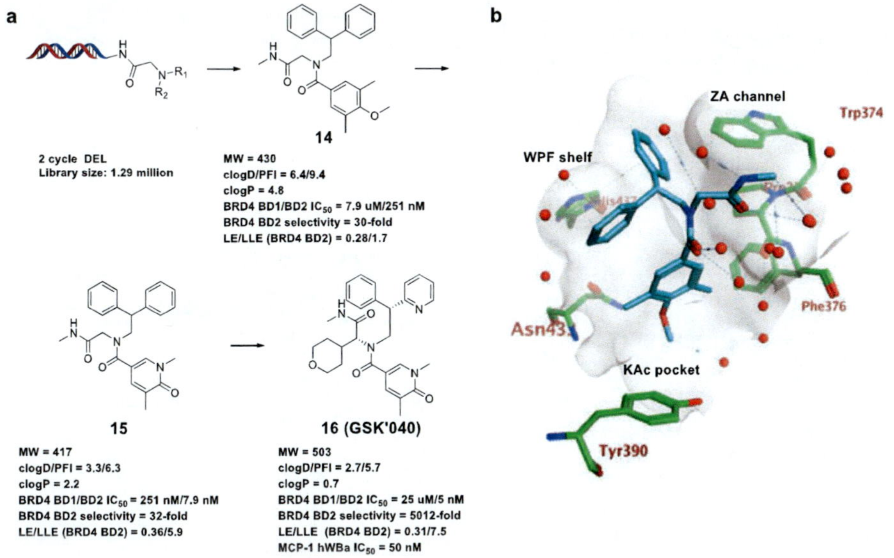

Fig. 11 (**a**) Hit-to-lead optimization for BRD4 BD2, (**b**) Crystal structure of **14** (cyan) in BRD2 BD2 [53]

Another example of substituting a lipophilic methoxyphenyl KAc mimetic with the dimethyl pyridone (DMP) was recently reported by Rianjongdee et al. [53]. In this example compound **14** was identified as a BRD4 BD2-selective inhibitor from a 2-cycle glycine-based DEL (Fig. 11a). A co-crystal structure of compound **14** bound to BRD2 BD2 was solved and is shown in Fig. 11b. The methoxyphenyl group sits in the KAc pocket, acting as a KAc mimetic [74]. Again the DMP was used as a KAc mimetic replacement for the more lipophilic 2-methoxy-1,3-dimethylphenyl group. Introduction of the DMP group both increased the BD2 potency by 1.5-log units while also lowering the clogD from 6.4 (**14**) to 3.3 (**15**); this led to a corresponding increase in the LE/LLE from 0.28/1.7 for compound **14**, to 0.36/5.9 for compound **15**. Further optimization was achieved through the introduction of a nitrogen group to one of the phenyl groups to further reduce clogD, and introduction of a tetrahydropyran group at the α-position to improve BD2-selectivity giving an unprecedented selectivity (5000-fold) against BRD4 BD1. The resulting lead molecule **16 (GSK’040)** possesses excellent selectivity against other bromodomains and good physicochemical properties (Fig. 11a). The lead molecule **16** exhibits oral bioavailability in rat models and represents a novel chemotype, serving as a potent BD2-selective probe for the understanding of BET biology.

α1-Antitrypsin is a major circulating protease inhibitor and its key function is regulation of the proteolytic effects of neutrophil elastase within the lung. Severe α1-antitrypsin deficiency leads to early onset emphysema. In their effort to discover an orally available small molecule treatment for α1-antitrypsin deficiency, Liddle et al. [61] identified a small molecule compound **17** that binds and inhibits Z α1-antitrypsin polymerization via a DEL screen (Fig. 12a). Compound **17** inhibited Z

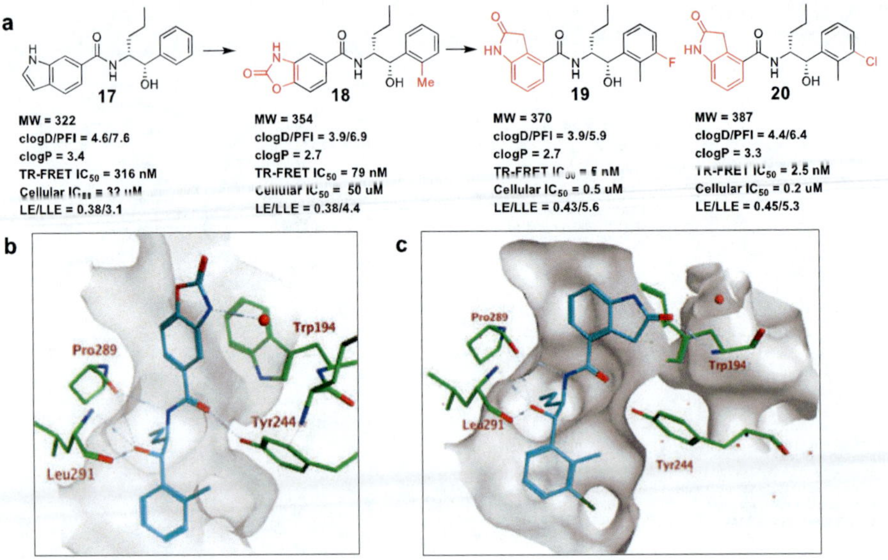

Fig. 12 (**a**) Hit-to-lead optimization for Z α1-antitrypsin with major changes highlighted in red, (**b**) Crystal structure of **18** (cyan) bound to apo α1-antitrypsin, (**c**) Crystal structure of **20** (cyan) bound to apo α1-antitrypsin [61]

α1-antitrypsin polymerization with an IC_{50} of 316 nM in the time-resolved fluorescence energy transfer (TR-FRET) polymerization assay and represents a highly ligand efficient (LE = 0.38) hit molecule found directly from DELs. The cellular activity of **17** to block Z α1-antitrypsin polymerization was modest ($IC_{50} = 32$ μM) and lead optimization effort was undertaken to lower lipophilicity and increase cellular activity. Replacing the indole ring with a benzoxazolone and the addition of 2-methyl group to modulate cell permeability led to compound **18**. The benzoxazolone **18** showed improved potency ($IC_{50} = 79$ nM) in the TR-FRET polymerization assay and more importantly a ten-fold drop in lipophilicity compared to the hit molecule **17** which resulted in a higher LLE of 4.4. A high-resolution co-crystal structure of **18** bound to apo α1-antitrypsin was solved (Fig. 12b). Several key polar interactions were observed. The hydroxyl group H-bonds to Leu291 backbone, the amide carbonyl forms an H-bond to Tyr244 OH and the amide NH H-bonds to the Pro289 carbonyl. The benzoxazolone bicycle stacks with the indole ring of Trp194 and the NH makes a water-bridged interaction with the backbone NH of Trp194. The phenyl ring and the propyl chain occupy two hydrophobic pockets.

The crystal structure suggested that substituting the 2- and 3-positions of the phenyl ring with small lipophilic groups may help to increase affinity, while modulation of the physicochemical properties may be achieved by introduction of polarity to the benzoxazolone portion of the molecule. Further optimization on both regions resulted in a 2-oxoindole lead molecule **19** with more than 10-fold improvement in potency in the TR-FRET assay. Examination of the crystal structure of α1-antitrypsin complexes with a 2-oxoindole analog **20** suggested the improved

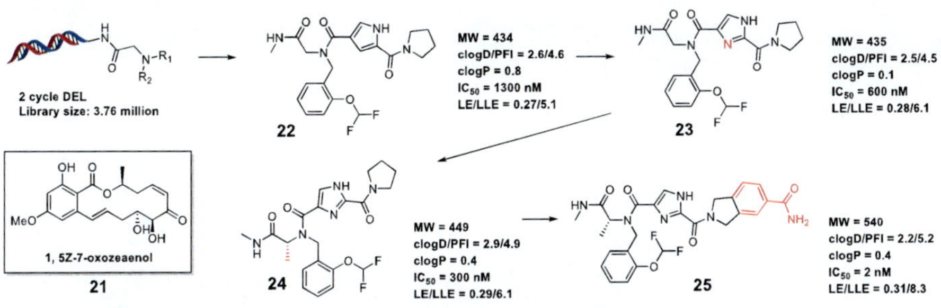

Fig. 13 Hit-to-lead optimization for TAK1, with major changes highlighted in red [24]

hydrogen bond interaction of the carbonyl group with Trp194 may be partially responsible for the increased potency compared to benzoxazolone **18** (Fig. 12c). Lead compound **19** is a highly ligand efficient (LE/LLE = 0.43/5.6) corrector of Z α1-antitrypsin misfolding. It demonstrated cellular activity (IC_{50} = 0.5 μM) and selectivity, while demonstrating an excellent in vitro profile, moderate bioavailability in rat and importantly is predicted to have high oral bioavailability in human.

SBDD has also been successfully applied in the context of transforming growth factor β-activated kinase 1 (TAK1) inhibitors [24]. Screening a mixture of 21 different DELs against TAK1–TAB1 (TAK1 binding protein 1) fusion protein at different protein concentrations, with and without a known inhibitor 5Z-7-oxozeaenol (**21**) identified a cluster of structurally related disynthons (combination of 2 building blocks) from a 2-cycle library with a total of 3.76 million different compounds (Fig. 13). These disynthons showed no enrichment in the selection with TAK1-TAB1 and a saturating concentration of 5Z-7-oxozeaenol, consistent with the hypothesis that both bind on the same site on the target protein. The off-DNA hit **22** based on the most enriched disynthon showed an IC_{50} value of 1.3 μM in the LanthaScreen biochemical kinase inhibition assay. A scaffold-hop to the corresponding imidazole resulted in compound **23** with improved biochemical potency and high kinase selectivity against a panel of 468 kinases. Introduction of a methyl group to constrain the rotation of the methyl amide group at the DNA attachment point led to a more potent compound **24**. Further optimization was achieved from structural information. A crystal structure of compound 23 with TAK1 showed it bound to the hinge in a type-I fashion and showed the imidazole and the carbonyl to bind as a donor–acceptor pair with the backbone of Ala107. The pyrrolidine was pointing to the back. There is a conformational change induced by compound **23** on the hinge region which may be responsible for the observed high kinase selectivity. Overlay of the co-crystal structures of compound **23** with an existing TAK inhibitor **26** [75] suggested that potency jump could be obtained by targeting the Lys63-Asp175 pair (Fig. 13a). A docking study showed this can be achieved by growing from the pyrrolidine ring and isoindoline-5-carboxamide was found to be a good fit for this purpose.

The resulting compound **25** showed >100-fold jump in biochemical potency resulting in a dramatic increase in LLE by 2.2, while superior kinase selectivity was

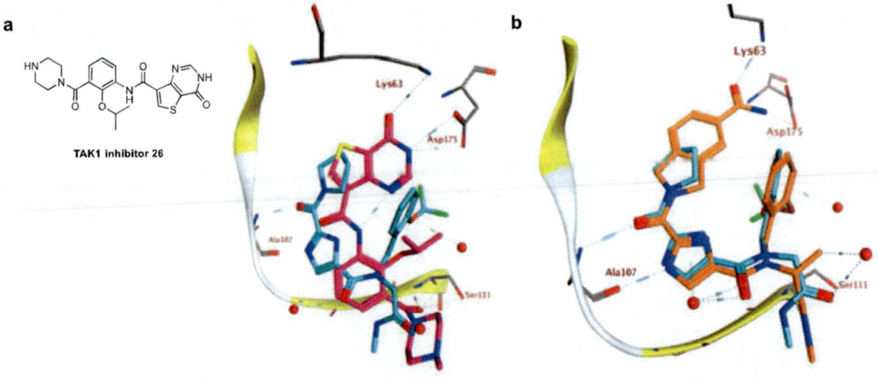

Fig. 14 (**a**) Structure of existing TAK1 inhibitor **26**, and overlay of the structures of TAK1 inhibitor **23** (cyan) and **26** (pink) bound in the ATP binding site of TAK1, (**b**) Overlay of the structures of **23** (cyan) and **25** (orange) bound in the ATP binding site of TAK1 [76]

maintained. A crystal structure of compound **25** bound to TAK1 confirmed the binding mode (Fig. 14b) with the isoindoline carboxamide showed a dual interaction with Lys63 as well as the Asp175. The α-methyl group helped to orientate the glycine carbonyl to make a hydrogen bond interaction with the Ser111 at the edge of the pocket, which was not observed for compound **23** (Fig. 14b, colored in cyan) without the α-methyl group.

5.3 Preorganization of Compound into Bound Conformation

Preorganization of the compound into the bound conformation via cyclization is another common technique to increase binding affinity and LLE; by decreasing ligand strain in the bound conformation to maximize entropic contributions to binding [77]. Johannes and co-workers described the discovery and optimization of a series of non-natural peptide myeloid cell leukemia 1 (Mcl-1) inhibitors derived from a tripeptide DEL [58]. The off-DNA compounds **27** and **28** were found to be 1.49 and 1.99 μM Mcl-1 inhibitors, respectively (Fig. 15a). The terminal methyl carboxamide is the DNA attachment site in the library. A crystal structure of compound **27** (cyan) bound to Mcl-1 was obtained (Fig. 15b). Compound **27** was confirmed to bind to the BH3 binding groove of Mcl-1 much like the previously reported Mcl-1 inhibitor 6-chloro-3-[3-(4-chloro-3,5-dimethyl-phenoxy)propyl]-1H-indole-2-carboxylic acid [78]. The key primary amine forms H-bonds to Ser255 and the backbone carbonyl of His252. In contrast to the indole acid, binding of DEL hit **27** does not involve a direct H-bond with Arg263 and thus represented a novel binding mode.

From the crystal structure of compound **27**, the authors noticed that the bound β-turn conformation positioned the two homophenylalanine groups in proximity. It

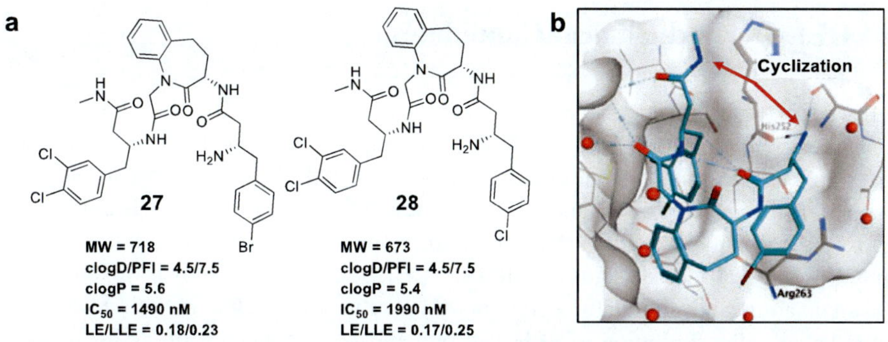

Fig. 15 (**a**) DEL hits for Mcl-1, (**b**) crystal structure of compound **27** (cyan) bound to Mcl-1 [58]

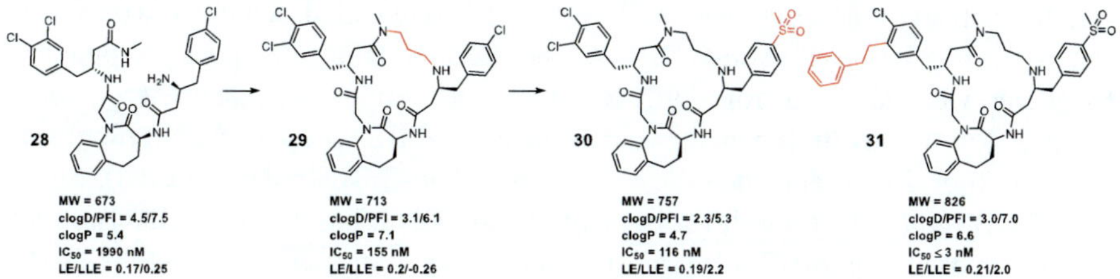

Fig. 16 Hit-to-lead optimization for Mcl-1 [58]

was hypothesized that additional potency could be achieved if these two ends of the molecule were linked into a macrocycle to preorganize the structure toward the bound conformation. Linking of the terminal primary amine with the methyl carboxamide was prioritized since this linkage (1) created no new stereocenters, (2) was more synthetically accessible, and (3) placed the linker in a more solvent exposed region where the likelihood of disruptive clashes was low. The methyl group at the newly formed amide nitrogen was required to force the amide to adopt the cis conformation that closely matches the bound conformation of **27**. The resulting macrocycle **29** was a 155 nM inhibitor of Mcl-1, which represents an over 12-fold improvement in binding potency relative to the uncyclized **28** (Fig. 16). To reduce the high clogP the chloride group was replaced with a more hydrophilic sulfone group (red) resulting in a 2.4-unit reduction of clogP without sacrificing potency (**30**). Further potency improvement was achieved by extending the dichlorophenyl moiety to fully exploit the back hydrophobic pocket. Installing the phenethyl group (red) led to compound **31** with an IC_{50} value of ≤3 nM in Mcl-1 FRET assay and EC_{50} of 3.78 μM in the MV4–11 caspase induction assay. In summary using SBDD, the binding potency of this series was improved around 500-fold without severely compromising the molecular weight or lipophilicity of the series [58].

6 DEL-Derived Clinical Candidates

DEL hits have been successively optimized and advanced to the clinic for various disease targets (e.g., sEH [16, 18], RIP1 [35, 59], and ATX [17]). Epoxide hydrolases are enzymes that convert epoxides to diols by hydrolysis. In mammals sEH is primarily responsible for the metabolism of arachidonic acid derivatives known as epoxyeicosatrienoic acids (EETs) into their corresponding less active dihydroxyeicosatrienoic acids (DHETs). EETs have been shown to mediate vasorelaxation and stimulate secretion of vascular endothelial growth factor by endothelial cells. Inhibition of sEH can prevent the degradation of EETs, restore vascular tone, and reduce loss of endothelial cells. Thus, sEH inhibitors could have beneficial effects in chronic obstructive pulmonary disease (COPD).

In their effort to identify novel inhibitors of sEH, Ding et al. screened two closely related triazine-based libraries: DEL-A and DEL-B (Fig. 17) [16]. A putative chemotype was identified from DEL-B, which was defined by a disynthon combination of 4-(aminomethyl) benzoic acid as the cycle 4 BB (colored in light blue) and cis-4-amino-cyclohexanecarboxylic acid as the cycle 2 BB (colored in red). Since cycle 1 BB (colored in purple) was linked to the DNA tag and more likely to be solvent exposed, for off-DNA confirmation it was truncated to a methyl group leading to compound **32** which exhibited an IC_{50} of 24 nM in the biochemical assay. Screening DEL-A identified several structurally related cycle-4 BBs including 2-(trifluoromethyl)phenylmethanamine. As the carboxylic acid group in **32** was a liability for cell permeability, it was replaced with 2-(trifluoromethyl)phenylmethanamine leading to compound **33** with similar activity but significantly higher clogP and lower LLE (5.4 and 2.2 for **32** vs. 6.6 and 0.83 for **33**, respectively). The clogP was successfully reduced by replacing the non-selected cycle 3(*R*)-2-phenylpropan-1-amine group with a more polar 1-methylpiperazine group and concurrently replacement of the 1,4-cyclohexane with 1,3-cyclohexane as the linker (compound **34**) resulting in a higher LLE (4.0) because of improved potency and reduced clogP. By utilizing the on-DNA SAR

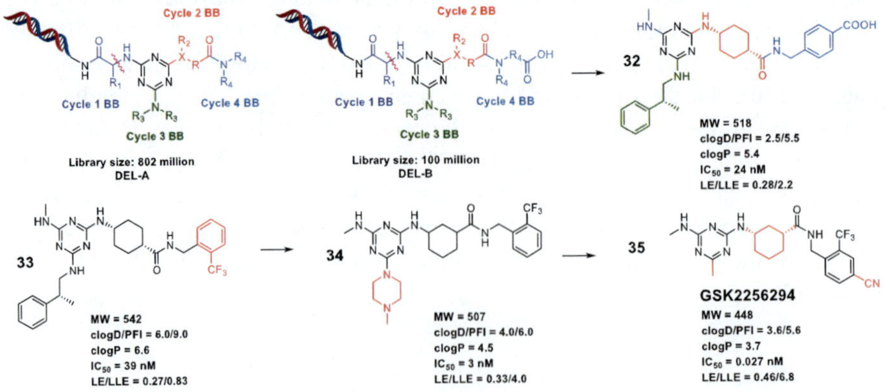

Fig. 17 Hit-to-lead optimization for sEH [16, 18]

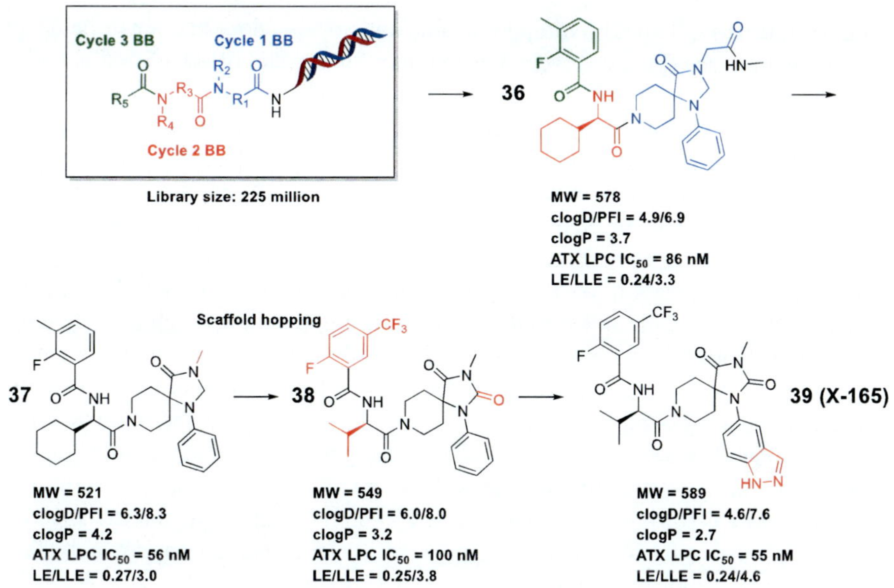

Fig. 18 Hit-to-lead optimization for ATX [17]

information, the authors successfully improved the pharmacokinetic (PK) profile of the DEL hit. Further modifications to improve potency and oral bioavailability were achieved by reducing the 1-methylpiperazine to a smaller methyl group and the addition of a cyano group at the 4-position of the cycle-4 2-(trifluoromethyl)phenyl) methanamine group. The resulting compound **GSK2256294** (**35**) is a potent, tight-binding inhibitor of recombinant human sEH with an IC_{50} of 27 pmol and an LLE of 6.8, while inhibiting sEH activity in human whole blood with an IC_{50} of 6.83 nM [79]. **GSK2256294** is highly selective and displays good oral bioavailability in rat and dog (94 and ~100%, respectively). **GSK2256294** further progressed to phase IIa studies for COPD. In clinical studies, **GSK2256294** demonstrated high levels of sEH inhibition with single dose of 6–20 mg and an effect on EET-mediated vasodilation in human resistance arteries on overweight smokers, as measured by blood flow in the forearm.

ATX is an ecto-enzyme that hydrolyzes lysophosphatidylcholine (LPC) to lysophosphatidic acid (LPA) [80]. The inflammatory signaling molecule LPA has been associated with a number of human diseases including idiopathic pulmonary fibrosis [81, 82]. Recently, ATX inhibitor **GLPG1690** has demonstrated efficacy in improving lung function in phase II studies. In their effort to identify new ATX inhibitors, Cuozzo and co-workers [17] screened a single 3-cycle DEL of 225 million compounds and identified a series of potent inhibitors. Optimization of this series led to the discovery of a clinical candidate **X-165** (**39**) which was approved by the FDA for a Phase I clinical trial (Fig. 18). The initial off-DNA hit (**36**) was based on a chemotype defined by an unusual spirocyclic amino acid (2-(4-oxo-1-phenyl-1,3,8-triazaspiro[4.5]decan-3-yl)acetic acid) at cycle 1. Other on-DNA SAR indicated

R-α-amino acids with small hydrophobic side chains at cycle 2 and small halogenated aromatic carboxylates at cycle 3 were preferred. Several off-DNA compounds were prepared in which the DNA linkage point was truncated to N-methylamide to confirm activity. The off-DNA hit compound **36** showed an IC_{50} of 86 nM in the ATX LPC biochemical assay. Further truncation of the DNA linkage to remove the amide group resulted in compound **37** with comparable activity and improved ligand efficiency (LE). Systematic SAR studies exploring additional spiro moieties were undertaken, in addition to replacing the cyclohexyl group with a smaller isopropyl at cycle 2 and optimization of the cycle 3 aromatic moiety to the 2-fluoro-5-trifluoromethylphenyl group led to the hydantoin compound **38** which has one unit drop in clogP and 0.8 unit improvement in LLE. Finally clinical candidate compound **X-165** (**39**) was discovered by replacing the phenyl group at the 1-position with a more polar indazole group which further reduced the clogP and improved the LLE. Compound **X-165** showed low nanomolar potency for LPA production in plasma both in vitro and in vivo as well as good PK properties. In addition, Compound **39** (**X-165**) demonstrated efficacy in both prophylactic and therapeutic models of fibrosis and safe in IND-enabling studies, which led to its approval by the FDA for Phase I studies in humans [17].

RIP1 is an upstream kinase that has been shown to regulate inflammation through both scaffolding and kinase-specific functions. Inhibition of RIP1 activation is likely to have broad therapeutic potential for multiple inflammatory diseases [83, 84]. Harris et al. described the identification of a series of benzoxazepinone inhibitors with complete monokinase selectivity for RIP1 from screening GSK's collection of DELs (Fig. 2) [35, 59]. After lead optimization focusing on improving the pharmacokinetic and developability profile of the series, a clinical candidate **GSK2982772** (**51,** Fig. 20b) was discovered which was advanced to phase IIa clinical studies for psoriasis, rheumatoid arthritis, and ulcerative colitis. Affinity selections against the RIP1 kinase domain (1–375) were carried out and a series of putative binders were identified from a three-cycle DEL of approximately 7.7 billion members (Fig. 2). The on-DNA disynthon hits were defined by an enantiopure atypical amino acid building block [(S)-3-amino-5-methyl-4-oxo-2,3,4,5-tetrahydrobenzo[b][1,4]oxazepine-7-carboxylic acid], as BB2 and three distinct but structurally related amine caps as BB3. There was no preference for BB1, suggesting that BB1 contributes very little to the binding (Fig. 2). Based on the on-DNA SAR information, off-DNA representative exemplars of the BB2 & BB3 combinations were designed with complete truncation of both BB1 and the carboxyl linker for BB2 (Fig. 19). The benzoxazepinones **40–43** were found to be potent biochemically as off-DNA compounds. Among them **GSK'481** (**40**) not only showed excellent biochemical activity (FP $IC_{50} = 10$ nM) but also exhibited excellent translation in the U937 cellular assay with an $IC_{50} = 10$ nM. **40** also showed complete selectivity against >450 off-target kinases. A co-crystal structure of **40** bound to RIP1 revealed that it is a type-III class of kinase inhibitor with no interactions with the hinge residues observed.

During the lead optimization process, systematic SAR studies of **40** were conducted with the goal of optimizing three key parameters: lipophilicity, solubility, and oral exposure in preclinical species while maintaining its favorable in vitro

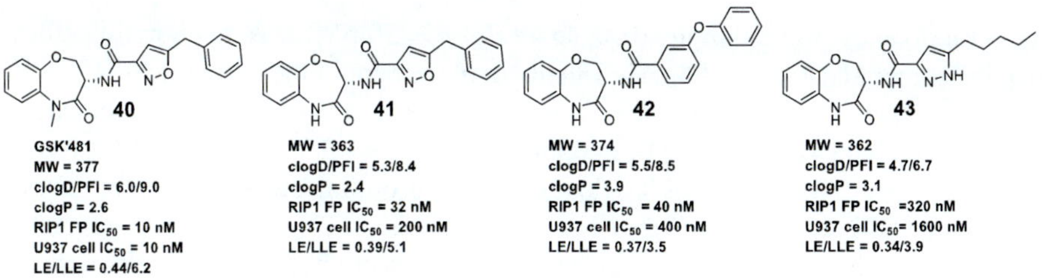

Fig. 19 Initial DEL-derived hits from screen against RIP1 [35, 59]

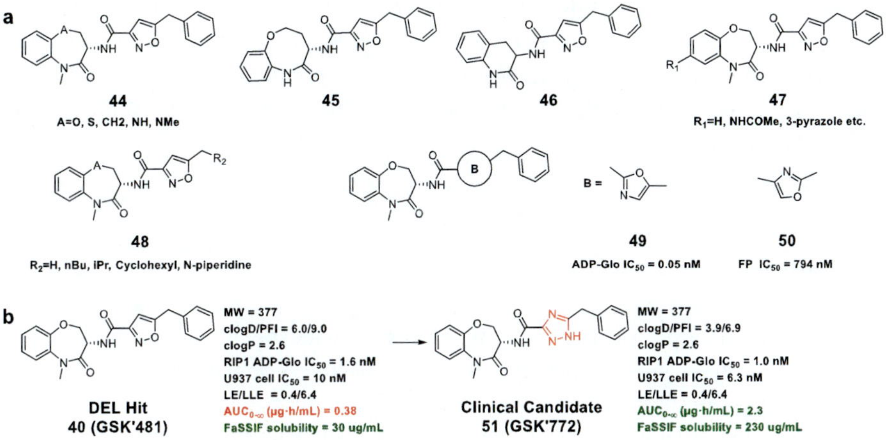

Fig. 20 (**a**) Lead optimization for RIPK1, (**b**) Identification of clinical candidate **51** (**GSK'772**) from DEL-derived hit **40** (**GSK'481**), resulting in the two-log unit improvement in logD and sevenfold improvement in oral exposure in rats [59]

profile. It was noted that compound **40**, although very potent and selective, suffers from high lipophilicity (clogD = 6.0), low fasted state simulating intestinal fluid (FaSSIF) crystalline solubility (30 μg/mL), and a suboptimal pharmacokinetic profile such as low oral exposure, high clearance (69 mL/min/kg), and high volume of distribution in rat. Extensive modification of the benzoxazepinone was carried out (Fig. 20a). Replacement of the heteroatom in the ring (**44**), expanding, contracting of the ring size (**45–46**) and substitution at the aryl ring of the benzoxazepinone ring (**47**), etc. did not yield a satisfactory molecule that met all the above desired criteria [59].

The attention was then turned to the benzyl group, which resides in an allosteric hydrophobic pocket at the back of the ATP binding site as indicated in the co-crystal structure of **40** bound to RIP1. SAR around this region was quite narrow and clearly favored lipophilic groups. Replacement of the phenyl ring with aliphatic groups resulted in a loss of activity (**48**). In the end, replacement of the isoxazole heterocycle had the greatest overall impact on improving the developability profile of this series. Various 5 and 6 member rings systems were explored and found to have significant effect on potency. For example, oxazole **49** ($IC_{50} = 50$ pM) was much more potent than its isomer **50** ($IC_{50} = 794$ nM). Finally, replacement of the isoxazole of **40** with a triazole led to benzoxazepinone **51** which was found to have optimal combination of in vitro potency ($IC_{50} = 1$ nM), lipophilicity (clogD = 3.9), and rat oral exposure ($AUC_{0\text{-}\infty}$ 2.3 μg h/mL at 2 mg/kg dose). This lead optimization culminated in the selection of **GSK2982772** (**51**) to progress into development (Fig. 20b). Compound **51** maintains the excellent potency and kinase selectivity of the hit **40** with a significantly improved developability profile as demonstrated by a two-log improvement in clogD and seven-fold improvements in oral exposure in rats of **51** compared to **40** as well as a much improved FaSSIF solubility (230 μg/mL) (Fig. 20b) [59].

Benzoxazepinone **51** exhibits highly favorable physicochemical and ADME properties. It demonstrated excellent activity in preventing TNF-induced necrotic cell death and was also able to reduce spontaneous production of cytokines from human ulcerative colitis explants in preclinical studies. It was advanced into phase 1 clinical trials in 2015.

7 Conclusions

DEL screening is a powerful strategy that has seen widespread use for the identification of bioactive, small molecules for drug discovery and more recently crop protection [85]. While triaging the output of a DEL screen can be overwhelming due to the sheer volume of data, various data visualization methods have been developed to facilitate this process. DEL selection data can also be combined with the output of other screening methods such as HTS, fragment-based screening, and virtual screening to inform the hit triaging or optimization process to maximize success [86]. In this chapter we have covered trends in DEL hit-to-lead optimization in the pharmaceutical industry and recommended approaches for improving potency and properties of these small molecules. As the field continues to evolve with new on-DNA chemistry and targeting more lead-like libraries (with fewer cycles of chemistry) DEL hit optimization tactics may continue to evolve as well. In addition, machine learning on DEL data offers opportunities for the discovery of diverse chemical series which may be unattainable as of yet due to current limitations of DEL-compatible chemistry [87].

Property Calculations All property calculations were done to make the analysis consistent across different publications.

clogP: Was calculated using commercial BioByte clogP, version 5.4

clogD: Was calculated using an in-house model based on chromlogD data

PFI: Was calculated by adding the clogD value to the number of aromatic rings

RotB: Was calculated using an in-house set of definitions using the ChemAxon JCHEM library and SMARTS to identify rotatable bonds

Ethical and Conflicts of Interest Statements All authors are employed by GSK, working on a DEL platform. No funding was received to assist with the preparation of this manuscript.

References

1. Clark MA, Acharya RA, Arico-Muendel CC, Belyanskaya SL, Benjamin DR, Carlson NR et al (2009) Design, synthesis and selection of DNA-encoded small-molecule libraries. Nat Chem Biol 5(9):647–654
2. Kuai L, O'Keeffe T, Arico-Muendel C (2018) Randomness in DNA encoded library selection data can be modeled for more reliable enrichment calculation. SLAS Discov 23(5):405–416
3. Kleiner RE, Dumelin CE, Liu DR (2011) Small-molecule discovery from DNA-encoded chemical libraries. Chem Soc Rev 40(12):5707–5717
4. Buller F, Mannocci L, Scheuermann J, Neri D (2010) Drug discovery with DNA-encoded chemical libraries. Bioconjug Chem 21(9):1571–1580
5. Faver JC, Riehle K, Lancia Jr DR, Milbank JBJ, Kollmann CS, Simmons N et al (2019) Quantitative comparison of enrichment from DNA-encoded chemical library selections. ACS Comb Sci 21(2):75–82
6. Richter H, Satz AL, Bedoucha M, Buettelmann B, Petersen AC, Harmeier A et al (2019) DNA-encoded library-derived DDR1 inhibitor prevents fibrosis and renal function loss in a genetic mouse model of Alport syndrome. ACS Chem Biol 14(1):37–49
7. Encinas L, O'Keefe H, Neu M, Remuiñán MJ, Patel AM, Guardia A et al (2014) Encoded library technology as a source of hits for the discovery and Lead optimization of a potent and selective class of bactericidal direct inhibitors of mycobacterium tuberculosis InhA. J Med Chem 57(4):1276–1288
8. Fu GK, Hu J, Wang PH, Fodor SP (2011) Counting individual DNA molecules by the stochastic attachment of diverse labels. Proc Natl Acad Sci U S A 108(22):9026–9031
9. Deng H, Zhou J, Sundersingh FS, Summerfield J, Somers D, Messer JA et al (2015) Discovery, SAR, and X-ray binding mode study of BCATm inhibitors from a novel DNA-encoded library. ACS Med Chem Lett 6(8):919–924
10. Satz AL (2016) Simulated screens of DNA encoded libraries: the potential influence of chemical synthesis fidelity on interpretation of structure-activity relationships. ACS Comb Sci 18(7):415–424
11. Buller F, Zhang Y, Scheuermann J, Schafer J, Buhlmann P, Neri D (2009) Discovery of TNF inhibitors from a DNA-encoded chemical library based on diels-alder cycloaddition. Chem Biol 16(10):1075–1086
12. Satz AL (2015) DNA encoded library selections and insights provided by computational simulations. ACS Chem Biol 10(10):2237–2245
13. Chan AI, McGregor LM, Liu DR (2015) Novel selection methods for DNA-encoded chemical libraries. Curr Opin Chem Biol 26:55–61
14. Buller F, Steiner M, Scheuermann J, Mannocci L, Nissen I, Kohler M et al (2010) High-throughput sequencing for the identification of binding molecules from DNA-encoded chemical libraries. Bioorg Med Chem Lett 20(14):4188–4192
15. Cuozzo JW, Centrella PA, Gikunju D, Habeshian S, Hupp CD, Keefe AD et al (2017) Discovery of a potent BTK inhibitor with a novel binding mode by using parallel selections with a DNA-encoded chemical library. Chembiochem 18(9):864–871
16. Ding Y, Belyanskaya S, DeLorey JL, Messer JA, Joseph Franklin G, Centrella PA et al (2021) Discovery of soluble epoxide hydrolase inhibitors through DNA-encoded library technology (ELT). Bioorg Med Chem 41:116216
17. Cuozzo JW, Clark MA, Keefe AD, Kohlmann A, Mulvihill M, Ni H et al (2020) Novel autotaxin inhibitor for the treatment of idiopathic pulmonary fibrosis: a clinical candidate discovered using DNA-encoded chemistry. J Med Chem 63(14):7840–7856

18. Belyanskaya SL, Ding Y, Callahan JF, Lazaar AL, Israel DI (2017) Discovering drugs with DNA-encoded library technology: from concept to clinic with an inhibitor of soluble epoxide hydrolase. Chembiochem 18(9):837–842
19. Soutter HH, Centrella P, Clark MA, Cuozzo JW, Dumelin CE, Guie MA et al (2016) Discovery of cofactor-specific, bactericidal mycobacterium tuberculosis InhA inhibitors using DNA-encoded library technology. Proc Natl Acad Sci U S A 113(49):E7880–E78E9
20. Clark MA (2010) Selecting chemicals: the emerging utility of DNA-encoded libraries. Curr Opin Chem Biol 14(3):396–403
21. Yang H, Medeiros PF, Raha K, Elkins P, Lind KE, Lehr R et al (2015) Discovery of a potent class of PI3Kalpha inhibitors with unique binding mode via encoded library technology (ELT). ACS Med Chem Lett 6(5):531–536
22. Wu Z, Graybill TL, Zeng X, Platchek M, Zhang J, Bodmer VQ et al (2015) Cell-based selection expands the utility of DNA-encoded small-molecule library technology to cell surface drug targets: identification of novel antagonists of the NK3 tachykinin receptor. ACS Comb Sci 17(12):722–731
23. Foley TL, Burchett W, Chen Q, Flanagan ME, Kapinos B, Li X et al (2021) Selecting approaches for hit identification and increasing options by building the efficient discovery of actionable chemical matter from DNA-encoded libraries. SLAS Discov 26(2):263–280
24. Veerman JJN, Bruseker YB, Damen E, Heijne EH, van Bruggen W, Hekking KFW et al (2021) Discovery of 2,4-1H-imidazole carboxamides as potent and selective TAK1 inhibitors. ACS Med Chem Lett 12(4):555–562
25. Encinas L, O'Keefe H, Neu M, Remuinan MJ, Patel AM, Guardia A et al (2014) Encoded library technology as a source of hits for the discovery and lead optimization of a potent and selective class of bactericidal direct inhibitors of mycobacterium tuberculosis InhA. J Med Chem 57(4):1276–1288
26. Lewis HD, Liddle J, Coote JE, Atkinson SJ, Barker MD, Bax BD et al (2015) Inhibition of PAD4 activity is sufficient to disrupt mouse and human NET formation. Nat Chem Biol 11(3): 189–191
27. Salamon H, Klika Skopic M, Jung K, Bugain O, Brunschweiger A (2016) Chemical biology probes from advanced DNA-encoded libraries. ACS Chem Biol 11(2):296–307
28. Disch JS, Evindar G, Chiu CH, Blum CA, Dai H, Jin L et al (2013) Discovery of thieno[3,2-d] pyrimidine-6-carboxamides as potent inhibitors of SIRT1, SIRT2, and SIRT3. J Med Chem 56(9):3666–3679
29. Franzini RM, Nauer A, Scheuermann J, Neri D (2015) Interrogating target-specificity by parallel screening of a DNA-encoded chemical library against closely related proteins. Chem Commun (Camb) 51(38):8014–8016
30. Chen Q, Cheng X, Zhang L, Li X, Chen P, Liu J et al (2020) Exploring the lower limit of individual DNA-encoded library molecules in selection. SLAS Discov 25(5):523–529
31. Taylor DM, Anglin J, Park S, Ucisik MN, Faver JC, Simmons N et al (2020) Identifying oxacillinase-48 carbapenemase inhibitors using DNA-encoded chemical libraries. ACS Infect Dis 6(5):1214–1227
32. Deng H, Zhou J, Sundersingh F, Messer JA, Somers DO, Ajakane M et al (2016) Discovery and optimization of potent, selective, and in vivo efficacious 2-aryl benzimidazole BCATm inhibitors. ACS Med Chem Lett 7(4):379–384
33. Satz AL, Hochstrasser R, Petersen AC (2017) Analysis of current DNA encoded library screening data indicates higher false negative rates for numerically larger libraries. ACS Comb Sci 19(4):234–238
34. Hackler AL, FitzGerald FG, Dang VQ, Satz AL, Paegel BM (2020) Off-DNA DNA-encoded library affinity screening. ACS Comb Sci 22(1):25–34
35. Harris PA, King BW, Bandyopadhyay D, Berger SB, Campobasso N, Capriotti CA et al (2016) DNA-encoded library screening identifies benzo[b][1,4]oxazepin-4-ones as highly potent and monoselective receptor interacting protein 1 kinase inhibitors. J Med Chem 59(5):2163–2178

36. Wellaway CR, Amans D, Bamborough P, Barnett H, Bit RA, Brown JA et al (2020) Discovery of a bromodomain and extraterminal inhibitor with a low predicted human dose through synergistic use of encoded library technology and fragment screening. J Med Chem 63(2): 714–746
37. Fernandez-Montalvan AE, Berger M, Kuropka B, Koo SJ, Badock V, Weiske J et al (2017) Isoform-selective ATAD2 chemical probe with novel chemical structure and unusual mode of action. ACS Chem Biol 12(11):2730–2736
38. Franzini RM, Randolph C (2016) Chemical space of DNA-encoded libraries. J Med Chem 59(14):6629–6644
39. Xia B, Franklin GJ, Lu X, Bedard KL, Grady LC, Summerfield JD et al (2021) DNA-encoded library hit confirmation: bridging the gap between on-DNA and off-DNA chemistry. ACS Med Chem Lett 12(7):1166–1172
40. Ratnayake AS, Flanagan ME, Foley TL, Hultgren SL, Bellenger J, Montgomery JI et al (2021) Toward the assembly and characterization of an encoded library hit confirmation platform: bead-assisted ligand isolation mass spectrometry (BALI-MS). Bioorg Med Chem 41:116205
41. Su W, Ge R, Ding D, Chen W, Wang W, Yan H et al (2021) Triaging of DNA-encoded library selection results by high-throughput resynthesis of DNA-conjugate and affinity selection mass spectrometry. Bioconjug Chem 32(5):1001–1007
42. Satz AL, Brunschweiger A, Flanagan ME, Gloger A, Hansen NJV, Kuai L et al (2022) DNA-encoded chemical libraries. Nat Rev Methods Primers 2(1):3
43. Shi Y, Wu YR, Yu JQ, Zhang WN, Zhuang CL (2021) DNA-encoded libraries (DELs): a review of on-DNA chemistries and their output. RSC Adv 11(4):2359–2376
44. Schultes S, de Graaf C, Haaksma EEJ, de Esch IJP, Leurs R, Krämer O (2010) Ligand efficiency as a guide in fragment hit selection and optimization. Drug Discov Today Technol 7(3):e157–ee62
45. Lipinski CA, Lombardo F, Dominy BW, Feeney PJ (2001) Experimental and computational approaches to estimate solubility and permeability in drug discovery and development settings. Adv Drug Deliv Rev 46(1–3):3–26
46. Veber DF, Johnson SR, Cheng HY, Smith BR, Ward KW, Kopple KD (2002) Molecular properties that influence the oral bioavailability of drug candidates. J Med Chem 45(12): 2615–2623
47. Shultz MD (2013) Setting expectations in molecular optimizations: strengths and limitations of commonly used composite parameters. Bioorg Med Chem Lett 23(21):5980–5991
48. Hann MM, Oprea TI (2004) Pursuing the leadlikeness concept in pharmaceutical research. Curr Opin Chem Biol 8(3):255–263
49. Satz AL (2018) What do you get from DNA-encoded libraries? ACS Med Chem Lett 9(5): 408–410
50. Eidam O, Satz AL (2016) Analysis of the productivity of DNA encoded libraries. Med Chem Commun 7(7):1323–1331
51. Reiher CA, Schuman DP, Simmons N, Wolkenberg SE (2021) Trends in hit-to-lead optimization following DNA-encoded library screens. ACS Med Chem Lett 12(3):343–350
52. Ding Y, O'Keefe H, DeLorey JL, Israel DI, Messer JA, Chiu CH et al (2015) Discovery of potent and selective inhibitors for ADAMTS-4 through DNA-encoded library technology (ELT). ACS Med Chem Lett 6(8):888–893
53. Rianjongdee F, Atkinson SJ, Chung CW, Grandi P, Gray JRJ, Kaushansky LJ et al (2021) Discovery of a highly selective BET BD2 inhibitor from a DNA-encoded library technology screening hit. J Med Chem 64(15):10806–10833
54. Disch JS, Duffy JM, Lee ECY, Gikunju D, Chan B, Levin B et al (2021) Bispecific estrogen receptor α degraders incorporating novel binders identified using DNA-encoded chemical library screening. J Med Chem 64(8):5049–5066
55. Gentile G, Merlo G, Pozzan A, Bernasconi G, Bax B, Bamborough P et al (2012) 5-Aryl-4-carboxamide-1,3-oxazoles: potent and selective GSK-3 inhibitors. Bioorg Med Chem Lett 22(5):1989–1994

56. Lee ECY, McRiner AJ, Georgiadis KE, Liu J, Wang Z, Ferguson AD et al (2021) Discovery of novel, potent inhibitors of hydroxy acid oxidase 1 (HAO1) using DNA-encoded chemical library screening. J Med Chem 64(10):6730–6744
57. Kazmierski WM, Xia B, Miller J, De la Rosa M, Favre D, Dunham RM et al (2020) DNA-encoded library technology-based discovery, Lead optimization, and prodrug strategy toward structurally unique indoleamine 2,3-dioxygenase-1 (IDO1) inhibitors. J Med Chem 63(7):3552–3562
58. Johannes JW, Bates S, Beigie C, Belmonte MA, Breen J, Cao S et al (2017) Structure based design of non-natural peptidic macrocyclic Mcl-1 inhibitors. ACS Med Chem Lett 8(2): 239–244
59. Harris PA, Berger SB, Jeong JU, Nagilla R, Bandyopadhyay D, Campobasso N et al (2017) Discovery of a first-in-class receptor interacting protein 1 (RIP1) kinase specific clinical candidate (GSK2982772) for the treatment of inflammatory diseases. J Med Chem 60(4): 1247–1261
60. Gilmartin AG, Faitg TH, Richter M, Groy A, Seefeld MA, Darcy MG et al (2014) Allosteric Wip1 phosphatase inhibition through flap-subdomain interaction. Nat Chem Biol 10(3): 181–187
61. Liddle J, Pearce AC, Arico-Muendel C, Belyanskaya S, Brewster A, Brown M et al (2021) The development of highly potent and selective small molecule correctors of Z alpha1-antitrypsin misfolding. Bioorg Med Chem Lett 41:127973
62. Petersen LK, Blakskjær P, Chaikuad A, Christensen AB, Dietvorst J, Holmkvist J et al (2016) Novel p38α MAP kinase inhibitors identified from yoctoReactor DNA-encoded small molecule library. Med Chem Commun 7(7):1332–1339
63. Rohm S, Berger BT, Schroder M, Chaikuad A, Winkel R, Hekking KFW et al (2019) Fast iterative synthetic approach toward identification of novel highly selective p38 MAP kinase inhibitors. J Med Chem 62(23):10757–10782
64. Young RJ, Green DV, Luscombe CN, Hill AP (2011) Getting physical in drug discovery II: the impact of chromatographic hydrophobicity measurements and aromaticity. Drug Discov Today 16(17–18):822–830
65. Johnson TW, Gallego RA, Edwards MP (2018) Lipophilic efficiency as an important metric in drug design. J Med Chem 61(15):6401–6420
66. Beutner GL, Young IS, Davies ML, Hickey MR, Park H, Stevens JM et al (2018) TCFH-NMI: direct access to N-acyl Imidazoliums for challenging amide bond formations. Org Lett 20(14): 4218–4222
67. Keseru GM, Makara GM (2009) The influence of lead discovery strategies on the properties of drug candidates. Nat Rev Drug Discov 8(3):203–212
68. Wigglesworth MJ, Murray DC, Blackett CJ, Kossenjans M, Nissink JW (2015) Increasing the delivery of next generation therapeutics from high throughput screening libraries. Curr Opin Chem Biol 26:104–110
69. Song M, Hwang GT (2020) DNA-encoded library screening as core platform technology in drug discovery: its synthetic method development and applications in DEL synthesis. J Med Chem 63(13):6578–6599
70. Konstantinidou M, Li J, Zhang B, Wang Z, Shaabani S, Ter Brake F et al (2019) PROTACs- a game-changing technology. Expert Opin Drug Discov 14(12):1255–1268
71. Young RJ, Leeson PD (2018) Mapping the efficiency and physicochemical trajectories of successful optimizations. J Med Chem 61(15):6421–6467
72. Brown DG, Bostrom J (2018) Where do recent small molecule clinical development candidates come from? J Med Chem 61(21):9442–9468
73. Leeson PD, Young RJ (2015) Molecular property design: does everyone get it? ACS Med Chem Lett 6(7):722–725
74. Umehara T, Nakamura Y, Jang MK, Nakano K, Tanaka A, Ozato K et al (2010) Structural basis for acetylated histone H4 recognition by the human BRD2 bromodomain. J Biol Chem 285(10): 7610–7618

75. Muraoka T, Ide M, Morikami K, Irie M, Nakamura M, Miura T et al (2016) Discovery of a potent and highly selective transforming growth factor β receptor-associated kinase 1 (TAK1) inhibitor by structure based drug design (SBDD). Bioorg Med Chem 24(18):4206–4217
76. Muraoka T, Ide M, Morikami K, Irie M, Nakamura M, Miura T et al (2016) Discovery of a potent and highly selective transforming growth factor beta receptor-associated kinase 1 (TAK1) inhibitor by structure based drug design (SBDD). Bioorg Med Chem 24(18): 4206–4217
77. Chang CE, Chen W, Gilson MK (2007) Ligand configurational entropy and protein binding. Proc Natl Acad Sci U S A 104(5):1534–1539
78. Friberg A, Vigil D, Zhao B, Daniels RN, Burke JP, Garcia-Barrantes PM et al (2013) Discovery of potent myeloid cell leukemia 1 (Mcl-1) inhibitors using fragment-based methods and structure-based design. J Med Chem 56(1):15–30
79. Podolin PL, Bolognese BJ, Foley JF, Long E, Peck B, Umbrecht S et al (2013) In vitro and in vivo characterization of a novel soluble epoxide hydrolase inhibitor. Prostaglandins Other Lipid Mediat 104–105:25–31
80. Nakanaga K, Hama K, Aoki J (2010) Autotaxin – an LPA producing enzyme with diverse functions. J Biochem 148(1):13–24
81. Oikonomou N, Mouratis MA, Tzouvelekis A, Kaffe E, Valavanis C, Vilaras G et al (2012) Pulmonary autotaxin expression contributes to the pathogenesis of pulmonary fibrosis. Am J Respir Cell Mol Biol 47(5):566–574
82. Montesi SB, Mathai SK, Brenner LN, Gorshkova IA, Berdyshev EV, Tager AM et al (2014) Docosatetraenoyl LPA is elevated in exhaled breath condensate in idiopathic pulmonary fibrosis. BMC Pulm Med 14(1):5
83. Berger SB, Kasparcova V, Hoffman S, Swift B, Dare L, Schaeffer M et al (2014) Cutting edge: RIP1 kinase activity is dispensable for normal development but is a key regulator of inflammation in SHARPIN-deficient mice. J Immunol 192(12):5476–5480
84. Ofengeim D, Yuan J (2013) Regulation of RIP1 kinase signalling at the crossroads of inflammation and cell death. Nat Rev Mol Cell Biol 14(11):727–736
85. Bomgardner MM (2020) Start-ups see an opening for new crop protection chemistry. Chem Eng News.
86. Leveridge M, Chung CW, Gross JW, Phelps CB, Green D (2018) Integration of lead discovery tactics and the evolution of the Lead discovery toolbox. SLAS Discov 23(9):881–897
87. McCloskey K, Sigel EA, Kearnes S, Xue L, Tian X, Moccia D et al (2020) Machine learning on DNA-encoded libraries: a new paradigm for hit finding. J Med Chem 63(16):8857–8866

Top Med Chem (2022) 40: 259–276
https://doi.org/10.1007/7355_2022_152

Published online: 22 October 2022

A Perspective on 30 Years of DNA-Encoded Chemistry

Barry A. Morgan

Contents

Abstract The history of DNA-encoded chemistry and its evolution into a universally accepted and broadly practised technology for ligand discovery in early stage small-molecule is outlined by a contributor with more than 20 years experience developing the technology. Interviews with the enterpreneures and scientists who made key contributions to DEL technology are included in this personal account of what has been described "In the Pipeline" as "the second coming of combinatorial chemistry, but this time it works!".

Keywords DNA-encoded chemistry, DNA-encoded library technology, Hit discovery

B. A. Morgan (✉)
HitGen Inc., Chengdu, China

Institute for Molecular Medicine, University of Texas Health Sciences, Houston, TX, USA
e-mail: barry.morgan@hitgen.com

1 Prologue

Over the past two decades, it has been my good fortune to observe the evolution of DNA-encoded library (DEL) technology from a controversial concept to a valued methodology practiced broadly in industry and academia. I have been particularly lucky to work with hundreds of dedicated scientists as a colleague or a collaborator, and several of these have suggested that I should write an account of those years. I made a start on such an endeavor when I recorded interviews with eight contributors [1–8] to the development of DEL and presented these interviews at a COVID19 compliant "virtual happy hour" as an entertainment entitled "Great Moments in DEL History" at an NIH sponsored workshop in September 2021. In this current account [9], I will reference those interviews to provide a historical perspective on DEL technology, and I will elaborate on that perspective to access the current impact of DEL on early-stage small molecule drug discovery. I will conclude with some thoughts on the future. But let's start at the beginning...

Sometime "at the end of the day, for no particular reason" in 1992 [1–8] (maybe 1991, the exact date is not recorded) Nobel Laureate Sidney Brenner walked over to Richard Lerner's presidential office suite at the Scripps Research Institute in La Jolla and, as was their habit, pondered on a range of topics. On this occasion however, an interruption by a somewhat "agitated young scientist" bemoaning the outcome of his [10] reaction of the day prompted a philosophical discussion on the inequities regarding the recording of biologic and chemical endeavors. Fully recognizant of the fundamental principles of molecular biology, Brenner and Lerner considered and conceived of the proposition that a series of chemical transformations could, in a conceptual parallel to protein synthesis, also be encoded by DNA, and proceeded to illustrate the concept in their seminal paper [11] "Encoded combinatorial chemistry" published in June of that year. While Brenner and Lerner chose to assemble the encoding oligonucleotide by Carruthers-style base-by-base organic synthesis [12], with consequent significant damage to the encoded product at the final strong base deprotection, this issue was resolved in a brief 1995 publication [13] by Kinoshita and Nishigaki who noted that enzymatic ligation was very effective for DNA assembly using preformed "codon" blocks. Some years past without further developments, then in the first years of the new millennium, Harvard professor David Liu [2] elaborated on a concept he had envisaged in a seminar during his doctoral studies [14], describing how DNA-encoding could not only record the presumptive product of a series of organic chemical transformations, but also may facilitate the reaction kinetics. He described this work, entitled "DNA-templated synthesis" in an elegant series of publications [15], and also founded Ensemble Therapeutics [16] to explore the commercial potential of the concept.

2 Praecis and the "Birth" of DEL

After some 20 years of experience with peptide-based drug discovery, I joined Praecis Pharmaceuticals in early 2001 with the mission to expand their chemistry capabilities in that direction. Praecis was anticipating FDA approval, with associated return-on-investment for the licensed-in GnRH antagonist Abarelix for treatment of prostate cancer. In late 2002, Rick Wagner was appointed to head research and CEO Malcolm Gefter gave us an ambitious remit [17] to propose a new direction for early-stage small molecule internal discovery. Given Praecis' ongoing program using phage display technology and noting the recent publications on DNA-templated methodology, we were attracted to the concept of Brenner and Lerner, and Rick and I decided to champion DNA-encoded combinatorial chemistry as the "new direction." Our brief proposal to this end was accepted, and funding provided by CEO Gefter. Rick's interview for "Great moments in DEL History" includes a lighthearted discussion of our decision process [4]. A side-project ongoing at that time in the chemistry group provided an alternative to DNA-templated synthesis. Given Praecis' interest in prostate cancer we were attempting to achieve intracellular delivery of small oligonucleotides by attachment to cell-penetrating peptides [18] such as TAT and penetratin, and synthetic studies showed that good yields of conjugate could be achieved by amide bond formation using a 10–50-fold excess of the peptide in aqueous media [19]. While the goal of intracellular delivery proved untenable with these conjugates, the ability to synthesize the conjugates by simple acylation of amine-containing oligonucleotide derivatives seemed to offer a pragmatic and economical alternative to DNA-templated synthesis. Further studies with a range of carboxylic acid substrates provided additional validation. The next challenge was to decide on the nature of the oligonucleotide component. Given my lack of experience with oligonucleotide chemistry and biophysics, I attended a timely symposium and workshop on Nucleic Acid Chemistry and Biology in Cambridge UK where, while browsing through a poster session, I became aware of stem-loop structures in tRNA and pondered whether the loop could be formed from a simpler structural linker connecting short DNA-duplexes appropriate for codon ligation [20]. On my return to Praecis, rudimentary modeling approaches starting from an appropriate DNA-duplex indicated that the loop could be constructed from commercially available phosphoramidite components to provide a structure which we christened "the Headpiece." Contract synthesis at Biosearch Technologies [21] provided the Headpiece which allowed organic chemistry elaboration via the amine group and attachment of duplex "codons" via ligation using T4 ligase. While there are alternative approaches [9] to assembling DNA-encoded libraries, Headpiece-based methodology has become ubiquitous for DEL design and is the standard component for new reaction development in DEL, allowing significant expansion of the chemistry repertoire available for DEL design [22, 23].

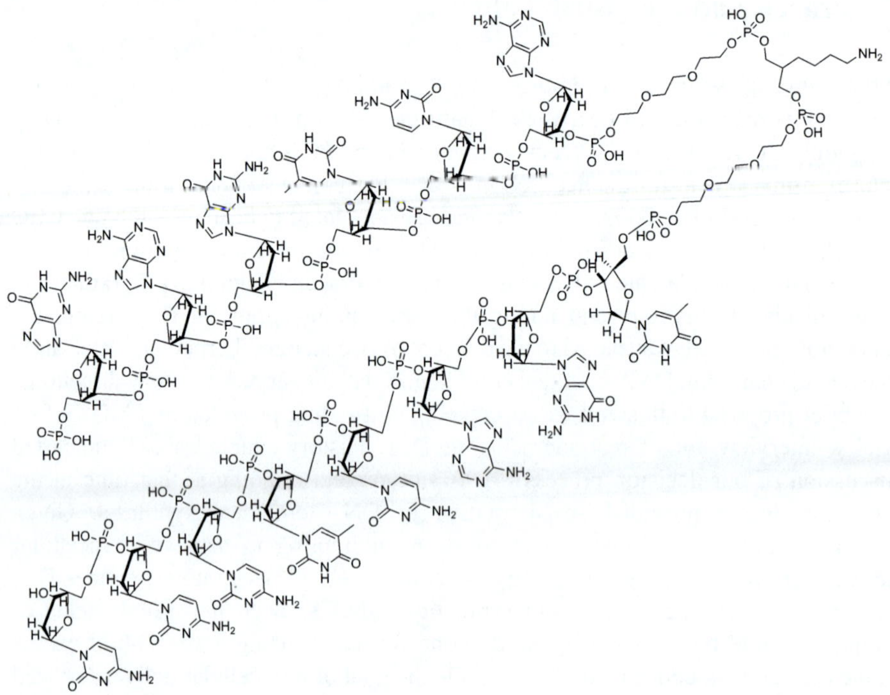

With our Headpiece design established and pragmatic chemistry partially validated, I began recruitment for leadership of the DEL synthesis team and was indeed fortunate to interview Matt Clark as a session at MIT! Matt joined Praecis in early 2004 and rapidly assembled a team which reduced our concept for DEL assembly to practice, completing the synthesis of "DEL 16" which we describe in our 2007 publication [24]. Matt's recollection of our MIT meeting and subsequent arrival at Praecis is included in his "Great moments in DEL History" interview [5]. It should be noted that in deference to CEO Malcolm Gefter's wishes we named the Praecis technology "Direct Select." Malcolm's well-founded trust in the remarkable power of screening by selection [25] and his response to my more naïve concern is covered in Malcolm's "Great moments in DEL History" interview [3].

It was at this time that a further key development required for the success of DEL technology became available. A colleague googling DNA sequencing became aware of the groundbreaking approach developed by 454 Life Sciences using emulsion PCR [26]. We immediately arranged an expedition down I-95 to view their machine, and quickly set up a proof-of-concept study for sequencing DEL DNA which worked spectacularly, allowing Praecis to confidently purchase one of the first commercially available machines.

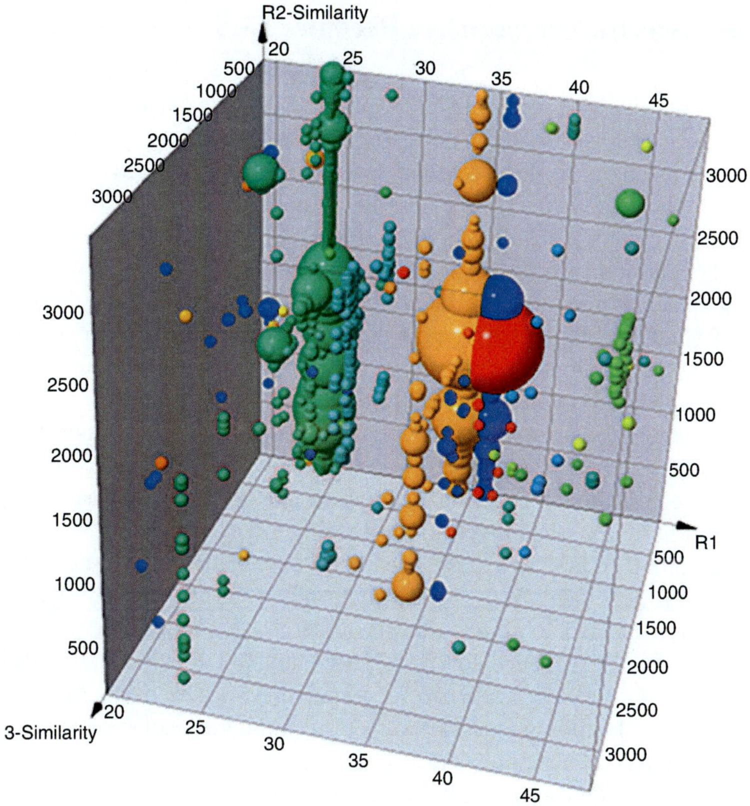

Now that we needed to sequence and examine millions of sequences, we also needed to develop methods for translating these data into the sets of structures they encoded [27] and presenting the outcome in a sufficiently facile manner that would be intuitive to our team. I had some familiarity with the multi-dimensional data visualization capabilities of Spotfire, so the "DEL Cubic Plot," and the concept of linear and planar "features" for identifying structurally related DEL chemotypes [28] became common practice.

Now we had DEL 16 [29, 30], a selection protocol loosely based on phage display, a means of interrogating the selection output by sequencing and a simple but effective tool for data display and interpretation. After some successful proof-of-concept studies on two kinase targets [24], we felt we were ready to look for some return on investment by seeking agreements with pharmaceutical companies.

3 GSK Steps in: Direct Select Becomes ELT

By 2005 high-throughput screening (HTS) was the well-established and reasonably successful methodology for small molecule “hit” discovery [31]. It was practiced at well over 50 locations in the USA and Europe, with many large pharmaceutical companies replicating their HTS operations at multiple sites. The main emerging issue was the economics: acquiring a curated HTS compound collection, typically of around one million structures, and establishing and maintaining a well-equipped facility to screen it required significant capital investment totaling hundreds of millions of USD. In addition, an ongoing cost per screen of the order of $500,000 and lengthy cycle time limited its deployment. We felt that the costs and timelines of DEL technology offered a disruptive [32] alternative to HTS. Our early presentations at symposia, and site visits to potential collaborators were not encouraging, meeting understandable skepticism, and in a few cases a less than favorable comparison to DNA-templated methodology. One example I like to refer to was at a SLAS conference in St Louis, when my abstract claiming interrogation of a ”10^{9}″ member library, was initially edited to a “109″ member library. This changed in 2005 when, once more, we benefitted from an unanticipated event. GlaxoSmithKline (now formally GSK) decided to reorganize its research operations into disease-class focused Centers of Excellence for Drug Discovery [33] or “CEDDs,” including one entitled the Center of Excellence for External Drug Discovery (the CEEDD) with the remit of identifying external assets or technologies suitable for investment [34]. After a series of meetings Praecis entered into a pilot agreement with the CEEDD where we would apply Direct-Select technology to four targets identified by the CEEDD following internal review at GSK. This agreement was signed in the Spring of 2006, and before the end of the Summer Praecis had identified highly potent, selective antagonists for the first two targets [35, 36]! One of these hits, for the target soluble epoxide hydrolase (sEH) was rapidly evolved to a clinical candidate [37] within less than a year of the initial Direct Select selection. This development attracted considerable interest at the CEEDD, resulting in CEEDD leader Maxine Gowan proposing the acquisition of Praecis by GSK to R&D head “Tashi” Yamada. Max recalls the internal debate at GSK leading to the December 2006 decision to acquire Praecis for $54.8 M in her “Great Moments in DEL History” interview [6]. Praecis became GSK Waltham [38] on Valentine’s Day 2007, “Direct Select” became “Encoded Library Technology” (ELT), and GSK invited me to lead the newly acquired site and promote ELT within GSK.

It is of worthy of note that ELT did not identify useful ligands for the original targets three and four, providing a success outcome of 50% from a library of a little under one billion encoded syntheses. Suffice it to say, I sometimes ponder the outcome if the target order had been reversed. We will return in a later section to discuss current success metrics.

During my tenure leading ELT at GSK, it was my pleasure to visit all the GSK research sites in the USA and EU, to form many scientific alliances and friendships and learn a great deal about small molecule drug discovery. We examined well over

100 targets. During a presentation at the Spring 2022 ACS meeting a GSK colleague noted that ELT had identified leads that led to four entities reaching clinical trials, and four additional advanced assets and an overall success of around 60% [39]. Perhaps that explains why a former head of business development at GSK used to remind me that in his view the Praecis acquisition was "the best deal" GSK completed. I provide a partial list [40] of published work of ELT outcomes from GSK and their collaborators, including perhaps the most referenced poster child: the selective RIP1 kinase inhibitor GSK2982772 [41]. The only change made in the original DEL hit shown on left was to change the isoxazole to a 1,2,4-triazole which provided sufficient improvement in oral exposure to provide the clinical candidate.

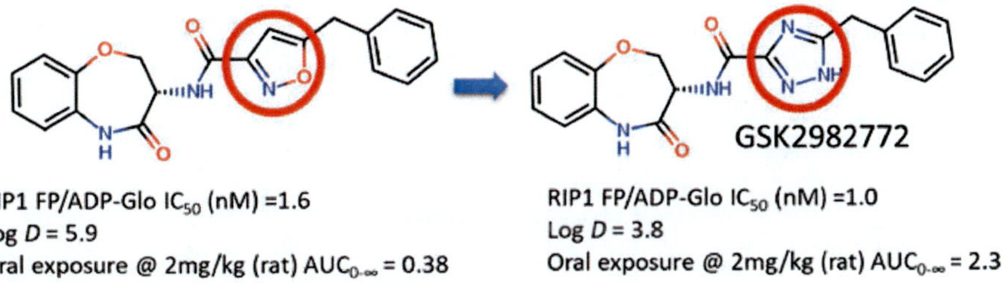

4 DEL Proliferates

The acquisition of Praecis for its DEL technology did not go unnoticed by the pharmaceutical industry, and several companies expressed interest in gaining access to the technology. Nuevolution, based in Copenhagen, announced several partnerships before its acquisition by Amgen [42] in 2019. X-Chem was founded [43] by former Praecis R&D head Rick Wagner in 2009, recruiting several Praecis/GSK scientists and announcing agreements with Roche [44] and AstraZeneca. In 2020 Rick formed ZebiAI, focused on the application of machine learning in an alliance with X-Chem [45], and Matt Clark assumed the position of X-Chem CEO. ZebiAI was acquired by Relay Therapeutics in 2021 [46]. We comment on ML and DEL below.

5 HitGen Enters the Game

DEL technology had been developed by companies based in the USA and EU but given the strength of the contract research sector in Asia, it was not too surprising to hear of the founding of HitGen, based in Chengdu, China by Jin Li in 2012 [47]. I had met Jin when I visited Astra Zeneca and at several of the International Encoded Library Symposia. Jin discusses the formation of HitGen in his "Great Moments in

DEL History" interview [8]. When I left GSK, Jin invited me to join him at HitGen, which I did, initially as a consultant, then becoming Chief Scientific Officer in 2016, and forming the USA-based subsidiary HitGen Pharmaceuticals Inc.

Over the past 5 years HitGen has assembled a DEL pool of more than 150 libraries encompassing greater than one trillion encoded syntheses. We have carried out campaigns with over 100 companies, foundations, and academic institutions, examining more than 200 targets, and identifying ligands meeting collaborator requirements for around 70% of these cases. We recently presented [48, 49] an analysis of the 380 species licensed by our collaborators 2016–2020 in terms of novelty as defined by Wills and Lipkus in terms of scaffold and shape [50]. On the left we show an example for the anti-inflammatory drug CELEBREX. Wills and Lipkus propose this classification as a measure of innovation and apply it to the 1,089 NMEs approved by FDA through 2020 yielding the distribution shown above.

Class	Definition	Total Count*
Pioneer	An NME whose shape and scaffold were not used in any previously approved drug.	511 (47%)
Settler	An NME whose shape was previously used but its scaffold was not used in any previously approved drug.	201 (19%)
Colonist	An NME whose shape and scaffold were used in a previously approved drug.	377 (35%)

Increasing structural novelty

* New molecular entities (NMEs) approved by FDA 1940-2020

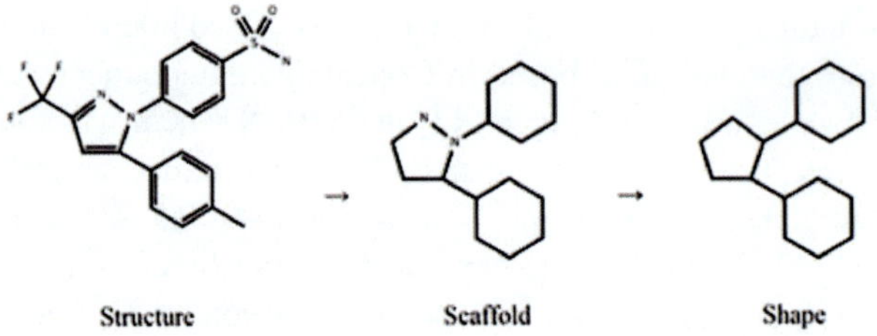

DEL libraries are generally designed with a focus on adapting and concatenating 2–4 organic transformations with a "diverse" substrate availability to an aqueous reaction medium [20]. In the years since the publication of a DEL build based on SNAr substitution on a triazine scaffold, DEL-compatible chemistry transformations have been expanded [51] to broadly mirror conventional medicinal chemistry [52, 53]. While sub-structures associated with "PAINS"[54] and genotoxicity are generally excluded, there is modest attention paid to chemotypic precedent or perceived "privileged structures" in the assembly of a DEL catalog. Consequently, we were interested in comparing the 380 HitGen DEL species that have been validated (that is, shown to be a ligand by biochemical and/or biophysical analysis) AND licensed by our collaborators in terms of the "Pioneers," "Settlers," and "Colonists" paradigm defined by Wills and Lipkus. We chose "licensed species"

as it cannot disclose structures without partner approval [55], and a licensing event represents the best metric of short-term partner-accessed return on investment [56, 57] for their commitment to a DEL campaign. Above on left we show HitGen Licensed Species graded in these terms. Note that vs. Self" means that the scaffold and shape, as defined by Wills and Lipkus, of each emerging HitGen Licensed species was compared to all earlier HitGen Licensed species and identified as a Pioneer, Settler, or Colonist. While only 47% of NMEs qualify as Pioneers, notably 85% of HitGen Licensed species are Pioneers, perhaps not too surprising given >1,200 libraries including a total of >1 trillion encoded syntheses in the current HitGen pool. Next, we checked against the 1,089 NMEs approved by FDA. Here the scaffold and shape, as defined by Wills and Lipkus, of each emerging HitGen licensed species was compared to all small-molecule NMEs approved by FDA, and defined as a Pioneer, Settler or Colonist. Remarkably 80% of all HitGen Licensed species are Pioneers in this context, compared to 47% when the NMEs were compared to "self"! While the NME "self" numbers are perhaps understandable as many were by design based on existing drugs [58], it is reasonable to conclude that selection from a DEL pool can provide significant novelty, supporting an earlier suggestion that multi-valent scaffolds may be the "currency" of diversity in DEL technology [59]. HitGen's success with DEL, now encompassing several hundred clients and collaborators, has prompted several Asia-based CROs to enter the field. Notably Wuxi Aptec [60], appointing several Praecis/GSK alumni to lead the science, now offers a range of DEL products.

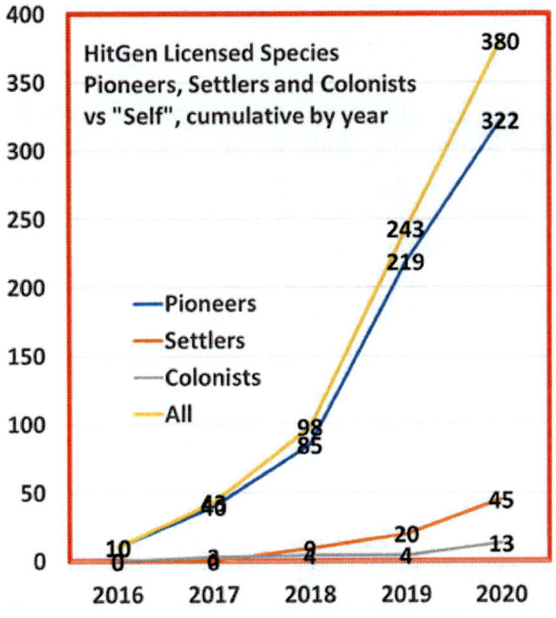

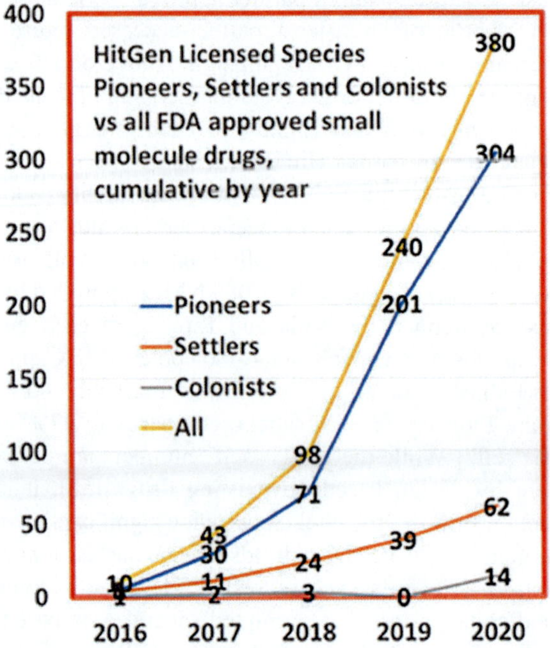

6 Adventures in Academia

Like most disruptive ideas, with the exception perhaps of contributions at Praecis, DEL was conceived in academia. While I have already noted the work of Brenner and Lerner, and of the Liu group, the pioneering contributions of the Neri group at ETH have been significant [61]. I became aware of Professor Neri's dedication to DNA-encoded library technology when he announced that the first International Symposium [62] on DNA-encoded Libraries would be held at the ETH in Zürich in 2006. I met Dario at the 2nd Symposium in 2010 and visited his lab at ETH and Philochem, a company which he founded as CEO and CSO and took public (as an affiliate of parent Philogen) in 2021[63]. Dario reviewed his interest in DNA-encoding in his interview for the "Great Moments in DEL History" [7]. The International Symposium on DNA-Encoded Libraries is now held annually (pandemics excepted), sited in cycles at Zürich, Boston, and Chengdu. In November 2022 it will be held in Cambridge MA [64].

As DEL technology gained acceptance, academic scientists in various disciplines took notice. Organic chemists dedicated time to adapting the conditions of their favored reactions to aqueous media [65, 66], while others chose to develop methods that significantly eliminated water [67] or identify ways of modifying the DNA to resist constraints on reaction conditions such as acidic pH. Most new drug targets emerge initially from academic investigators [68], and DEL was soon recognized as potentially "leveling the playing field," in that the investment required to identify ligands for novel proteins using DEL technology could be more manageable than attempting to match pharma HTS activities. HitGen recognized this need and has acknowledged collaborations with many academic centers of biologic excellence, including the University of Texas, the University of North Carolina, and the Scripps Research Institute. We also applaud the vision of not-for-profit entities such as the Bill and Melinda Gates Foundation, where we have focused on collaborations with academic laboratories sponsored by the BMGF focused on TB and malaria [69]. We are also aligned with the vision of foundations focused on rare diseases [70], which are frequently sponsored by activists seeking solutions to these untreated conditions.

7 DNA-Encoded Chemistry in 2022: Where Are We, and Some Thoughts on the Future...

I have already noted the transition of DEL to a widely practiced technology for ligand discovery over the past 5 years. Seventeen [71] of the top 20 pharmaceutical companies by revenue in 2021 have announced significant investment in DEL either internally or with a partner such as HitGen or X-Chem. DEL is also an appealing "hit" discovery approach for the Biotech sector, where nascent companies, frequently emerging from academia are typically focused on a novel mechanism or specific target, view the economics and timelines of DEL as the modality of choice to potentially allow rapid early-stage/ pre-clinical validation. Above I list many of HitGen's announced collaborators to illustrate the broad acceptance of the technology [47].

While initial collaborations were largely target-based, where our partner identified specific proteins of interest, several major pharmaceutical companies have committed to acquiring an exclusive DEL pool [72–74], frequently recruiting scientists with DEL chemistry or selection biology experience to establish the technology in house. It is a particular pleasure to note that many of these recruits gained their experience in the laboratories of Praecis Pharmaceuticals.

While the initial DEL design focused on conventional small-molecule concepts identifying reversible ligands followed up by biochemical and biophysical characterization, we have seen expansion recently into electrophile pools [75], fragment-like species [76], enablement for RNA-based targets [77], and approaches where the direct readout of the function of the potential ligand is the focus [78].

Electrophile DEL, where a pool of encoded syntheses terminating in electrophiles is interrogated in a single round modified selection protocol has gained more interest in recent years due to the increased acceptance of irreversible ligands in oncology and the presence of nucleophilic sulfur in many virology targets. We recommend consideration of this mode when mechanistic or well-placed sulfur nucleophiles are identified on a target protein.

The millimolar range affinity of initial conventional fragment "hits" was considered to be a potential issue for differentiating true fragment-like ligands from background DNA "noise," yet the power of DNA-encoding had attraction in terms of expanded sets of encoded fragment and minimal target requirements. In a proof-of-concept collaboration study with our colleagues at affiliate Vernalis Research, HitGen has recently shown that incorporating a photoactivatable diazirine moiety between the encoding DNA and a set of around 100,000 linker/fragment species can identify known and novel ligands for PAK4. The approach has been extended to identify fragments for 2-epimerase, a more challenging bacterial target.

In recent years the level of interest in identifying small molecule modifiers of RNA function has increased culminating with FDA approval of Risdiplam for the treatment of spinal muscular atrophy. Addressing a concern that target RNA/encoding DEL DNA interaction may give rise to complications in interpretation of selection outcome, HitGen has developed an approach to identify small molecule ligands for RNA targets. A combination of algorithms examining selected encoding DNA output for known and novel consensus sequences and synthetic RNA "patches" has been found to be useful in distinguishing between target RNA/DEL DNA interaction and small molecule "hits" for the E. coli riboswitch, allowing identification of high affinity small molecule ligands.

One of the advantages inherent in the utility of ligands emerging from DEL technology is the assurance that the structures of these ligands can accommodate attachment of ancillary elements such as fluorophores or solubilizing moieties at the site of linkage to the DNA. The emergence of Proteolysis Targeting Chimeras (PROTACs) to address "undruggable" targets [79] by linking a target ligand to a ligand for an appropriate E3 ligase has attracted significant attention. Success then requires that both target and E3 ligase ligands be linked to in a manner that still allows each ligand to maintain its original affinity and that the resulting heterodimeric species enables significant degradation via the ubiquitin- proteasome system. The utility of ligands discovered by DEL for this purpose has been highlighted [80].

Screening by selection identifies ligands for target proteins providing little information on the effect of those ligands on target function, leading in some interest in developing methodologies that allow functional selection (sometimes described as "phenotypic DEL"). While the results of a microfluidics-based study have been published, and several companies have been founded, the impact and generality of such an endeavor remain to be resolved.

HitGen also pioneered "OpenDEL," where a relatively small collection of encoded syntheses can be provided as a kit for consumers wishing to test the technology [81], which brings us to consider the issues limiting DEL success. From the early years at Praecis to some data discussed this month with a HitGen collaborator, the two challenges we most frequently encounter are the sometimes-unattractive (not drug-like) physicochemical properties of DEL hits and the perplexing situation when enriched, reproducible DEL features do not result in biochemically or biophysically characterizable ligands.

While physicochemical properties are to a degree in the eye of the beholder and can be ameliorated by limiting the number of cycles and choice of reagents in a DEL design, 64% of HitGen licensed species came from 4-cycle libraries.

Enriched features, and by that I refer to species that reproducibly demonstrate read counts and/or PolyO scores [82] significant above background DNA "noise" vary in their prospects of delivering a validated (that is a hit that matches or exceeds collaborator-defined values in a biochemical or biophysical assay) "hit." Although there are sufficient exceptions to require checking, ratios >20 above background provide a > 70% chance of validation, while ratios <3 yield a 20% hit likelihood. Because almost all DEL features are remarkably reproducible, it has been my bias to assume there is "something" in that set of encoded syntheses that is indeed a ligand for some component of the folded ensemble presented for selection. However, that component unfortunately may not be relevant in the chosen biochemical assay. Consequently, we recommend that a biophysical assay be in place for hit validation. In recent years, HitGen has developed a series of procedures for on-DNA syntheses (both following the original library build protocol [83] and in collaboration with our Pfizer colleagues [84], including the addition of a cleavable linker [85] to remove DNA post synthesis) of individual encoded syntheses followed by characterization of the product mix by AS/MS methods. While there are reasonably data to suggest this approach significantly improves the percentage of active off-DNA follow-up syntheses [86], it is not universally practiced.

Finally, let us return to the utility of applying machine learning (ML) algorithms to DEL selection outcomes. The Google X-Chem publication [87] and subsequent claims [88] suggesting that ML eliminated the need for feature validation by synthesis (either off- or on-DNA), allowing ML interrogation of proprietary or commercially available small molecule collections to yield "hits" have attracted significant attention. While we find that validated DEL hits with good (sub-micromolar) potency can indeed yield additional chemotypes, the use of unvalidated selection data is of significantly less value and, in a limited number of cases did not provide a useful outcome. However, we continue to monitor ML approaches to the analysis of DEL outcomes.

In summary, I've attempted to provide an admittedly personal perspective on the progression of DNA-encoded chemistry from a concept to a now broadly practiced technology for small molecule ligand discovery. While challenges remain, it has clearly impacted the economics and timelines of early-stage endeavors in industry and academia. I would note two potential, likely related areas for advancement: a better understanding of the dynamics of folded ensembles of target constructs presented for selection, and how that understanding may relate to the highly reproducible yet biophysically and biochemically uncharacterized features that continue to frustrate even seasoned practitioners.

Acknowledgments Over the past 20 years, it has been my honor to work with many visionary entrepreneurs and outstanding scientists. In addition to those noted in the text and references/comments, I would like to thank Dr. David Israel (my colleague over 20 years at Praecis, GSK and HitGen), Dr. Nils Hansen (my colleague at Praecis, discoverer of the "Yoctoreactor" and founder of Vipergen), Dr. Alex Shaginian (my colleague at GSK and HitGen), Dr. Ghotas Evindar (my colleague at Praecis and GSK), Dr. Bryan King (who led the DEL team at GSK Upper Providence and the RIP1 DEL campaign), Dr. Graham Simpson (who advocated for DEL technology at GSK Stevenage), Dr. Bob Herzberg (my GSK manager and mentor to GSK culture), Dr. Gwenn Hansen (my DEL collaborator during our tenure at Baylor College of Medicine, Prof. Zhiqiang An (my colleague and sponsor at UT Health Sciences, Houston), Drs. Jinqiao Wan and Guansai Liu (Library Chemistry leaders at HitGen), Dr. Dengfeng Dou (Target Campaign leader at HitGen), Dr. Xuemin Cheng (Scientific Computing leader at HitGen), Dr. Matt Tremblay (collaborator and sponsor at the Scripps Research Institute, La Jolla), Dr. Jimmi Seitzberg (collaborator at LEO Pharma), Drs. Sylvie Sakata, Mark Flanagan and Justin Montgomery (collaborators at Pfizer Inc.), Dr. Zhicai Shi (collaborator at MSD and Janssen, and persistent advocate for me to write this account), and Drs. Adam Weinglass and Abbas Walji (collaborators at MSD). Finally I'd like to thank Dr. Damian Young for his persistence and assistance in the preparation of this monograph.

Ethical Statements The author is employed by HitGen Inc., and working on the DEL platform of the company.

He did not receive funding for writing this chapter.

This chapter does not contain any studies with human participants or animals performed by any of the authors.

References

1. Greatest Moments in DEL History: Prof. Richard Lerner, former President of Research, the Scripps Research Institute
2. Prof. David Liu, Harvard and the Broad Institute
3. Prof. Malcolm Gefter, founder and CEO, Praecis Pharmaceuticals and Professor Emeritus, MIT
4. Dr. Rick Wagner Research Director, Praecis Pharmaceuticals, founder and CEO X-Chem
5. Dr. Matt Clark, Director of DNA-encoded chemistry, Praecis Pharmaceuticals, currently CEO X-Chem
6. Dr. Maxine Gowen, Director, CEEDD GSK
7. Prof. Dario Neri; ETH Zurich, founder and CEO Philochem
8. Dr. Jin Li, Director of Global Compound Sciences, Astra Zeneca, founder and CEO HitGen Inc. The video interviews listed in references 1-8 will be made available on the NCATS website
9. This account is essentially a personal recollection of the conception, development and practice of DNA-"recorded" library technology at Praecis Pharmaceuticals, GSK and HitGen Inc., where the DNA simply serves as a barcode of the synthesis of the attached reaction conditions and components. Other DNA-encoded methodologies are referenced and may be reviewed at https://en.wikipedia.org/wiki/DNA-encoded_chemical_library
10. The identity of this "agitated young scientist" referred to by Richard in his "Great moments in DEL history" interview as perhaps the catalyst for the Brenner and Lerner conception of

DNA-encoded chemistry remains unknown. Information leading to his (we are unformed the scientist is male) identity will be welcomed by the author
11. Brenner S, Lerner RA (1992) Encoded combinatorial chemistry. Proc Natl Acad Sci 89:5381–5383
12. It is interesting to reflect that Sidney Brenner, given his contributions to the development of the fundamentals of molecular biology did not choose to assemble the encoding DNA by an enzymatic approach
13. Kinoshita Y, Nishigaki K (1995) Enzymatic synthesis of code regions for encoded combinatorial chemistry (ECC). Nucleic Acids Symp Ser:201–202
14. David recounts his development of the concept of DNA-templated synthesis in his "Great Moments in DEL History" interview
15. Kleiner RE, Dumelin CE, Liu DR (2011) Small-molecule discovery from DNA-encoded chemical libraries. Chem Soc Rev 40:5707–5717
16. Cambridge's Ensemble Therapeutics Quietly Shuts Down After 13 Years. In: BioSpace. https://www.biospace.com/article/unique-cambridge-s-ensemble-therapeutics-quietly-shuts-down-after-13-years/. Accessed 27 Aug 2022
17. Perhaps anticipating funding from Abarelix approval. Although FDA approval as Plenaxis was secured, sales were less than anticipated: Wikipedia (2022) https://en.wikipedia.org/wiki/Abarelix. Accessed 29 Aug 2022
18. Derakhshankhah H, Jafari S (2018) Cell penetrating peptides: a concise review with emphasis on biomedical applications. Biomed Pharmacother 108:1090–1096
19. Kavarana M, personal communication (2002) Praecis Pharmaceuticals, Waltham
20. We decided to use preformed duplex DNA "codons" with cycle-specific 2-base overhangs for ligation, although we were unaware of the Kinoshita publication at this time. I acknowledge the contributions of David Israel and Nils Hansen to this concept
21. Biosearch Technologies is still a major supplier of oligonucleotide materials for the assembly of DEL libraries. DNA encoded library (DEL) components. In: LGC Biosearch Technologies. https://www.biosearchtech.com/products/oligos-probes-and-primers/dna-encoded-library-del-oligos. Accessed 26 Aug 2022
22. Kölmel DK, Zhu H, Flanagan ME et al (2021) Employing photocatalysis for the design and preparation of DNA-encoded libraries: a case study. Chem Rec 21:616–630
23. Liu W, Huang W, Lin Q et al (2021) Development of DNA-compatible hydroxycarbonylation reactions using chloroform as a source of carbon monoxide. Bioorg Med Chem 38:116118
24. Clark MA, Acharya RA, Arico-Muendel CC et al (2009) Design, synthesis and selection of DNA-encoded small-molecule libraries. Nat Chem Biol 5:647–654
25. A description of the development of phage display, including selection procedures, is to be found in the Nobel Lecture of Prof. George Smith: Nobel Prize Outreach AG. (2019) https://www.nobelprize.org/prizes/chemistry/2018/smith/lecture/. Accessed 27 Aug 2022
26. 454 Life Sciences was founded by Jonathan Rothberg and was originally known as 454 Corporation, a subsidiary of CuraGen. For their method for low-cost gene sequencing, 454 Life Sciences was awarded the Wall Street Journal's Gold Medal for Innovation in the Biotech-Medical category in 2005. Without the development of sequencing technologies such as 454, and subsequently Illumina, I would not be writing this review
27. The procedures for deconvoluting the negative ion mass spectra of our DNA conjugates, recording the codon/structure encoding relationship, the curation, analysis and translation of the raw sequence 454 sequence data in a manner ready to be displayed in Spotfire were all designed and coded by Jeff Messer at Praecis
28. The Cubic Plot survives as the most intuitive graphical presentation of DEL selection data, and planar and linear sets of structurally related encoded syntheses are still referred to as "features"
29. DEL 16 was a 4-cycle library based on a non-encoded 2,4,6-trichloro-1,3,5-triazine scaffold, displacing each chlorine sequentially with amine nucleophiles via SNAr chemistry. We were aware of the design and a successful hit-finding outcome from an ECLiPS library built at

Pharmacopeia Inc: Leftheris K, Ahmed G, Chan R, et al (2004) The discovery of orally active Triaminotriazine aniline amides as inhibitors of p38 MAP kinase. J Med Chem 47:6283–6291
30. Despite a somewhat arbitrary bias (I am not aware of any published structure-class based impediment or toxicity) against the triazine core by a number of medicinal chemists, triazine-based libraries were an early boon in popularizing DEL technology
31. Pereira DA, Williams JA (2007) Origin and evolution of high throughput screening. Br J Pharmacol 152:53–61
32. Bower JL, Christensen CM (1995) Disruptive technologies: catching the wave. Harvard Business Review
33. GlaxoSmithKline: Reorganizing Drug Discovery (A) ^ 605074. In: HBR Store. https://store.hbr.org/product/glaxosmithkline-reorganizing-drug-discovery-a/605074. Accessed 24 Aug 2022
34. GSK to create seventh centre of excellence for drug discovery. http://test.pharmabiz.com/news/gsk-to-create-seventh-centre-of-excellence-for-drug-discovery-18771. Accessed 28 Aug 2022
35. Ding Y, Belyanskaya S, DeLorey JL et al (2021) Discovery of soluble epoxide hydrolase inhibitors through DNA-encoded library technology (ELT). Bioorg Med Chem 41:116216
36. Deng H, O'Keefe H, Davie CP et al (2012) Discovery of highly potent and selective small molecule ADAMTS-5 inhibitors that inhibit human cartilage degradation via encoded library technology (ELT). J Med Chem 55:7061–7079
37. Lazaar AL, Yang L, Boardley RL et al (2016) Pharmacokinetics, pharmacodynamics and adverse event profile of GSK2256294, a novel soluble epoxide hydrolase inhibitor. Br J Clin Pharmacol 81:971–979
38. PharmaTimes (2006) GSK to acquire Praecis pharmaceuticals. In: PharmaTimes https://www.pharmatimes.com/news/gsk_to_acquire_praecis_pharmaceuticals_995468. Accessed 28 Aug 2022
39. Ding Y (2022) Advances in DNA-encoded library technology (ELT): from on-DNA chemistry to preclinical candidate. San Diego, CA
40. A comprehensive review of DEL outcomes at GSK may be found in Chapter 7 "From DEL Selections to Validated Hits to Clinical Leads" of this book by authors Lisa Marcaurelle, Westley Tear and Gang Yao
41. Harris PA, Berger SB, Jeong JU et al (2017) Discovery of a first-in-class receptor interacting protein 1 (RIP1) kinase specific clinical candidate (GSK2982772) for the treatment of inflammatory diseases. J Med Chem 60:1247–1261
42. Amgen announces a recommended public cash offer to the shareholders of Nuevolution. In: Amgen. https://www.amgen.com/newsroom/press-releases/2019/05/amgen-announces-a-recommended-public-cash-offer-to-the-shareholders-of-nuevolution. Accessed 26 Aug 2022
43. X-Chem | Our History in DEL-Driven Drug Discovery. In: X-Chem. https://www.x-chemrx.com/history/. Accessed 28 Aug 2022
44. X-Chem (2022) https://www.x-chemrx.com/wp-content/uploads/2021/02/X-Chem-and-Roche-Enter-Into-Drug-Discovery-Collaboration-and-License-Agreement.pdf Accessed 26 Aug 2022
45. (2020) ZebiAI Therapeutics, Inc. and X-Chem Enter into Strategic Agreement to Enable Validation of Novel Therapeutic Targets and Acceleration of Drug Candidate Discovery. https://www.businesswire.com/news/home/20200113005795/en/ZebiAI-Therapeutics-Inc.-and-X-Chem-Enter-into-Strategic-Agreement-to-Enable-Validation-of-Novel-Therapeutic-Targets-and-Acceleration-of-Drug-Candidate-Discovery. Accessed 28 Aug 2022
46. In one of their first ever acquisitions, Relay bets $85M cash on a new AI-based screening approach. In: Endpoints News. https://endpts.com/in-one-of-their-first-ever-acquisition-relay-bets-85m-cash-on-a-new-ai-based-screening-approach/. Accessed 28 Aug 2022
47. HitGen (2022). https://www.hitgen.com/en/about.html. Accessed 28 Aug 2022
48. Morgan BA (2021) 29 years of DNA-encoded chemistry: managing serendipity in pre-clinical small molecule drug discovery
49. Morgan BA (2022) Does DNA-encoded library technology provide novelty?

50. Wills TJ, Lipkus AH (2020) Structural approach to assessing the innovativeness of new drugs finds accelerating rate of innovation. ACS Med Chem Lett 11:2114–2119
51. Fair RJ, Walsh RT, Hupp CD (2021) The expanding reaction toolkit for DNA-encoded libraries. Bioorg Med Chem Lett 51:128339
52. Brown DG, Boström J (2016) Analysis of past and present synthetic methodologies on medicinal chemistry: where have all the new reactions gone? J Med Chem 59:4443–4458
53. Dragovich PS, Haap W, Mulvihill MM et al (2022) Small-molecule Lead-finding trends across the Roche and Genentech research organizations. J Med Chem 65:3606–3615
54. Baell JB, Holloway GA (2010) New substructure filters for removal of pan assay interference compounds (PAINS) from screening libraries and for their exclusion in bioassays. J Med Chem 53:2719–2740
55. HitGen is committed to publish studies related to assets discovered during DEL collaborations, but only following agreement of the licensee, for example: Kung P-P, Bingham P, Burke BJ et al (2020) Characterization of specific N-α-acetyltransferase 50 (Naa50) inhibitors identified using a DNA encoded library. ACS Med Chem Lett 11:1175–1184
56. While we do not address the economics of early-stage small molecule drug discovery here see: Paul SM, Mytelka DS, Dunwiddie CT et al (2010) How to improve R&D productivity: the pharmaceutical industry's grand challenge. Nat Rev Drug Discov 9:203–214
57. O'Connell KE, Frei P, Dev KK (2014) The premium of a big pharma license deal. Nat Biotechnol 32:617–619
58. "The most fruitful basis for the discovery of a new drug is to start with an old drug." Attributed to James Black, winner of the 1988 Nobel Prize in Medicine: Raju TN (2000) The Nobel chronicles. 1988: James Whyte black, (b 1924), Gertrude Elion (1918-99), and George H Hitchings (1905-98). Lancet 355:1022
59. While "building blocks" (DEL components with a single reactive site, for example amines or carboxylic acids) are broadly commercially available, "scaffolds" (DEL components with >1 reactive site) are more limited in terms of commercial availability. Hence my suggestion in an early International Symposium on DNA-encoded Libraries that "scaffolds are the currency of diversity" in DEL design
60. WuXi AppTec Launches DNA Encoded Library (DEL) Service Package "DELight" to Expedite Early Drug Discovery. https://www.wuxiapptec.com/news/wuxi-news/3428. Accessed 28 Aug 2022
61. Gironda-Martínez A, Donckele EJ, Samain F, Neri D (2021) DNA-encoded chemical libraries: a comprehensive review with successful stories and future challenges. ACS Pharmacol Transl Sci 4:1265–1279
62. 1st International Symposium on DNA-Encoded Libraries, held at ETH Zurich 2006
63. (2020) Our co-founder and Chief Scientific Officer, Prof. Dario Neri will join Philogen in a full time capacity. In: Philogen Spa https://www.philogen.com/2020/08/20/our-co-founder-and-chief-scientific-officer-prof-dario-neri-will-join-philogen-in-a-full-time-capacity/. Accessed 28 Aug 2022
64. The 10th International Symposium on DNA-Encoded Chemical Libraries has been postponed from its original 2020 date and will now be held at MIT, Cambridge MA on November 3rd–4th 2022
65. Monty OBC, Nyshadham P, Bohren KM et al (2020) Homogeneous and functional group tolerant ring-closing metathesis for DNA-encoded chemical libraries. ACS Comb Sci 22:80–88
66. Phelan JP, Lang SB, Sim J et al (2019) Open-air alkylation reactions in photoredox-catalyzed DNA-encoded library synthesis. J Am Chem Soc 141:3723–3732
67. Flood DT, Zhang X, Fu X et al (2020) RASS-enabled S/P-C and S-N bond formation for DEL synthesis. Angew Chem Int Ed Engl 59:7377–7383
68. Since 1953, 62 Nobel prizes in physiology or medicine have been awarded to a total of 150 recipients. Only two of those were awarded to a total of 4 scientists who carried out their prize-winning work in industrial laboratories while 85% of the drugs approved by FDA 1998-2007 were discovered in pharma or biotech companies. Kneller R (2010) the importance of new

companies for drug discovery: origins of a decade of new drugs. Nat Rev Drug Discov 9:867–882
69. Ltd H (2018) HitGen launching initiative to discover new medicines to combat tuberculosis and malaria. In: GlobeNewswire News Room https://www.globenewswire.com/fr/news-release/2018/08/03/1546776/0/en/HitGen-Launching-Initiative-to-Discover-New-Medicines-to-Combat-Tuberculosis-and-Malaria.html. Accessed 28 Aug 2022
70. HitGen and the PACS1 Syndrome Research Foundation Announce an Agreement to Discover Ligands That Address Issues Related to the Mutated PACS1 Gene Product – CMOCRO. In: Domain Name. http://www.cmocro.com/. Accessed 28 Aug 2022
71. Fierce Pharma (2022) https://www.fiercepharma.com/special-reports/top-20-pharma-companies-2021-revenue Accessed 26 Aug 2022
72. (2017) HitGen and Pfizer Enter Research Collaboration and License Agreement to Build and Screen Novel DNA-Encoded Libraries. https://www.businesswire.com/news/home/20170418005640/en/HitGen-and-Pfizer-Enter-Research-Collaboration-and-License-Agreement-to-Build-and-Screen-Novel-DNA-Encoded-Libraries. Accessed 29 Aug 2022
73. Hale C (2018) Merck, HitGen expand DNA library screening collaboration. In: Fierce Biotech. https://www.fiercebiotech.com/medtech/merck-hitgen-expand-dna-library-screening-collaboration. Accessed 29 Aug 2022
74. HitGen Inc. Announces Strategic Collaboration with Janssen to Design and Build Novel DNA-Encoded Libraries | MarketScreener. https://www.marketscreener.com/quote/stock/HITGEN-INC-119082779/news/HitGen-Inc-Announces-Strategic-Collaboration-with-Janssen-to-Design-and-Build-Novel-DNA-Encoded-Lib-34049815/. Accessed 26 Aug 2022
75. Zhu Z, Grady LC, Ding Y et al (2019) Development of a selection method for discovering irreversible (covalent) binders from a DNA-encoded library. SLAS Discov 24:169–174
76. Ma H, Murray JB, Luo H et al (2022) PAC-FragmentDEL – photoactivated covalent capture of DNA-encoded fragments for hit discovery. RSC Med Chem
77. Chen Q, Li Y, Lin C et al (2022) Expanding the DNA-encoded library toolbox: identifying small molecules targeting RNA. Nucleic Acids Res 50:e67
78. MacConnell AB, Price AK, Paegel BM (2017) An integrated microfluidic processor for DNA-encoded combinatorial library functional screening. ACS Comb Sci 19:181–192
79. The multiple sessions on PROTACS and Molecular Glues at the 2022 Fall ACS Meeting, where virtually all the presenters used "undruggable" in their introductions speak to the resources committed to this concept in industry and academia
80. Bond MJ, Crews CM (2021) Proteolysis targeting chimeras (PROTACs) come of age: entering the third decade of targeted protein degradation. RSC Chem Biol 2:725–742
81. How could an open access DNA-encoded library benefit your research? In: Endpoints Webinars. https://webinars.endpts.com/how-could-an-open-access-dna-encoded-library-benefit-your-research/. Accessed 29 Aug 2022
82. A proprietary HitGen algorithm that compares feature intensities. It is described in the supplementary materials to reference 65
83. Foley TL, Burchett W, Chen Q et al (2021) Selecting approaches for hit identification and increasing options by building the efficient discovery of actionable chemical matter from DNA-encoded libraries. SLAS Discov 26:263–280
84. Montgomery JI (2022) Triage strategies for DEL screens: Toward Tractable Chemical Matter
85. Xia B, Franklin GJ, Lu X et al (2021) DNA-encoded library hit confirmation: bridging the gap between on-DNA and off-DNA chemistry. ACS Med Chem Lett 12:1166–1172
86. A rigorous adherence to only following up on-DNA syntheses with a positive AS/MS outcome will always give an improved % of "hits" but can increase timelines
87. McCloskey K, Sigel EA, Kearnes S et al (2020) Machine learning on DNA-encoded libraries: a new paradigm for hit finding. J Med Chem 63:8857–8866
88. Machine Learning On Top of DNA Encoded Libraries. https://www.science.org/content/blog-post/machine-learning-top-dna-encoded-libraries. Accessed 26 Aug 2022